Carbon Dioxide Reduction through Advanced Conversion and Utilization Technologies

Electrochemical Energy Storage and Conversion

Series Editor: Jiujun Zhang
National Research Council Institute for Fuel Cell Innovation Vancouver,
British Columbia, Canada

Recently Published Titles

Graphene: Energy Storage and Conversion Applications
Zhaoping Liu and Xufeng Zhou

Lithium-Ion Batteries: Fundamentals and Applications
Yuping Wu

Lead-Acid Battery Technologies: Fundamentals, Materials, and Applications
Joey Jung, Lei Zhang, and Jiujun Zhang

Solar Energy Conversion and Storage: Photochemical Modes
Suresh C. Ameta and Rakshit Ameta

Electrolytes for Electrochemical Supercapacitors
Cheng Zhong, Yida Deng, Wenbin Hu, Daoming Sun, Xiaopeng Han, Jinli Qiao, and Jiujun Zhang

Electrochemical Reduction of Carbon Dioxide: Fundamentals and Technologies
Jinli Qiao, Yuyu Liu, and Jiujun Zhang

Metal–Air and Metal–Sulfur Batteries: Fundamentals and Applications
Vladimir Neburchilov and Jiujun Zhang

Photochemical Water Splitting: Materials and Applications
Neelu Chouhan, Ru-Shi Liu, and Jiujun Zhang

Electrochemical Supercapacitors for Energy Storage and Delivery: Fundamentals and Applications
Aiping Yu, Victor Chabot, and Jiujun Zhang

Electrochemical Polymer Electrolyte Membranes
Jianhua Fang, Jinli Qiao, David P. Wilkinson, and Jiujun Zhang

Electrochemical Energy: Advanced Materials and Technologies
Pei Kang Shen, Chao-Yang Wang, San Ping Jiang, Xueliang Sun, and Jiujun Zhang

High-Temperature Electrochemical Energy Conversion and Storage: Fundamentals and Applications
Yixiang Shi, Ningsheng Cai, and Jiujun Zhang

Redox Flow Batteries: Fundamentals and Applications
Huamin Zhang, Xianfeng Li, and Jiujun Zhang

Carbon Nanomaterials for Electrochemical Energy Technologies: Fundamentals and Applications
Shuhui Sun, Xueliang Sun, Zhongwei Chen, Jinli Qiao, David P. Wilkinson, and Jiujun Zhang

Proton Exchange Membrane Fuel Cells
Zhigang Qi

Hydrothermal Reduction of Carbon Dioxide to Low-Carbon Fuels
Fangming Jin

Lithium-Ion Supercapacitors: Fundamentals and Energy Applications
Lei Zhang, David P. Wilkinson, Zhongwei Chen, Jiujun Zhang

Advanced Bifunctional Electrochemical Catalysts for Metal-Air Batteries
Yan-Jie Wang, Rusheng Yuan, Anna Ignaszak, David P. Wilkinson, Jiujun Zhang

Carbon Dioxide Reduction through Advanced Conversion and Utilization Technologies
Yun Zheng, Bo Yu, Jianchen Wang, Jiujun Zhang

https://www.crcpress.com/Electrochemical-Energy-Storage-and-Conversion/book-series/CRCELEENESTO

Carbon Dioxide Reduction through Advanced Conversion and Utilization Technologies

Yun Zheng
Bo Yu
Jianchen Wang
Jiujun Zhang

CRC Press is an imprint of the
Taylor & Francis Group, an **informa** business

CRC Press
Taylor & Francis Group
6000 Broken Sound Parkway NW, Suite 300
Boca Raton, FL 33487-2742

First issued in paperback 2020

ISBN-13: 978-1-138-09529-8 (hbk)
ISBN-13: 978-0-367-77986-3 (pbk)

Library of Congress Cataloging-in-Publication Data

Names: Zheng, Yun (Chemical engineer), author.
Title: Carbon dioxide reduction through advanced conversion and utilization technologies / Yun Zheng, Bo Yu, Jianchen Wang, Jiujun Zhang.
Description: Boca Raton : Taylor & Francis, CRC Press, 2019. | Series: Electrochemical energy storage and conversion | Includes bibliographical references and index.
Identifiers: LCCN 2018057440| ISBN 9781138095298 (hardback : alk. paper) | ISBN 9781315104171 (ebook)
Subjects: LCSH: Carbon dioxide--Separation. | Electrolytic reduction. | Carbon dioxide--Recycling.
Classification: LCC TP244.C1 Z44 2019 | DDC 660.028/6--dc23
LC record available at https://lccn.loc.gov/2018057440

Visit the Taylor & Francis Web site at
http://www.taylorandfrancis.com

and the CRC Press Web site at
http://www.crcpress.com

Contents

Preface

Continuous increases in the consumption of global fossil fuels has led to a rapid increase in CO_2 concentration in our atmosphere. As one of the major contributors to undesired possible climate change and environmental destruction, excessive CO_2 emission could disturb the harmony between nature and humans. Developing efficient methods to reduce CO_2 and converting it into useful fuels and chemicals are obviously one of the solutions for both reducing greenhouse effects and energy shortages. Currently, the technologies of CO_2 conversion and utilization are not mature and efficient enough in meeting the requirements and overcoming the challenges for practical implementation and commercialization. A book focusing on advanced technologies of CO_2 conversion and the associated energy storage is necessary and could have a great impact on the efforts to accelerate the progress of R&D.

This book is a comprehensive introduction that focuses on CO_2 conversion technologies to produce useful fuels and chemicals. It is a chapter-by-chapter breakdown of the key topics. The background of CO_2 reduction through advanced conversion and utilization technologies is introduced in Chapter 1. Then the fundamentals of CO_2 are demonstrated in Chapter 2, including the molecular structure of CO_2, the thermodynamics and kinetics of CO_2 conversion, and the cause analysis for low activation/conversion rates of CO_2. Typical CO_2 conversion technologies such as enzymatic, mineralization, photocatalysis, and thermocatalytic are classified and summarized as non-electrochemical methods in Chapters 3 and 4. The electrochemical conversion approaches are divided into low- and high-temperature electrochemical processes; the former is introduced in Chapter 5, and the latter is summarized thoroughly in Chapters 6 through 14.

Particularly, as a research topic of our group, the advanced materials and technology of high temperature co-electrolysis of H_2O and CO_2 to produce sustainable fuels using solid oxide cells (SOCs) are reviewed in detail in these chapters. To be specific, the introduction and fundamentals of CO_2/CO_2 co-electrolysis using SOCs are introduced in Chapter 6, and the research status is reviewed in Chapter 7. As key topics in SOCs, the materials and their related detection technique are summarized in Chapters 8 and 9. Advanced fabrication methods, advanced structure, and a significant phenomenon are also introduced in Chapters 11 through 13, respectively. As an obstacle to further develop SOCs technique, degradation issues are emphasized in Chapter 14.

In addition to analyzing advanced materials and technology used for CO_2 conversion, an economic analysis of CO_2 conversion into useful fuels and chemicals is given in Chapter 15. The technical and application challenges are analyzed and summarized in Chapter 16, with proposed possible research directions for CO_2 conversion technologies to overcome the challenges in future research and applications.

This book provides a comprehensive and clear picture of the fundamentals, technologies, and advanced materials for CO_2 conversion into useful fuels and chemicals. We hope it will help our readers further understand CO_2 reduction through advanced conversion and utilization technologies. We sincerely hope this book will ignite more interest in CO_2 issues.

Acknowledgments

The authors would like to acknowledge the earnest contributions from the Group for High Temperature Electrolysis, the Institute of Nuclear and New Energy Technology (INET), Tsinghua University (Chapter 3: Chenhuan Zhao, Chapter 4: Yifeng Li, Chapter 10: Tong Wu, Chapter 11: Yifeng Li, Chapter 13: Wenqiang Zhang), and the great efforts of Jianxin Zhu, Wangxv Yue and Junwen Cao in editing this book.

The authors would also like to acknowledge the support from the National Natural Science Foundation of China (No. 91645126, 21273128, 51425403, and 21677162), the National Key Research and Development Program of China (2017YFB013001), the "Program for Changjiang Scholars and Innovative Research Team in University" (IRT13026), and the "Tsinghua-MIT-Cambridge" Low Carbon Energy University Alliance Seed Fund Program (201LC004).

Authors

Yun Zheng is currently studying for a PhD from the Institute of Nuclear and New Energy Technology (INET) at Tsinghua University, China. He received his BS and MS degrees in chemistry from China Three Gorges University in 2012 and Ocean University of China in 2015, respectively. From 2009 to 2012, he studied in the Engineering Research Center of Eco-environment in Three Gorges Reservoir Region, Ministry of Education, China, for investigations of photocatalysis and photoelectrocatalysis of carbonaceous contaminants using doped semiconductor and organometallic complex materials. His research topics at the Key Laboratory of Marine Chemistry Theory and Technology, Ministry of Education, China, from 2012 to 2015, were electrodialysis and membrane bioreactor. His current research interests are electrochemical energy storage and conversion, especially for the research of high-temperature electrolysis of CO_2/H_2O to produce sustainable fuels using solid oxide electrolytic cells (SOECs).

Dr. Bo Yu is an associate professor at Institute of Nuclear and New Energy Technology (INET) at Tsinghua University, China. Dr. Yu obtained her BS and MSc in physical chemistry from Northeastern University in 1997 and 2000, respectively. She received her PhD from Tsinghua University in 2004 and joined the Nuclear Science and Engineering Department at Massachusetts Institute of Technology (MIT) as a visiting researcher in 2012. Dr. Yu has been responsible for the research and development of nuclear hydrogen or syngas production through high-temperature electrolysis of CO_2/H_2O at INET since 2005. Her research interests are electrochemical energy storage and conversion, with some focus on solid oxide cell (SOFC/SOEC) technologies, surface/interface electrochemistry, theoretical and experimental analyses of electrode kinetics, electrocatalysis, and applied electrochemistry. She is the author or co-author of more than 100 peer-reviewed articles, over 30 patents, and several book chapters. Dr. Yu is a Board Member of Nuclear Hydrogen Division of International Associate Hydrogen Energy (IAHE). She is also an Associated Editor of the *International Journal of Energy Technology & Policy*.

Dr. Jianchen Wang is a professor at the Institute of Nuclear and New Energy Technology (INET) at Tsinghua University, China. He earned his BS degree in physical chemistry and master's degree in applied chemistry from Tsinghua University in 1987 and 1992, respectively. He has worked at INET since 1987. He obtained a position as an associate professor in 1998 and took a position as a professor in 2004 at INET. He is a deputy director of Committee of nuclear chemistry and radiochemistry, Chinese chemical society. His research interests are in the areas of nuclear chemistry, and separation chemistry in actinides and other nuclear fission elements. He participated in setting up the Chinese partitioning process for high-level liquid waste (HLLW), including the trialkyl phosphine oxides (TRPOs) process to separate actinides, the crown ether process to separate strontium, and the calixcrown process to separate cesium from HLLW. He was awarded the second prize of the State Technological Invention of China for his contribution to researching TRPO process. He is the author or co-author of more than 60 peer-reviewed articles and over 30 patents in these areas. In recent years, his research area also includes electrochemistry, including high-temperature co-electrolysis (HTCE) of H_2O and CO_2 by using solid oxide electrolytic cells (SOECs) and the disintegration of graphite matrix from spent fuel of high-temperature gas-cooled reactors (HTGRs) by electrochemical technology.

Dr. Jiujun Zhang is a professor at Shanghai University and Principal Research Officer (Emeritus) at National Research Council of Canada (NRC). Dr. Zhang received his BS and MSc in electrochemistry from Peking University in 1982 and 1985, respectively, and his PhD in electrochemistry from Wuhan University in 1988. After completing his PhD, he took a position as an associate professor at the Huazhong Normal University for two years. Starting in 1990, he carried out three terms of postdoctoral research at the California Institute of Technology, York University, and the University of British Columbia. Dr. Zhang has over 30 years of R&D experience in theoretical and applied electrochemistry, including over 18 years of fuel cell R&D (including six years at Ballard Power Systems and 12 years at NRC-IFCI (before 2011; NRC-EME after 2011), and three years of electrochemical sensor experience. Dr. Zhang holds several adjunct professorships, including one at the University of Waterloo, one at the University of British Columbia, and one at Peking University. Dr. Zhang has co-authored more than 400 publications, including 230 refereed journal papers with approximately

22,000 citations, 18 edited or co-authored books, 41 book chapters, as well as 110 conference and invited oral presentations. He also holds over 10 US/EU/WO/JP/CA patents and 11 US patent publications, and he has produced more than 90 industrial technical reports. Dr. Zhang serves as the editor or as an editorial board member for several international journals as well as editor for a book series (Electrochemical Energy Storage and Conversion, CRC Press). Dr. Zhang is an active member of the Electrochemical Society (ECS), the American Chemical Society (ACS), and the Canadian Institute of Chemistry (CIC); a Fellow Member of the International Society of Electrochemistry (ISE); as well as a board committee member of the International Academy of Electrochemical Energy Science (IAOEES).

1 Introduction to CO_2 Reduction through Advanced Conversion and Utilization Technologies

1.1 GLOBE ENERGY STATUS, CHALLENGES, AND PERSPECTIVES

Energy is one of the most important strategic resources for social development and national economy. According to the data from the International Energy Agency (IEA), as shown in Figure 1.1, global energy needs in the new policies scenario will still increase by 30% between today and 2040, although it will rise more slowly than in the past. This increased energy need is basically the equivalent of adding another China and India to today's global demand.[1] By comparison, the largest portion of this energy demand growth (about 30%) comes from India, and its global energy consumption will rise to 11% by 2040. Southeast Asia is also an important contributor in global energy demand, and its growth rate will be twice of China. In short, about two-thirds of global energy growth results from the developing countries in Asia, and some other countries and regions including the Middle East, Africa and Latin America.[1]

With respect to the specific energy resources, in the last nearly two decades, global consumption of fossil fuels such as coal, oil, and gas have still increased significantly (a typical example of China is shown in Figure 1.2),[1] thus increasing the levels of CO_2 in our atmosphere greatly and leading to undesired climate change and environmental destruction.[2] Meanwhile, the limited reserves of coal and gas on the earth will gradually dry up with over-exploitation. Accordingly, as proposed by the IEA, energy consumption should be changed dramatically in the future with reference to new policies scenario; specifically, decreased use of coal and oil, rapid increase in the use of renewables and enhanced energy efficiency are necessary to deal with above issues. The development of advanced energy conversion and storage technologies should therefore be explored in conjunction with the exploration of clean, renewable alternative energy sources, especially solar radiation, wind, and so on.[3–8]

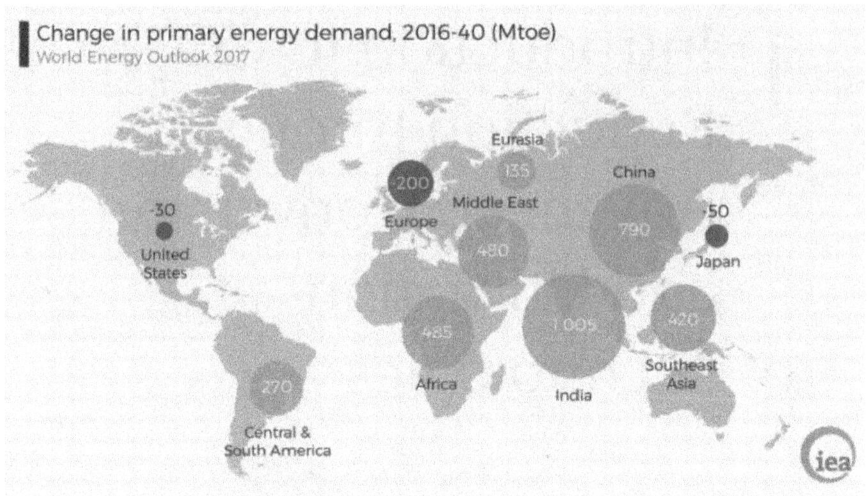

FIGURE 1.1 Growing global energy demand from 2016 to 2040. (From *World Energy Outlook 2017*, International Energy Agency, Paris, France, 2018.)

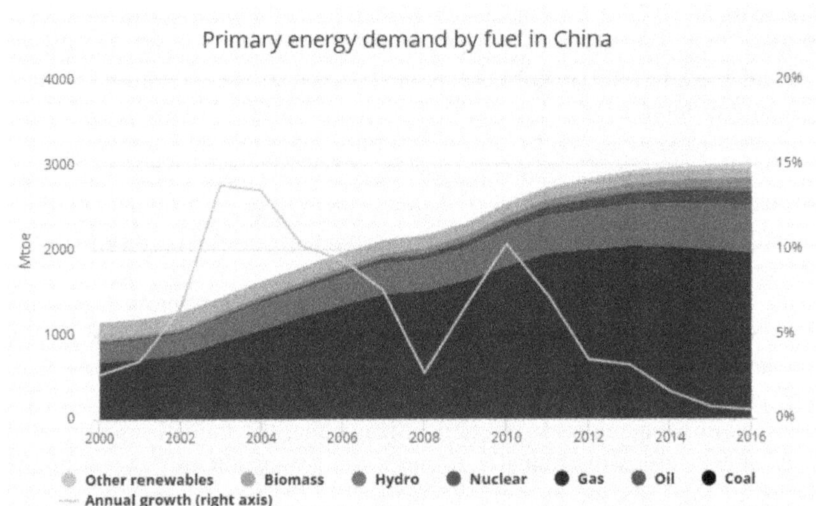

FIGURE 1.2 Primary energy demand by fuel in China from 2000 to 2016. (From *World Energy Outlook 2017*, International Energy Agency, Paris, France, 2018.)

1.2 CO_2 EMISSION AND REDUCING

Continuously increasing global consumption of fossil fuels has led to the rapid growth of CO_2 concentrations in our atmosphere.[9,10] As one of the major contributors to undesired climate change and environmental destruction, excessive CO_2 emissions have greatly affected our world despite it being a necessary material for life and many industrial processes.[10–12] Despite their recent flattening, global energy-related CO_2 emissions will still increase slightly through 2040 in the new policies scenario. This outcome is far from enough to avoid severe impacts of climate change; thus, it is critical to develop efficient methods for reducing CO_2 emissions and exploring green and sustainable energy resources.[13,14] For this purpose, several treatment systems to reduce CO_2, including more efficient electric energy production, better energy utilization effciency, fuel shift (from coal to gas), carbon capture and storage, and conversion have been proposed by researchers (more details are shown in Table 1.1).[14] Based on important factors such as cost, energy resource distribution, and implementation time, the reduction and conversion of CO_2 into useful fuels/chemicals for utilization is an attractive solution that reduces greenhouse effects and provides effective energy sources.[11,15]

TABLE 1.1
Strategies for Reducing CO_2 Emission

Strategies	Features	Limitations
Higher efficiency in electric energy production	Use of innovative technologies usually. Efficiency improvements from <30% to 53% (IGCC)[a]	High costs and long implementation times
Utilization of energy with better efficiency	Requires more conscientious attitude towards energy usage. Methods such as driving more fuel-efficient vehicles, using electrical appliances more effciently	Difficulties for unified plan and control
Fuel shift	From coal to gas. Coal (950) > oil > gas (497)[b]	Distribution of resources
Carbon capture and storage (CCS)	Large potential in CO_2 storage. Storage in natural fields[c]. Not been accepted by many countries	Security and economic (costly, uncertainty, intensive energy requirements)
CO_2 conversion and utilization	Recycles carbon. Reduces the extraction of fossil-C. Avoids emission of CO_2	

Source: Aresta, M. et al., *Chem. Rev.*, 114, 1709–1742, 2014.

[a] IGCC: integrated gasification combined cycle.

[b] Emissions of CO_2 per kWh of electric energy generated, g/kWh.

[c] For example, depleted oil and gas fields, aquifers, and coal beds.

1.3 MAIN APPROACHES OF CO_2 CONVERSION AND UTILIZATION

Generally, the conversion of CO_2 can be divided into two reaction categories (Figure 1.3). One uses less energy supply to convert CO_2 into high-carbon organic chemicals with its carbon in an oxidation state of +4 or +3; such conversion reactions also tend to happen at temperature below 273 K. The other uses greater energy to reduce CO_2 into low-carbon chemicals/fuels with its carbon in the reductive states of +2 or lower. Typical products in the two categories are shown in Figure 1.3. In terms of global market demands, fuels face significantly higher demand than do chemicals (12–14 times)[14,16] and the manufacture of fuels convert significantly greater volumes of CO_2 (Gt/y) than that of chemicals (>300 Mt/y).

Regarding the specific approaches for CO_2 conversion and utilization (see Figure 1.4), there are three main approaches: chemical/catalytic conversion (chemical process), enzymatic conversion (biological/biochemical process), and technological utilization (physical process). Here, technological utilization is not closely related to CO_2 conversion because of its physical nature, including compressing, recycling, phase transition, and so on. However, it is important for the practical usage of CO_2 which includes cereal preservation (bactericide), additives to beverages, food packaging/conversion, dry cleaning, extraction, mechanical industries, fire extinguishers, air-conditioning as well as water treatment.[13,14]

CO_2 enzymatic conversions (such as photosynthesis)[10,17–20] and mineralization[21–27] are the most common methods for the conversion/fixation of CO_2. CO_2 mineralization, from CO_2 to inorganic or organic hydro-carbonate (HCO_3^-)/carbonate(CO_3^{2-}) ($\Delta G < 0$), can fixate and store CO_2 effectively, but it is inefficient. Enzymatic conversions are biological/biochemical processes that occur in organisms such as animals, plants or microorganisms where the conversion mechanisms in plants are much more complex and difficult to mimic than in bacteria.[14]

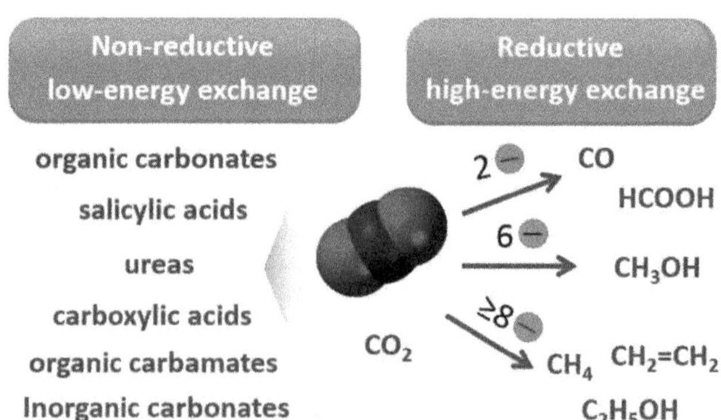

FIGURE 1.3 Synthetic processes for some products from CO_2 in two main categories. (Antonio, J.M., *Green Chem.*, 17, 5114–5130, 2015. Reproduced by permission of The Royal Society of Chemistry.)

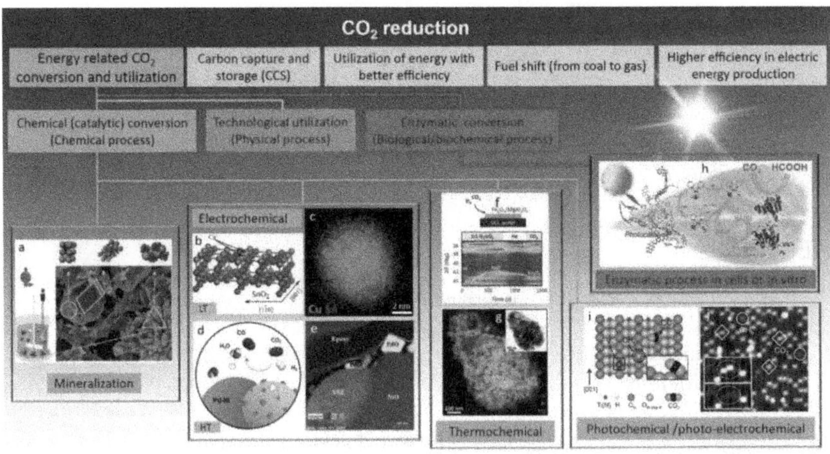

FIGURE 1.4 Main approaches for carbon dioxide reduction. (From Aresta, M. et al., *Philos. Trans. A Math. Phys. Eng. Sci.*, 371, 20120111, 2013; *Chem. Rev.*, 114, 1709–1742, 2014.) (a) Controlling the polymorphism of $CaCO_3$ to generate value-added mineral carbonation products (Reprinted with permission from Chang, R. et al., *ACS Sustain. Chem. Eng.*, 5, 1659–1667, 2017. Copyright 2017 American Chemical Society.); 2 × 3 supercells (b) and EELS elemental mapping (c) of the core/shell Cu/SnO_2 catalyst (Reprinted with permission from Li, Q. et al., *J. Am. Chem. Soc.*, 139, 4290–4293, 2017. Copyright 2017 American Chemical Society.); the reverse water gas shift (RWGS) reaction on Pd-Ni nanoalloys (d) and EDX analysis (e) of this nanoalloys on fuel-electrode substrate (Reprinted from *Appl. Catal. B: Environ.*, 200, Kim, S. et al., In situ nano-alloying Pd-Ni for economical control of syngas production from high-temperature thermo-electrochemical reduction of steam/CO_2, 265–273, Copyright 2017, with permission from Elsevier.); 2D in situ XRD map (f) and EDX analysis and corresponding STEM image (g) of $50Fe_2O_3/MgAl_2O_4$ used in super-dry reforming of CH_4 (From Buelens, L.C. et al., *Science*, 354, 449–452, 2016. Reprinted with permission of AAAS.); artificial photosynthesis for HCOOH generation from CO_2 (h) (Reprinted with permission from Yadav, R.K. et al., *J. Am. Chem. Soc.*, 134, 11455–11461, 2012. Copyright 2012 American Chemical Society.); behaviors of an oxygen vacancy defect (V_O), a bridging hydroxyl (OH_b), and an absorbed CO_2 molecule on reduced TiO_2 (110) surface (i) and STM image of this surface after CO_2 adsorption (j) (Reprinted with permission from Lee, J. et al., *J. Am. Chem. Soc.*, 133, 10066–10069, 2011. Copyright 2011 American Chemical Society.); LT, Low-temperature, HT, High-temperature.

Chemical conversions include approaches such as photocatalysis and photochemical catalysis[28–34] as well as thermochemical[35–39] and electrochemical[6,40–48] processes. Of these, electrochemical conversions are more attractive both in terms of energy efficiency and cost for several reasons: (1) the products that are achieved from chemical synthesis or energy storage can be applied to transportation fuels; (2) the conversion process is precise and controllable though reaction temperatures and electrode potentials; (3) the electrochemical conversion systems are modular, compact, highly efficient, on-demand, and scalable; (4) the electricity used in the process can be obtained from clean energy sources with zero CO_2 emissions such as solar, wind, water, nuclear, geothermal, tidal, and so on.[8,12,45] This book will provide detailed descriptions and discussions on the electrochemical conversion process. The more specific categories of electrochemical catalysis such as low-temperature and high-temperature electrochemical processes are classified and shown in Figure 1.4. As a typical high-temperature electrochemical process, a solid oxide electrolysis cell operated at high temperature will be the main focus and will be detailed in following some chapters especially.

REFERENCES

1. *World Energy Outlook 2017* (International Energy Agency, Paris, France, 2018).
2. Ebbesen, S. D., Knibbe, R., Mogensen, M. Co-electrolysis of steam and carbon dioxide in solid oxide cells. *Journal of the Electrochemical Society* **159**, F482–F489 (2012).
3. Sawyer, J. Man-made carbon dioxide and the "Greenhouse" effect. *Nature* **239**, 23–26 (1972).
4. Qiao, J., Liu, Y., Hong, F., Zhang, J. A review of catalysts for the electroreduction of carbon dioxide to produce low-carbon fuels. *Chemical Society Reviews* **43**, 631–75 (2014).
5. Kintisch, E. Report backs more projects to sequester CO_2 from coal. *Science* **315**, 1481 (2007).
6. Cyrille, C., Marc, R., Jean-Michel, S. Catalysis of the electrochemical reduction of carbon dioxide. *Chemical Society Reviews* **42**, 2423 (2013).
7. Chen, Y., Jing, Z., Miao, J., Zhang, Y., Fan, J. Reduction of CO_2 with water splitting hydrogen under subcritical and supercritical hydrothermal conditions. *International Journal of Hydrogen Energy* **41**, 9123–9127 (2015).
8. Gao, S. et al. Partially oxidized atomic cobalt layers for carbon dioxide electroreduction to liquid fuel. *Nature* **529**, 68–71 (2016).
9. Zheng, Y. et al. A review of high temperature co-electrolysis of H_2O and CO_2 to produce sustainable fuels using solid oxide electrolysis cells (SOECs): Advanced materials and technology. *Chemical Society Reviews* **46**, 1427–1463 (2017).
10. Liu, C., Colón, B. C., Ziesack, M., Silver, P. A., Nocera, D. G. Water splitting-biosynthetic system with CO_2 reduction efficiencies exceeding photosynthesis. *Science* **352**, 1210–1213 (2016).
11. Banerjee, A., Dick, G. R., Yoshino, T., Kanan, M. W. Carbon dioxide utilization via carbonate-promoted C-H carboxylation. *Nature* **531**, 215–219 (2016).
12. Christina, W. L., Jim, C., Matthew, W. K. Electroreduction of carbon monoxide to liquid fuel on oxide-derived nanocrystalline copper. *Nature* **508**, 504–507 (2014).
13. Aresta, M., Dibenedetto, A., Angelini, A. The use of solar energy can enhance the conversion of carbon dioxide into energy-rich products: Stepping towards artificial photosynthesis. *Philosophical Transactions A Mathematical, Physical, and Engineering Sciences* **371**, 20120111 (2013).
14. Aresta, M., Dibenedetto, A., Angelini, A. Catalysis for the valorization of exhaust carbon: From CO_2 to chemicals, materials, and fuels. Technological use of CO_2. *Chemical Reviews* **114**, 1709–1742 (2014).
15. Schwander, T., Borzyskowski, L., Burgener, S., Cortina, N. S., Erb, T. J. A synthetic pathway for the fixation of carbon dioxide *in vitro*. *Science* **354** (6314), 900–9004 (2016).
16. Antonio, J. M. Towards sustainable fuels and chemicals through the electrochemical reduction of CO_2: Lessons from water electrolysis. *Green Chemistry* **17**, 5114–5130 (2015).
17. Wagner, T., Ermler, U., Shima, S. The methanogenic CO_2 reducing-and-fixing enzyme is bifunctional and contains 46 [4Fe-4S] clusters. *Science* **354**, 114–117 (2016).

18. Lee, S. Y., Lim, S. Y., Seo, D., Lee, J., Chung, T. D. Light-driven highly selective conversion of CO_2 to formate by electrosynthesized enzyme/cofactor thin film electrode. *Advanced Energy Materials* **6**, 1502207 (2016).
19. Beller, M., Bornscheuer, U. T. CO_2 fixation through hydrogenation by chemical or enzymatic methods. *Angewandte Chemie International Edition* **53**, 4527–4528 (2014).
20. Schuchmann, K., Müller, V. Direct and reversible hydrogenation of CO_2 to formate by a bacterial carbon dioxide reductase. *Science* **342**, 1382–1385 (2013).
21. Ji, L. et al. CO_2 sequestration by direct mineralisation using fly ash from Chinese Shenfu coal. *Fuel Processing Technology* **156**, 429–437 (2017).
22. Chang, R., Choi, D., Kim, M. H., Park, Y. Tuning crystal polymorphisms and structural investigation of precipitated calcium carbonates for CO_2 mineralization. *ACS Sustainable Chemistry & Engineering* **5**, 1659–1667 (2017).
23. Yang, W., Zaoui, A. Mineralization of CO_2 in hydrated calcium Montmorillonite. *Physica A: Statistical Mechanics and its Applications* **464**, 191–197 (2016).
24. Yuen, Y. T., Sharratt, P. N., Jie, B. Carbon dioxide mineralization process design and evaluation: Concepts, case studies, and considerations. *Environmental Science and Pollution Research* **23**, 22309–22330 (2016).
25. Helwani, Z., Wiheeb, A. D., Kim, J., Othman, M. R. *In-situ* mineralization of carbon dioxide in a coal- fired power plant. *Energy Sources, Part A: Recovery, Utilization, and Environmental Effects* **38**, 606–611 (2016).
26. Xie, H. et al. Feedstocks study on CO_2 mineralization technology. *Environmental Earth Sciences* **75**, 615 (2016).
27. Bhardwaj, R., Van Ommen, J. R., Nugteren, H. W., Geerlings, H. Accelerating natural CO_2 mineralization in a fluidized bed. *Industrial & Engineering Chemistry Research* **55**, 2946–2951 (2016).
28. Yang, M., Xu, Y. Photocatalytic conversion of CO_2 over graphene-based composites: Current status and future perspective. *Nanoscale Horizons* **1**, 185–182 (2016).
29. Kan L., Bosi P., Tianyou P. Recent advances in heterogeneous photocatalytic CO_2 conversion to solar fuels. *ACS Catalysis* **6**, 7485–7527 (2016).
30. Ong, W., Tan, L., Chai, S., Yong, S. Graphene oxide as a structure-directing agent for the two-dimensional interface engineering of sandwich-like graphene-g-C_3N_4 hybrid nanostructures with enhanced visible-light photoreduction of CO_2 to methane. *Chemical Communications* **51**, 858–861 (2015).
31. Low, J., Yu, J., Ho, W. Graphene-based photocatalysts for CO_2 reduction to solar fuel. *The Journal of Physical Chemistry Letters* **6**, 4244–4251 (2015).
32. Kumar, P., Bansiwal, A., Labhsetwar, N., Jain, S. L. Visible light assisted photocatalytic reduction of CO_2 using a graphene oxide supported heteroleptic ruthenium complex. *Green Chemistry* **17**, 1605–1609 (2015).
33. Tu, W., Zhou, Y., Zou, Z. Photocatalytic conversion of CO_2 into renewable hydrocarbon fuels: State-of-the-art accomplishment, challenges, and prospects. *Advanced Materials* **26**, 4607–4626 (2014).
34. Ong, W., Tan, L., Chai, S., Yong, S., Mohamed, A. R. Self-assembly of nitrogen-doped TiO_2 with exposed {001} facets on a graphene scaffold as photo-active hybrid nanostructures for reduction of carbon dioxide to methane. *Nano Research* 7, 1528–1547 (2014).
35. Atsonios, K., Panopoulos, K. D., Kakaras, E. Thermocatalytic CO_2 hydrogenation for methanol and ethanol production: Process improvements. *International Journal of Hydrogen Energy* **41**, 792–806 (2016).
36. Li, Y., Chan, S. H., Sun, Q. Heterogeneous catalytic conversion of CO_2: A comprehensive theoretical review. *Nanoscale* 7, 8663–8683 (2015).

37. Pakhare, D., Spivey, J. A review of dry (CO_2) reforming of methane over noble metal catalysts. *Chemical Society Reviews* **43**, 7813–7837 (2014).
38. Grabow, L. C., Mavrikakis, M. Mechanism of methanol synthesis on Cu through CO_2 and CO hydrogenation. *ACS Catalysis* **1**, 365–384 (2011).
39. Ashcroft, A. T., Cheetham, A. K. Partial oxidation of methane to synthesis gas using carbon dioxide. *Nature* **352**, 225–226 (1991).
40. Ganesh, I. Electrochemical conversion of carbon dioxide into renewable fuel chemicals-The role of nanomaterials and the commercialization. *Renewable and Sustainable Energy Reviews* **59**, 1269–1297 (2016).
41. Kortlever, R., Shen, J., Schouten, K. J. P., Calle-Vallejo, F., Koper, M. T. M. Catalysts and reaction pathways for the electrochemical reduction of carbon dioxide. *The Journal of Physical Chemistry Letters* **6**, 4073–4082 (2015).
42. Jonathan, A., Manuel, A. Towards the electrochemical conversion of carbon dioxide into methanol. *Green Chemistry* **17**, 2304–2324 (2015).
43. Spinner, N. S., Vega, J. A., Mustain, W. E. Recent progress in the electrochemical conversion and utilization of CO_2. *Catalysis Science & Technology* **2**, 19–28 (2012).
44. Finn, C., Schnittger, S., Yellowlees, L. J., Love, J. B. Molecular approaches to the electrochemical reduction of carbon dioxide. *Chemical Communications* (*Camb*) **48**, 1392–1399 (2012).
45. Whipple, D. T., Kenis, P. J. A. Prospects of CO_2 utilization via direct heterogeneous electrochemical reduction. *The Journal of Physical Chemistry Letters* **1**, 3451–3458 (2010).
46. Whipple, D. T., Finke, E. C., Kenis, P. J. A. Microfluidic reactor for the electrochemical reduction of carbon dioxide: The effect of pH. *Electrochemical and Solid-State Letters* **13**, B109–B111 (2010).
47. Delacourt, C., Ridgway, P. L., Kerr, J. B., Newman, J. Design of an electrochemical cell making syngas ($CO + H_2$) from CO_2 and H_2O reduction at room temperature. *Journal of the Electrochemical Society* **155**, B42–B49 (2008).
48. Tanaka, K., Ooyama, D. Multi-electron reduction of CO_2 via Ru CO_2, C(O)OH, CO, CHO, and CH_2OH species. *Coordination Chemistry Reviews* **226**, 211–218 (2002).
49. Li, Q. et al. Tuning Sn-catalysis for electrochemical reduction of CO_2 to CO via the core/shell Cu/SnO_2 structure. *Journal of the American Chemical Society* **139**, 4290–4293 (2017).
50. Kim, S. et al. *In-situ* nano-alloying Pd-Ni for economical control of syngas production from high-temperature thermo-electrochemical reduction of steam/CO_2. *Applied Catalysis B: Environmental* **200**, 265–273 (2017).
51. Buelens, L. C., Galvita, V. V., Poelman, H., Detavernier, C., Marin, G. B. Super-dry reforming of methane intensifies CO_2 utilization via Le Chatelier's principle. *Science* **354**, 449–452 (2016).
52. Yadav, R. K. et al. A photocatalyst-enzyme coupled artificial photosynthesis system for solar energy in production of formic acid from CO_2. *Journal of the American Chemical Society* **134**, 11455–11461 (2012).
53. Lee, J., Sorescu, D. C., Deng, X. Electron-induced dissociation of CO_2 on TiO_2 (110). *Journal of the American Chemical Society* **133**, 10066–10069 (2011).

2 Fundamentals of CO_2 Structure, Thermodynamics, and Kinetics

2.1 MOLECULAR STRUCTURE OF CO_2

CO_2 is a linear triatomic molecule with a molecular weight of 44 Dalton, as shown in Figure 2.1. In the molecule, carbon (C) and oxygen (O) atoms are held together through bonds formed by sharing electrons, and these bonds possess strong electrical affinities. They possess a circular axial symmetry, a center of symmetry, and a horizontal plane of symmetry.[1] O atoms are better in grabbing the electron pair than that of C atoms and therefore the electrons are pulled partially away from the C atom, resulting in the C atom being in a relatively low energy state.[2]

The bond length between C and O in molecular CO_2 is 116 pm. This is shorter than that of typical double bonds (e.g., C=O in acetaldehyde, 124 pm) and close to that of triple bonds (e.g., C≡O in carbon monoxide, 112.8 pm). The bond energy of C=O in CO_2 is 803 kJ/mol,[3] which is much higher than that of O–H in H_2O molecules (463 kJ/mol). These properties seem to be ascribed to the presence of two σ bonds and two π bonds (π_3^4) in CO_2, as shown in Figure 2.2; two *sp*-hybridization orbitals are formed by the mixing of 2*s* and $2p_x$ orbitals in a CO_2 molecular.[4] Each *sp*-hybridization orbital of C and a $2p_z$ orbital of O then combine to form an σ bond, and each π bond (π_3^4) is formed with a $2p_y$ or $2p_z$ orbital, two 2*px* and a 2*py* orbital of O. The combination of high symmetry, low polarity and high bond energy leads to the high stability of CO_2.

There are three vibrational modes in CO_2: a bend vibration and two stretch vibrations (symmetric and anti-symmetric), as shown in Figure 2.3. The corresponding frequencies of CO_2 with various phases (g, s, and aq.) are listed in this figure. In addition to infrared (IR) spectroscopy, nuclear magnetic resonance (NMR) spectroscopy can be widely employed for diagnosis and even quantitative determinations of CO_2. ^{13}C NMR resonance of CO_2 in non-polar solvents such as benzene or toluene occurs around 124 ppm, and the resonance is close to 125 ppm when CO_2 is performed in aqueous solutions.[5,6] This can be used for the quantification of free CO_2 molecules.

2.2 THERMODYNAMICS OF CO_2

CO_2 is one of the most common final products in any combustion reaction, including chemical or biological.[7] CO_2 conversion is difficult from a thermodynamic standpoint. Figure 2.4 shows the standard Gibbs free-energy of formation ("B" represents

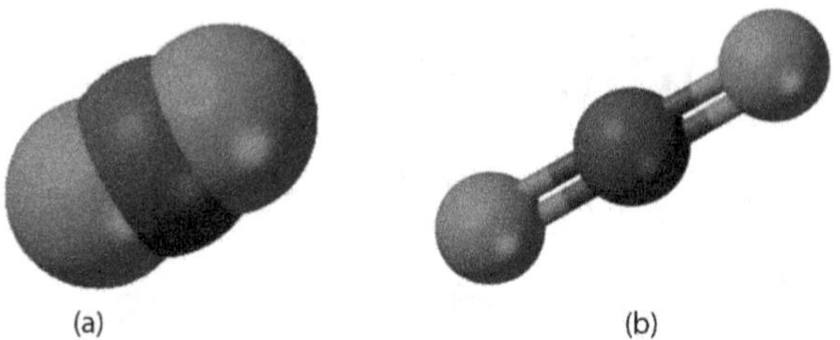

FIGURE 2.1 Structure of CO_2 molecule with space filling model (a) and ball-and-stick model (b). The light gray ball is for O atom and the dark gray one is for C; the bond length is 116 pm; the bond angle is 180°.

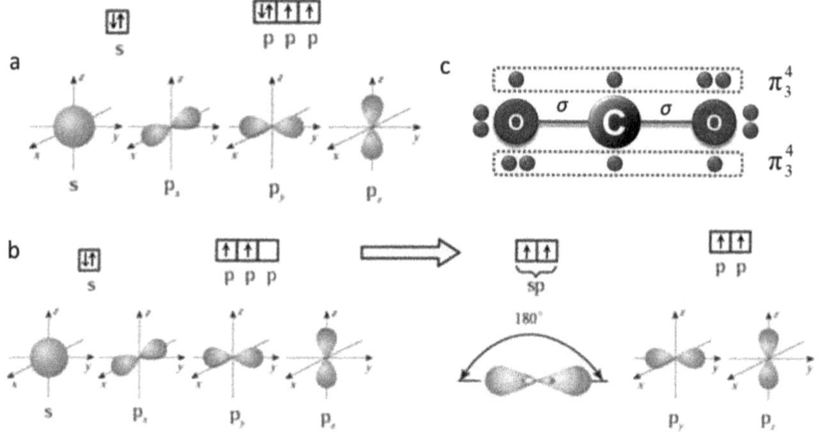

FIGURE 2.2 Hybridization form of oxide (a) and carbon (b) and the bonding style of the CO_2 molecule. (From Ishida, S. et al., *Nature*, 421, 725–727, 2003.)

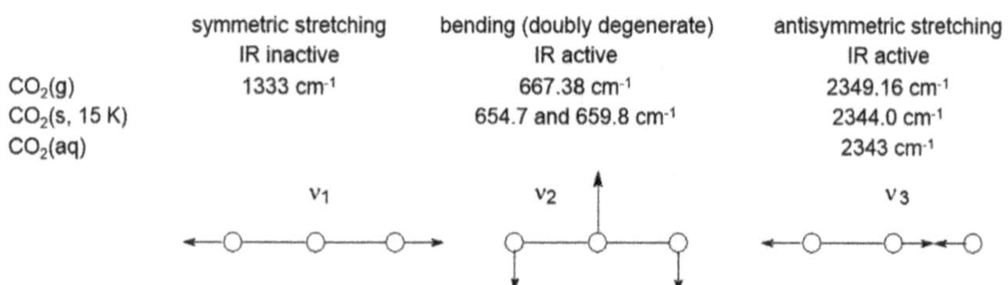

FIGURE 2.3 Normal vibration modes and corresponding frequencies of CO_2 with various phases (g, s, and aq.) in its ground state. (Reproduced from *J. Catal.*, 343, Aresta, M. et al., State of the art and perspectives in catalytic processes for CO_2 conversion into chemicals and fuels: The distinctive contribution of chemical catalysis and biotechnology, 2–45, Copyright 2016, with permission from Elsevier.)

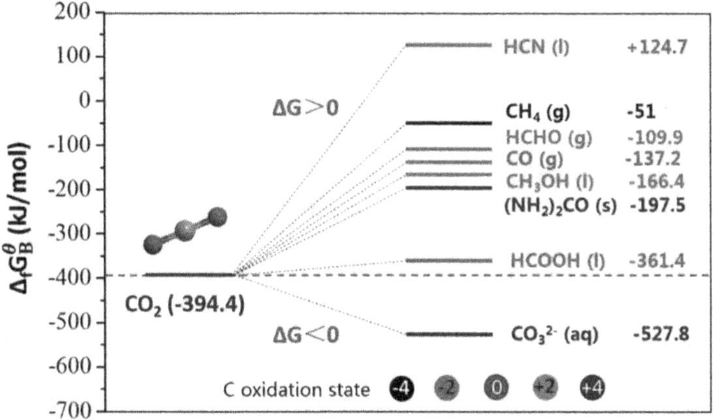

FIGURE 2.4 Formation Gibbs free energy of some C1 species. (Reprinted with permission from Aresta, M. et al., *Chem. Rev.*, 114, 1709–1742, 2014. Copyright 2014 American Chemical Society.)

one of the reactants or products in the reaction, "f" represents the "formation", and the "θ" indicates the standard state of 298.15 K and 101.325 kPa) of C1 species with different oxidation states of carbon. In terms of thermodynamics, it indicates that the CO_2 conversion reaction tends to be endergonic when the O/C ratio of one product is less than 2[7]. On the contrary, the reaction tends to be exergonic when CO_2 is kept in its +4 oxidation state with various O/C ratios in the products (such as carbonates or urea). In regard to specific conversion reactions as listed in Table 2.1, the Gibbs free energy change (ΔG) of these conversion reactions from CO_2 to other products

TABLE 2.1
Calculated Thermodynamic Parameters (ΔG^{θ}, ΔH^{θ}, and ΔS^{θ}) of Some CO_2 Conversion Reactions

Product	Net Reaction	ΔG^{θ} (kJ·mol^{-1})	ΔH^{θ} (kJ·mol^{-1})	ΔS^{θ} (J·mol^{-1}·K^{1})
Hydrogen	$H_2O \rightleftharpoons H_2\uparrow + 0.5O_2\uparrow$	237.17	285.83	163.30
Carbon monoxide	$CO_2 + H_2O \rightleftharpoons CO\uparrow + H_2O + 0.5O_2\uparrow$	257.38	283.01	86.55
Formic acid	$CO_2 + H_2O \rightleftharpoons HCOOH + 0.5O_2\uparrow$	269.86	254.34	–52.15
Formaldehyde	$CO_2 + H_2O \rightleftharpoons HCHO + O_2\uparrow$	528.94	570.74	140.25
Methanol	$CO_2 + 2H_2O \rightleftharpoons CH_3OH + 1.5O_2\uparrow$	701.87	725.97	80.85
Ethanol	$2CO_2 + 3H_2O \rightleftharpoons C_2H_5OH + 3O_2\uparrow$	1325.56	1366.90	138.75
Propanol	$3CO_2 + 4H_2O \rightleftharpoons C_3H_7OH + 4.5O_2\uparrow$	1962.94	2021.24	195.65
Methane	$CO_2 + 2H_2O \rightleftharpoons CH_4\uparrow + 2O_2\uparrow$	818.18	890.57	242.90
Ethane	$2CO_2 + 3H_2O \rightleftharpoons C_2H_6\uparrow + 3.5O_2\uparrow$	1468.18	1560.51	309.80
Ethylene	$2CO_2 + 2H_2O \rightleftharpoons C_2H_4\uparrow + 3O_2\uparrow$	1331.42	1411.08	267.30

Source: Yang, M. and Xu, Y., *Nanoscale Horiz.*, 1, 185–200, 2016.

as shown under a standard state is positive ($\Delta G^\theta > 0$), showing that the conversion reactions are non-spontaneous. Additionally, with respect to the standard enthalpy changes ($\Delta H^\theta > 0$) of various products listed in Table 2.1, the ΔH^θ values of all these conversion reactions are also positive, demonstrating that these CO_2 conversion processes are endothermic under a standard state.

Thermodynamic parameters (Gibbs free energy change [ΔG], enthalpy change [ΔH], and entropy change [ΔS]) are a function of temperature. ΔG of CO_2 conversion reactions at 298.15 K or even at other temperatures can be calculated with reference to some thermodynamic equations (Equations 2.1 through 2.6) and parameters of CO_2 (Table 2.2) and other reactants or products.[9] Even with exothermic reactions ($\Delta H < 0$), the results may be endergonic ($\Delta G > 0$) due to the effects of entropy (S) and temperature (T) in accordance with Equation (2.1). Therefore, negative thermodynamics can make the conversion unfeasible even at higher temperature.

$$\Delta G = \Delta H - T\Delta S \tag{2.1}$$

$$\Delta H\,(298.15\text{ K}) = \sum_{B} \nu_B \Delta_f H_B(298.15\text{ K}) \tag{2.2}$$

$$\Delta S\,(298.15\text{ K}) = \sum_{B} \nu_B S_B(298.15\text{ K}) \tag{2.3}$$

$$\Delta H\,(T) = \Delta H\,(298.15\text{ K}) + \int_{298.15\text{K}}^{T} \Delta C_p dT \tag{2.4}$$

$$\Delta S\,(T) = \Delta S\,(298.15\text{ K}) + \int_{298.15\text{K}}^{T} \Delta C_p dT \tag{2.5}$$

$$\Delta C_p = \sum_{B} \nu_B C_p(B) \tag{2.6}$$

where in Equations (2.1) through (2.6), ΔH is the enthalpy change (kJ·mol^{-1}); $\Delta_f H_B$ is the enthalpy of formation for B (kJ·mol^{-1}); S is the entropy (kJ·mol^{-1} K^{-1}); ΔS is the entropy change (kJ·mol^{-1} K^{-1}); T is the temperature (K); ΔG is the Gibbs free energy change (kJ·mol^{-1}); C_p is the heat capacity (J·mol^{-1}K^{-1}); and ΔC_p is the heat capacity change (J·mol^{-1} K^{-1}).

TABLE 2.2
Thermodynamic Parameters of CO_2 at 298.15 K and 101.325 kPa (Here the Definitions of Other Parameter Symbols Are Presented Above in the Text)

	$\Delta_f G^\theta$ (kJ·mol^{-1})	$\Delta_f H^\theta$ (kJ·mol^{-1})	S^θ (kJ·mol^{-1}·K^{1})	Heat Capacity (C_p) (J·mol^{-1}·K^{-1})
CO_2(g)	−394.4	−393.5	0.214	44.141 + 0.00904T − 853500/T^2

Source: Ihsan, B., ed., *Thermochemical Data of Pure Substances*, WILEY-VCH Verlag GmbH, New York, 1993.

For an equilibrium reaction, the ΔG^θ can be closely related to the standard equilibrium constant K_f^θ (Equation 2.7), where R is the molar gas constant ($J \cdot mol^{-1} \cdot K^{-1}$). In addition, the relationship between temperatures and ΔG^θ is needed to obtain the equilibrium constant at a reaction temperature. This relationship can be expressed with reference to Van't Hoff equation (Equation 2.8), where *p* represents the constant pressure. For CO_2 conversion, the heat of the reaction and the equilibrium constant can be calculated as shown in Table 2.3; it is seen that increasing pressures, for example, from 1 to 6 MPa, can increase the equilibrium conversion process.[10] However, increasing temperature in a certain range may slow down some equilibrium conversions (e.g., production of CH_4, C_2H_4, CH_3OH, C_2H_5OH), although increasing temperature can speed up other conversion processes such as the production of HCHO and HCOOH.[10] Therefore, higher temperatures are not always conducive to CO_2 conversion.

$$\Delta G^\theta = -RT \ln K_f^\theta \tag{2.7}$$

$$\left(\frac{\partial \ln K_f^\theta}{\partial T} \right)_P = \frac{\Delta H^\theta}{RT^2} \tag{2.8}$$

TABLE 2.3
Heat and Equilibrium Constants of Several CO_2 Conversion Reactions at Ideal State

Reaction	Tem. (°C)	ΔH^θ, (kJ/mol)	K_f^θ
$CO_2 + 4H_2 = CH_4 + 2H_2O$	25	−0.16486E+3	0.82755E+20
	100	−0.16819E+3	0.11260E+15
	200	−0.17211E+3	0.10359E+10
$2CO_2 + 6H_2 = C_2H_4 + 4H_2O$	25	−0.12791E+3	0.12223E+11
	100	−0.13230E+3	0.32014E+6
	200	−0.13720E+3	0.32913E+2
$CO_2 + 3H_2 = CH_3OH + H_2O$	25	−0.49449E+2	0.26804
	100	−0.52464E+2	0.43142E−2
	200	−0.55839E+2	0.10792E−3
$2CO_2 + 6H_2 = C_2H_5OH + 3H_2O$	25	−0.17314E+3	0.28230E+12
	100	−0.17828E+3	0.18262E+6
	200	−0.18383E+3	0.79786
$CO_2 + 2H_2 = HCHO + H_2O$	25	0.35739E+2	0.16345E−9
	100	0.41270E+2	0.36155E−8
	200	0.51494E+2	0.81597E−7
$CO_2 + H_2 = HCOOH$	25	0.14881E+2	0.25001E−7
	100	0.13797E+2	0.48226E−6
	200	0.12982E+2	0.68332E−6

Source: Gao, F. et al., *J. Nat. Gas. Chem.*, 10, 24–33, 2001.

2.3 KINETICS OF CO_2 CONVERSION

In CO_2 conversion reactions, the C atom in the molecule is normally first attacked. In an addition reaction, the nucleophile first attacks the C atom with a partially positive charge (Figure 2.5). The nucleophile can be a neutral species with a lone pair of electrons (e.g., an amine) and can possess a C-metal σ bond (e.g., a Grignard reagent) or an electron-rich π bond (e.g., a phenolate).[6] In such situations, the chemical reactivity of CO_2 depends on the polarization of O and C bonds, and the kinetics is dominated by both CO_2 and nucleophiles.

In terms of the chemistry of CO_2, the other key feature is its coordination to metals. The electron distribution and molecular geometry within the CO_2 molecule can be changed significantly via the coordination of CO_2 to a metal. Generally, the electrons in the bonding orbitals of CO_2 can act as an electron donor (referred to as μ_a–η^b, where "a" donates receptor number, and "b" donates donor number), weakening the bond between C and O, and leading to increased chemical reactivity. Several CO_2-metal complex geometries are shown in Figure 2.6.

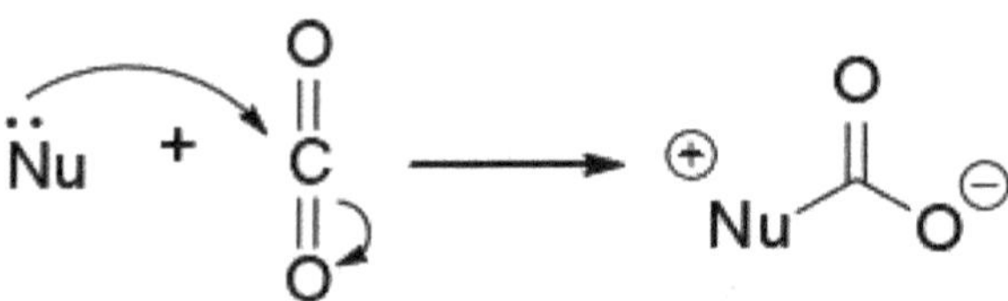

FIGURE 2.5 Addition reaction of carbon dioxide with nucleophiles. (From North, M., Chapter 1: What is CO_2? Thermodynamics basic reactions and physical chemistry, in Styring, P. et al. (Eds.), *Carbon Dioxide Utilisation: Closing the Carbon Cycle*, Elsevier, Oxford, UK, 2015.)

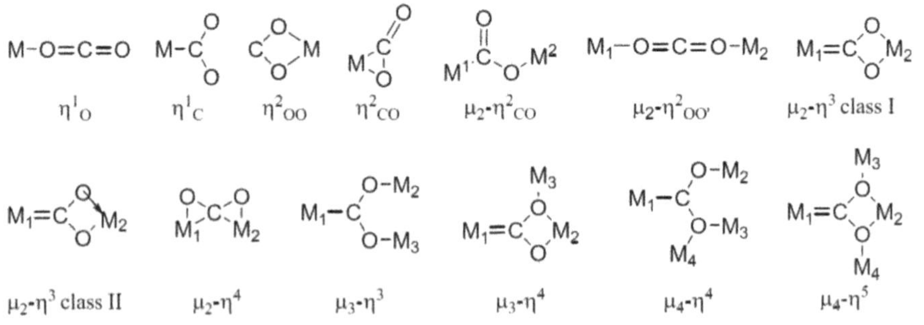

FIGURE 2.6 Metal complex geometries of CO_2. η^1_C is for the coordination of a metal to one C atom; η^1_O is for the coordination of a metal to one of the O atoms; the n in μ_n is for the number of metals; and the m in η^m is for the number of M-C/O bands. (Reprinted from North, M., Chapter 1: What is CO_2? Thermodynamics basic reactions and physical chemistry, in Styring, P. et al. (Eds.), *Carbon Dioxide Utilisation: Closing the Carbon Cycle*, Elsevier, Oxford, UK, Copyright 2015, with permission from Elsevier.)

From a thermodynamic standpoint, as mentioned above, a reaction is spontaneous when its Gibbs free energy change is favorable ($\Delta G > 0$). Nevertheless, some reactions may still be non-spontaneous due to high activation energies. Consequently, these CO_2 conversion reactions require a suitable catalyst. Catalysts can lower activation energies and allow or even accelerate the reaction at suitable temperatures. For example, in the production of ethylene carbonate from ethylene oxide and CO_2, even with remarkably favorable ΔG (−56 kJ/mol) and ΔH (−144 kJ/mol), a suitable catalyst is required for the reaction to occur. This can be interpreted with Figure 2.7 where four types of CO_2 conversion reactions (A–D) are defined. The reaction pathway energy diagrams are divided into two groups, also shown in Figure 2.7: (a) the processes of reactions A and B, both of which possess a negative Gibbs free energy change ($\Delta G < 0$), and (b) the processes of reactions C and D, both of which possess a positive Gibbs free energy change ($\Delta G > 0$).

In group (a) mentioned above, the activation energy (ΔG^*) of reaction B is much higher than that of A, thus, reaction A is expected to occur without the need of a catalyst whereas a catalyst is necessary for reaction B due to its significant activation energy. The synthesis of sodium salicylate from CO_2 and sodium phenolate is a good example of reaction A and the production of ethylene carbonate from CO_2, and ethylene oxide is a good example of reaction B. One thing to note here is that simply promoting the reaction temperature may not always be effective in overcoming the activation barrier as the ΔG will be positive if the ΔS of the reaction is negative.

In group (b), with regard to reaction C in Figure 2.7, the ΔG^* is slightly higher than $\Delta_r G$ (the Gibbs free energy change of CO_2 conversion reactions), thus catalysts are not needed, but the removal of the formed products is challenging. Therefore, the reaction equilibrium will shift positively or even not occur. Compared with

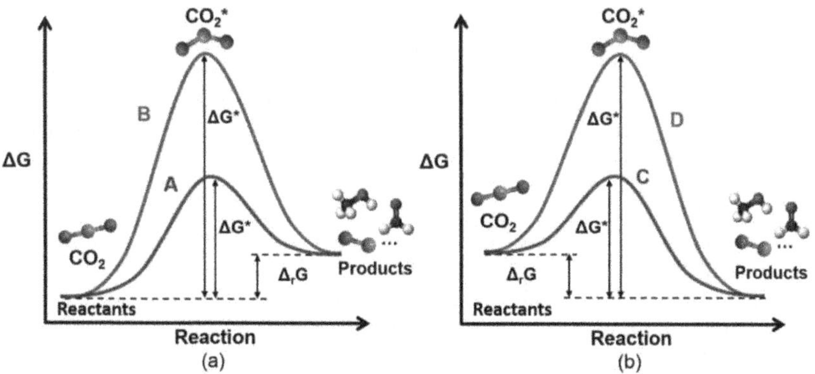

FIGURE 2.7 (a, b) Reaction pathway energy diagrams for reactions of carbon dioxide. Here $\Delta_r G$ is the Gibbs free energy change of CO_2 conversion reactions. (Reprinted from North, M., Chapter 1: What is CO_2? Thermodynamics basic reactions and physical chemistry, in Styring, P. et al. (Eds.), *Carbon Dioxide Utilisation: Closing the Carbon Cycle*, Elsevier, Oxford, UK, Copyright 2015, with permission from Elsevier.)

reaction C, the ΔG^* of reaction D is much higher than its $\Delta_r G$. This is the most challenging situation and unfortunately it is common in CO_2 conversions. Consequently, an effective catalyst for reducing ΔG^* is necessary, one that can facilitate the reaction at a lower and more suitable temperature, because lower reaction temperatures can contribute to a decrease in $\Delta_r G$ and an increase in equilibrium constant for the reaction. It is recognized that poor catalytic activity and/or insufficient stability of catalysts are the largest challenges in CO_2 conversions (especially in CO_2 electroreduction).[11] Thus, CO_2 conversion is also difficult from a kinetics standpoint.[12]

2.4 SUMMARY

CO_2 is thermodynamically and kinetically stable, making its conversions to other chemicals/fuels difficult. However, CO_2 can be attacked by nucleophiles and electron-donating reagents ascribe to the electron deficiency of carbonyl carbons.[13] Therefore, the conversion of CO_2 is achievable in practice despite technological barriers.[14] Typical approaches, especially for the production of fuels with high energy exchange, are detailed in following chapters.

REFERENCES

1. Armenise, I., Kustova, E. V. State-to-state models for CO_2 molecules: From the theory to an application to hypersonic boundary layers. *Chemical Physics* **415**, 269–281 (2013).
2. Rothman, L. S., Hawkins, R. L., Wattson, R. B. Energy levels, intensities, and linewidths of atmospheric carbon dioxide bands. *Journal of Quantitative Spectroscopy & Radiative Transfer* **48**, 573–566 (1992).
3. Stanbury, M. et al. Mn-carbonyl molecular catalysts containing a redox-active phenanthroline-5, 6-dione for selective electro- and photoreduction of CO_2 to CO or HCOOH. *Electrochimica Acta* **240**, 288–299 (2017).
4. Ishida, S., Iwamoto, T., Kabuto, C. A stable silicon-based allene analogue with a formally sp-hybridized silicon atom. *Nature* **421**, 725–727 (2003).
5. Aresta, M., Dibenedetto, A., Quaranta, E. State of the art and perspectives in ca talytic processes for CO_2 conversion into chemicals and fuels: The distinctive contribution of chemical catalysis and biotechnology. *Journal of Catalysis* **343**, 2–45 (2016).
6. North, M. Chapter 1: What is CO_2? Thermodynamics basic reactions and physical chemistry. In Styring, P., Quadrelli, A., Armstrong, K. (Eds.). *Carbon Dioxide Utilisation: Closing the Carbon Cycle* (Elsevier, Oxford, UK, 2015).
7. Aresta, M., Dibenedetto, A., Angelini, A. Catalysis for the valorization of exhaust carbon: From CO_2 to chemicals, materials, and fuels. Technological use of CO_2. *Chemical Reviews* **114**, 1709–1742 (2014).
8. Yang, M., Xu, Y. Photocatalytic conversion of CO_2 over graphene-based composites: Current status and future perspective. *Nanoscale Horizons* **1**, 185–2 (2016).
9. Ihsan, B. (ed.) *Thermochemical Data of Pure Substances* (WILEY-VCH Verlag GmbH, New York, 1993).
10. Cao, F. et al. Thermodynamic analysis of CO_2 direct hydrogenation reaction. *Journal of Natural Gas Chemistry* **10**, 24–23 (2001).
11. Qiao, J., Liu, Y., Hong, F., Zhang, J. A review of catalysts for the electroreduction of carbon dioxide to produce low-carbon fuels. *Chemical Society Reviews* **43**, 631–675 (2014).

12. Gao, S. et al. Partially oxidized atomic cobalt layers for carbon dioxide electroreduction to liquid fuel. *Nature* **529**, 68–71 (2016).
13. Sakakura, T., Choi, J., Yasuda, H. Transformation of carbon dioxide. *Chemical Reviews* **107**, 2365–2387 (2007).
14. Aresta, M., Dibenedetto, A., Angelini, A. The use of solar energy can enhance the conversion of carbon dioxide into energy-rich products: Stepping towards artificial photosynthesis. *Philosophical Transaction A Mathematical Physical and Engineering Sciences* **371**, 20120111 (2013).

3 Enzymatic and Mineralized Conversion Process of CO_2 Conversion

3.1 ENZYMATIC CONVERSION OF CO_2

As one of the most widespread processes in nature, enzymatic conversion of CO_2 is absolutely essential for life and biological evolution. Enzymatic conversion of CO_2 is advantageous for its high efficiency and non-polluting nature thanks to the specificity and selectivity of the biological reaction.[1–3] The Calvin cycle, the 3-hydroxypropionate cycle, and the reductive citric acid cycle are the most common approaches of the enzymatic conversion of CO_2 in cells. As the most important biosynthetic cycle in nature, the Calvin cycle is exists in plants, cyanobacteria, algae, and so on.[4] The process of converting hydro-carbon chemicals/fuels (C_nH_m) to CO_2 is exothermal and easily facilitated due to a negative Gibbs free-energy ($\Delta G < 0$), while the conversion of CO_2 to carbohydrates ($C_nH_mO_x$) via photosynthesis is energy intensive ($\Delta G > 0$). Figure 3.1a shows the principle of photosynthesis in nature. Typical photosynthesis consists of a light reaction from H_2O to O_2 by daylight and a dark reaction (Calvin cycle) from CO_2 to organic substance like glucose.[5] Significantly, the committed step for enzymatic conversion of CO_2 is the Calvin cycle. The Calvin cycle includes three stages, as shown in Figure 3.1b. First, the first key enzyme 1,5-bisphosphate ribulose bisphosphate carboxylaset accelerates the reaction from CO_2 and 1,5-bisphosphate ribulose to 3-phosphoglycerate. Then, 1,3-diphosphoglycerate is generated from 3-phosphoglycerate and a Pi, further producing 3-phosphate glyceraldehyde by the second key enzyme phosphoglyceral-dehyde dehydrogenase. Finally, 5-phosphate ribulose is produced from 5/6 of the generated 3-phosphate glyceraldehyde and further transformed into 1,5-bisphosphate ribulose, which forms the reactant with CO_2 in the next cycle. At the same time, the extra 3-phosphate glyceraldehyde transforms to organic substances such as sugar, fatty acids, and amino acids as shown in Figure 3.1b. Meanwhile, the nicotinamide adenine dinucleotide phosphate (NADPH) acts as a proton carrier and a hydrogen source, and the transformation of adenosine triphosphate (ATP) into adenosine diphosphate (ADP) provides energy for enzymatic conversion of CO_2.

Except for CO_2 conversion in cells, there are different enzymatic conversion approaches with other enzymes for producing chemicals/fuels, as listed in Table 3.1.[1] According to the types of catalytic enzyme, the conversion approaches are sorted into two large categories, namely, single enzyme and multienzyme systems.

In regard to CO_2 conversion with single enzyme catalytic, Reda et al. have developed a typical approach for formate production facilitated by an electro-active enzyme

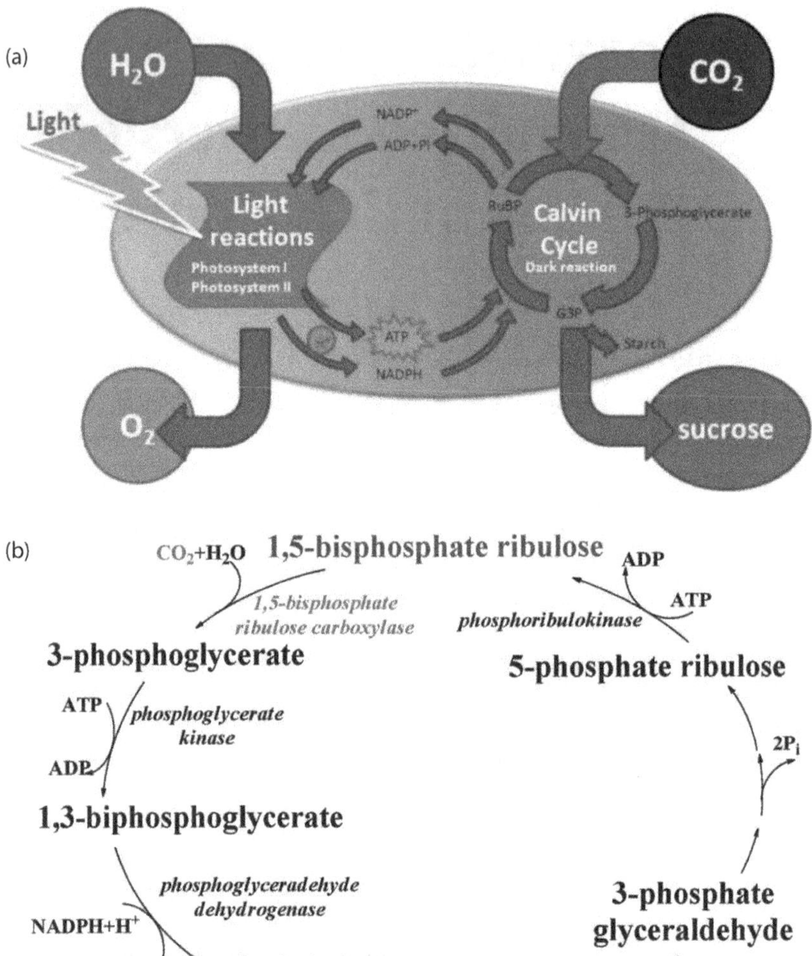

FIGURE 3.1 (a) Process of photosynthesis. (Reprinted from *Appl. Catal. B Environ.*, Lee, W. et al., A novel twin reactor for CO_2 photoreduction to mimic artificial photosynthesis, 132–133, 445–451, Copyright 2013, with permission from Elsevier.) (b) The Calvin cycle. (Shi, J. et al., *Chem. Soc. Rev.*, 44, 5981–6000, 2015, Reproduced by permission of The Royal Society of Chemistry.)

TABLE 3.1
Some Products from CO_2 Conversion with Different Enzymes In Vitro

Products	Specific Enzyme[a]	References
Formate (HCOO–)	Formate dehydrogenase (F_{ate}DH)	7–9
Formate (HCOO–)	Carbon dioxide reductase	10,11
Carbon monodioxide (CO)	Carbon monodioxide dehydrogenase (CODH)	12–16
Methane (CH_4)	Remodeled nitrogenase	17,18
Bicarbonate (or minerals) (HCO_3^-)	Carbonic anhydrase (CA)	19–23
Biodegradable chemicals (e.g., $CH_3COCOOH$)	Decarboxylases (e.g., pyruvate decarboxylase)	1,24
Methanol (CH_3OH)	Multiple dehydrogenases (e.g., F_{ate}DH + F_{ald}DH + ADH)	25,26

Source: Shi, J. et al., *Chem. Soc. Rev.*, 44, 5981–6000, 2015.
[a] F_{ald}DH, formaldehyde dehydrogenase; ADH, alcohol dehydrogenase.

(tungsten-containing formate dehydrogenase enzyme), as shown in Figure 3.2a.[27] The conversion reaction is reversible, where the process occurs on the formate dehydrogenase enzyme when two electrons transfer from the electrode to the tungsten active site. Yadav et al.[9] have proposed another approach with four stages for CO_2 conversion to formic acid in an artificial photosynthesis system facilitated by photocatalysts and enzymes simultaneously, as shown in Figure 3.2b. First, the photon absorption occurs as an exchange between located orbitals around a chromophore. Second, a rhodium (Rh) complex is reduced after receiving an electron. Third, the reduced rhodium complex helps to extract a proton, with some electrons and a hydride transferred to nicotinamide adenine dinucleotide (NAD^+) for reduced nicotinamide

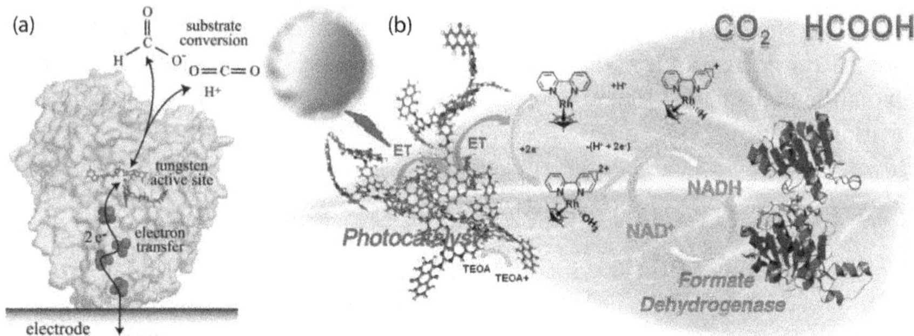

FIGURE 3.2 Schematics of (a) a typical approach for formate production facilitated by an electro-active enzyme (Reprinted with permission from Reda, T. et al., 2008. Reversible interconversion of carbon dioxide and formate by an electroactive enzyme. *Proc. Natl. Acad. Sci. USA*, 105, 10654–10658. Copyright 2008 National Academy of Sciences, U.S.A.) and (b) the artificial photosynthesis for HCOOH generation from CO_2. (Reprinted with permission from Yadav, R.K. et al., *J. Am. Chem. Soc.*, 134, 11455–11461, 2012 American Chemical Society.)

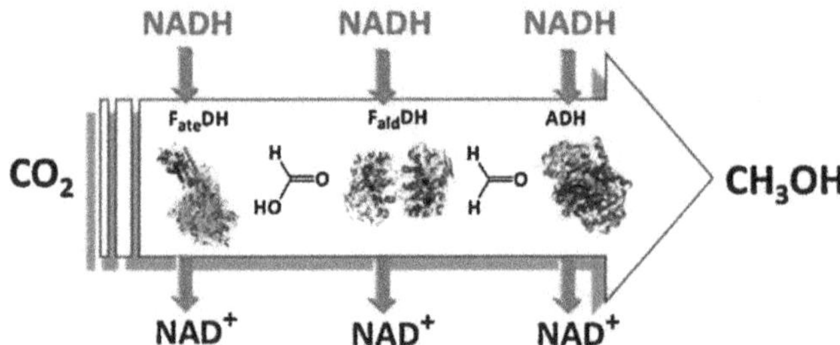

FIGURE 3.3 The schematic of overall process of CO_2 conversion to CH_3OH with enzymes T_{ate}DH-formate dehydrogenase, T_{ald}DH-formaldehyde dehydrogenase and ADH-alcohol dehydrogenase. (Reprinted from *J CO_2 Util.*, Aresta, M. et al., 3–4, The changing paradigm in CO_2 utilization, 65–73, Copyright 2013, with permission from Elsevier.)

adenine dinucleotide (NADH) generation. Finally, the generated NADH is used up for the CO_2 conversion to HCOOH with formate dehydrogenase.

Formate or HCOOH are the common products in single enzyme systems, while other products are CO, CH_4, HCO_3^-, and other chemicals with various enzymes in vitro, as listed in Table 3.1. The typical instance in the multienzyme system is the conversion from CO_2 to CH_3OH. This conversion process includes three steps with corresponding enzymes: the first step is the transformation from CO_2 into HCOOH, which is facilitated by formatdehydrogenase (T_{ate}DH enzymes). Second, formaldehydedehydro (T_{ald}DH enzymes) further change HCOOH into HCHO, followed by the transformation of HCHO into CH_3OH facilitated by alcohol dehydrogenase (ADH). As shown in Figure 3.3, NADH supplies electrons for the conversion reaction, and the overall conversion process is accompanied by three $2e^-$ transfers.

In this section, we have submitted a series of conversion processes for enzymatic conversions of CO_2. Although these processes are environmentally friendly and highly efficient methods to transform CO_2 into chemicals/fuels, it should be noted that more research about enzyme conversion process is needed to further promote the conversion effciency and facilitate the reaction kinetics, especially for industrialization.

3.2 MINERALIZATION PROCESS OF CO_2 CONVERSION

The mineralization of CO_2 is valued for the large-scale CO_2 storage and conversion of CO_2 to chemicals.[28] The complex conversion process consists of carbonation of silicide and formation of stable carbonate minerals.[29–31] The CO_2 mineralization is inspired by the natural reaction of CO_2 absorption to silicide. It has been reported by Seifritz[32] that CO_2 and $CaSiO_3$ can be converted together into Ca^{2+}, HCO_3^- and SiO_2 with the participation of water. In nature, CO_2 can be gathered by silicate minerals such as serpentine ($Mg_3Si_2O_5(OH)_4(s)$), olivine (forsterite, $Mg_2SiO_4(s)$), wollastonite ($CaSiO_3(s)$), talcum, pyroxenes and amphiboles and then transformed to solid products that are applied as material of construction, fire retardants, and so on.[33–36] The process of chemical transformation is[37,38]:

$$(Mg, Ca)_xSi_yO_{x+2y}H_{2z} + xCO_2 \rightarrow x\,(Mg, Ca)CO_3 + ySiO_2 + zH_2O \quad (3.1)$$

$$Mg_3Si_2O_5(OH)_4\,(s) + 3CO_2\,(g) \rightarrow 3MgCO_3\,(s) + 2SiO_2\,(s) + 2H_2O\,(l) \quad (3.2)$$

$$Mg_2SiO_4\,(s) + 2CO_2\,(g) \rightarrow 2MgCO_3\,(s) + SiO_2\,(s) \quad (3.3)$$

$$CaSiO_3\,(s) + CO_2\,(g) \rightarrow CaCO_3\,(s) + SiO_2\,(s) \quad (3.4)$$

According to the $\mathbf{\Delta_f G^\theta}$ of typical products shown in Table 3.2, the overall conversion reactions for the carbonation of silicate minerals process in nature are exothermic and occur spontaneously by thermodynamics ($\Delta G<0$).[40] Unfortunately, due to the limitation of three-phase contact, the reaction rates of this process are very slow and thus could not be applied in practical application directly and should be further advanced. It is necessary to develop economical industrial technology of CO_2 conversion which could accelerate the chemical reaction kinetics and improve the extracting concentrations of MgO/CaO from silicate minerals.

The most widely used methods for silicide carbonization include two kinds of technological routes, as shown in Figure 3.4. One is a direct (single-step) processes[41] in which the conversion reaction could be completed in one step directly. The other is an indirect (multistep) process[42] that involves two steps: (1) magnesium or calcium extraction step, and (2) carbonation.

The slow reaction kinetics at room temperature is a major challenge for the use of direct carbonation processes. The economic cost holds back the possibility of elevated temperatures and pressures.[38] Lackner et al.[43] reported that the carbonation conversion efficiency of serpentine particles was about 25%–30% at 500°C

TABLE 3.2
The Standard Gibbs Free-Energy of Formation ($\Delta_f G^\theta$) for Process Compounds

Species	$\Delta_f G^\theta$ (kJ/mol)
CO_2 (g)	−394.4
CO_2 (aq)	−386.2
CO_3^{2-} (aq)	−528.1
HCO_3^- (aq)	−587.1
H_2CO_3 (aq)	−623.4
$MgCO_3$ (s)	−1029.5
$CaCO_3$ (s)	−1128.8
Mg_2SiO_4 (s)	−2051.3
$Mg_3Si_2O_5(OH)_4$ (s)	−4034.0

Source: Ihsan, B., *Thermochemical Data of Pure Substances*, WILEY-VCH Verlag GmbH, New York, 1993.

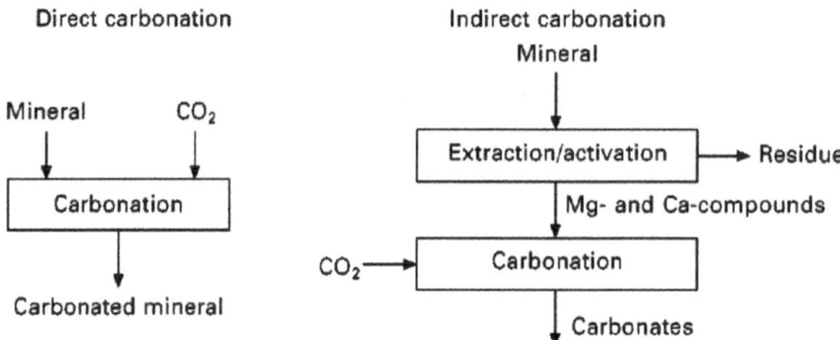

FIGURE 3.4 Schematic for the detailed process of direct and indirect carbonation process. (Reproduced from Zevenhoven, R. and Fagerlund, J., *Developments and Innovation in Carbon Dioxide (CO_2) Capture and Storage Technology*, pp. 433–462, Woodhead Publishing, Oxford, UK, with permission from Woodhead Publishing, Copyright 2010.)

and 3.4×10^7 Pa. The chemical kinetics of the direct aqueous carbonation process is much higher than gas-solid carbonation conversions.[44,45] By using $NaHCO_3$ and NaCl solutions, conversion reaction kinetics can be increased, which attracted much attention.[37] Further, the application of suitable reactor types has been proven to help conversion levels because of the existence of strong intra-particle mechanical effects, such as fluidized bed proposed by Bhardwaj et al. As shown in Figure 3.5,[31] by using

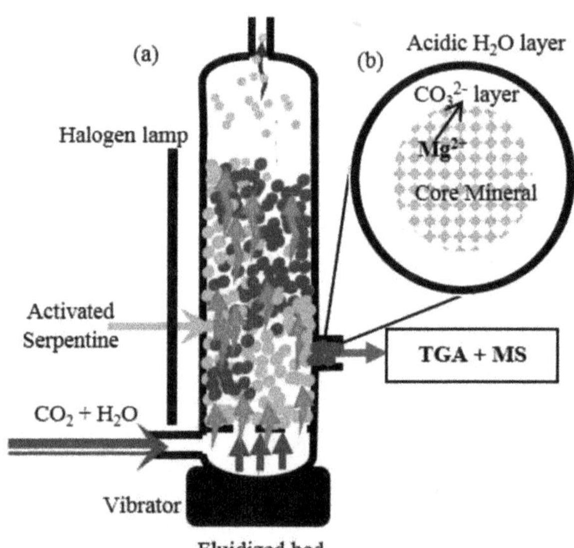

FIGURE 3.5 Schematic for (a) the process of CO_2 mineralization using fluidized bed and (b) the formation process of carbonate-silica product on the surface of serpentinite particles. TGA, thermogravimetric analysis; MS, mass spectrometry. (Reproduced with permission from Bhardwaj, R. et al., *Ind. Eng. Chem. Res.*, 55, 2946–2951, 2016. Copyright 2016 American Chemical Society.)

fluidized and serpentinite, the CO_2 conversion level and mineral conversion level can reach 40% and 50%, respectively (in 8 min at 9°C and 0.1 MPa).

The indirect (multistep) process consists of the pretreatments and subsequent aqueous carbonation process. The chemical, thermal, and mechanical pretreatments are introduced to improve aqueous rock dissolution and activate the particles, including extracting magnesium and calcium from minerals. Weak acids (such as citrates and oxalates), HCl and NaOH solutions are mostly used in the intermediate process.[33,34] For example, Kakizawa et al.[46] and Yuen et al.[40] have reported a process including two steps, with the dissolution of silicate by acetic acid and the carbonation of calcium acetate, as shown in Figure 3.6. The final products in the indirect approach is usually purer compared to the direct approach. There are still many challenges to using the indirect conversion routes, including the huge energy input and the unrecovered intermediate chemicals. More research to optimize the technical route is needed.

Note that a series of value-added mineral products via carbon mineralization could help to further decrease the total cost. For instance, Chang et al.[29] have successfully generated high value-added mineral carbonation products by controlling the polymorphism of $CaCO_3$. Xie et al.[47] have made advances in the mineralization of CO_2 and the separation of Ca^{2+}/Mg^{2+} ions from hard water to some high valuable carbonates. A novel, environmentally friendly route using industrial alkaline solid wastes and waste water as feedstocks for CO_2 mineralization has been proposed.[30,48] It has been proven by researchers that the carbonation of the waste and by-products could not only lower the energy cost but also reduce the pollution significantly.[49–51] Therefore, the carbonation of industrial waste or by-products has attracted more and more attention in academic research and technological development.[52–54]

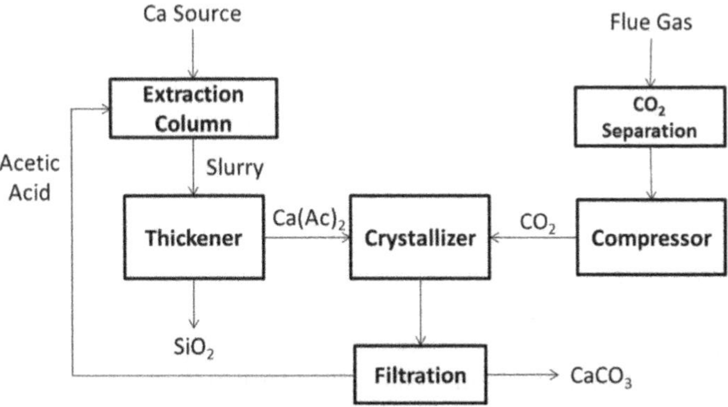

FIGURE 3.6 Schematic of indirect pressure carbonation process. (With kind permission from Springer Science+Business Media: *Environ. Sci. Pollut. Res.*, Carbon dioxide mineralization process design and evaluation, concepts, case studies, and considerations, 23, 2016, 22309–22330, Yuen, Y.T. et al.)

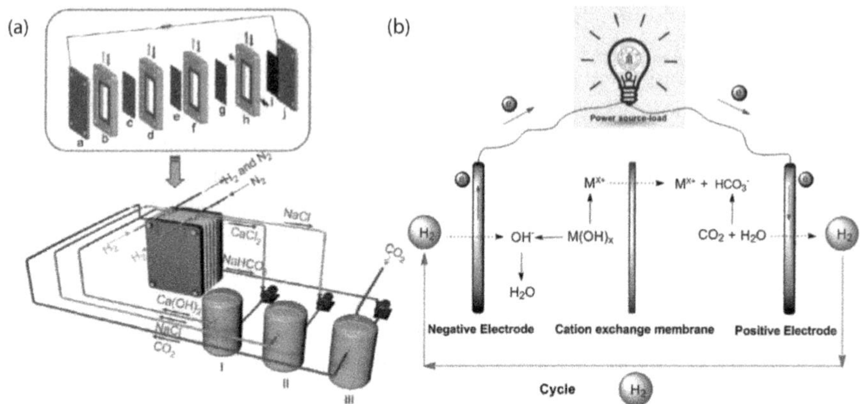

FIGURE 3.7 Schematics of (a) a CO_2 mineralization fuel cell system and the structure of a single cell and (b) a CO_2 mineralization cell with cation exchange membrane (With kind permission from Springer Science+Business Media: *Sci. China. Technol. Sci.*, Generation of electricity from CO_2 mineralization: Principle and realization, 57, 2014, 2335–2343, Xie, H. et al., Copyright 2014, Science China Press.)

Carbon mineralization is accompanied by electricity generation because the mineralization reactions are exothermic and occur spontaneously ($\Delta G < 0$).[55] Xie et al.[56,57] have applied exchange membranes to generate electricity by using industrial alkaline wastes containing $Ca(OH)_2$ and other industrial CO_2 pollutants as feedstock and producing value-added $NaHCO_3$ simultaneously (Equation 3.5, Figure 3.7a[57]). The output power density of this membrane conversion device in experimental scale is 5.5 W m^{-2}, which is much higher than microbial fuel cells. The thermodynamics investigation of this route shown in Figure 3.7b shows that the theoretical maximum electric power output could reach up to 416.67 kW h per 1 t CO_2 consumption (298.15 K, 0.1 MPa, saturated NaCl solution).[56] The results indicate the technique is quite promising and needs more research.

$$Ca(OH)_2 + 2CO_2 + 2NaCl \rightarrow 2NaHCO_3 + CaCl_{2,}\ \Delta_f G^{\theta} = -62.75\ \text{kJ/mol} \quad (3.5)$$

3.3 SUMMARY

The environmentally friendly process of carbon mineralization is an advantageous technique for sustainable, long-term development. The raw materials for CO_2 carbonation are cheap and easily available, while carbonate product is outstandingly thermal stable. However, the limitations of the conversion process, such as low carbonation rates and high energy penalties and costs, hinder the large-scale development of the technique and need to be further resolved.

REFERENCES

1. Shi, J. et al. Enzymatic conversion of carbon dioxide. *Chemical Society Reviews* **44**, 5981–6000 (2015).
2. Demars, B. O. L. et al. Impact of warming on CO_2 emissions from streams countered by aquatic photosynthesis. *Nature Geoscience* **9**, 758–761 (2016).
3. Zhou, H. et al. Biomimetic polymeric semiconductor based hybrid nanosystems for artificial photosynthesis towards solar fuels generation via CO_2 reduction. *Nano Energy* **25**, 128–135 (2016).
4. Johnson, M. P. An overview of photosynthesis. *Essays in Biochemistry* **60**, 255–273 (2016).
5. Armor, J. N. Addressing the CO_2 dilemma. *Catalysis Letters* **114**, 115–121 (2007).
6. Lee, W., Liao, C., Tsai, M., Huang, C., Wu, J. C. S. A novel twin reactor for CO_2 photoreduction to mimic artificial photosynthesis. *Applied Catalysis B: Environmental* **132– 133**, 445–451 (2013).
7. Thauer, R. K. A fifth pathway of carbon fixation. *Science* **318**, 1732–1733 (2007).
8. Castillo, R., Oliva, M., Martí, S., Moliner, V. A theoretical study of the catalytic mechanism of formate dehydrogenase. *The Journal of Physical Chemistry B* **112**, 10012–10022 (2008).
9. Yadav, R. K. et al. A photocatalyst-enzyme coupled artificial photosynthesis system for solar energy in production of formic acid from CO_2. *Journal of the American Chemical Society* **134**, 11455–11461 (2012).
10. Beller, M., Bornscheuer, U. T. CO_2 Fixation through hydrogenation by chemical or enzymatic methods. *Angewandte Chemie International Edition* **53**, 4527–4528 (2014).
11. Schuchmann, K., Müller, V. Direct and reversible hydrogenation of CO_2 to formate by a bacterial carbon dioxide Reductase. *Science* **342**, 1382–1385 (2013).
12. Bachmeier, A. et al. How light-harvesting semiconductors can alter the bias of reversible electrocatalysts in favor of H_2 production and CO_2 reduction. *Journal of the American Chemical Society* **135**, 15026–15032 (2013).
13. Appel, A. M. et al. Frontiers, opportunities, and challenges in biochemical and chemical catalysis of CO_2 fixation. *Chemical Reviews* **113**, 6621–6658 (2013).
14. Jeoung, J., Dobbek, H. Carbon dioxide activation at the Ni, Fe-cluster of anaerobic carbon monoxide dehydrogenase. *Science* **318**, 1461–1464 (2007).
15. Woolerton, T. W. et al. Efficient and clean photoreduction of CO_2 to CO by enzyme-modified TiO_2 nanoparticles using visible light. *Journal of the American Chemical Society* **132**, 2132–2133 (2010).
16. Chaudhary, Y. S. et al. Visible light-driven CO_2 reduction by enzyme coupled CdS nanocrystals. *Chemical Communications* **48**, 58–60 (2012).
17. Seefeldt, L. C., Rasche, M. E., Ensign, S. A. Carbonyl sulfide and carbon dioxide as new substrates, and carbon disulfide as a new inhibitor, of nitrogenase. *Biochemistry* **34**, 5382–5389 (1995).
18. Yang, Z. Y., Moure, V. R., Dean, D. R., Seefeldt, L. C. Carbon dioxide reduction to methane and coupling with acetylene to form propylene catalyzed by remodeled nitrogenase. *Proceedings of the National Academy of Sciences* **109**, 19644–19648 (2012).
19. Zhang, S., Lu, H., Lu, Y. Enhanced stability and chemical resistance of a new nanoscale biocatalyst for accelerating CO_2 absorption into a carbonate solution. *Environmental Science & Technology* **47**, 13882–13888 (2013).
20. Vinoba, M. et al. CO_2 absorption and sequestration as various polymorphs of $CaCO_3$ using sterically hindered amine. *Langmuir* **29**, 15655–15663 (2013).

21. Hwang, E. T., Gang, H., Chung, J., Gu, M. B. Carbonic anhydrase assisted calcium carbonate crystalline composites as a biocatalyst. *Green Chemistry* **14**, 2216–2220 (2012).
22. Zhang, Y., Zhang, L., Chen, H., Zhang, H. Selective separation of low concentration CO_2 using hydrogel immobilized CA enzyme based hollow fiber membrane reactors. *Chemical Engineering Science* **65**, 3199–3207 (2010).
23. Forsyth, C., Yip, T. W. S., Patwardhan, S. V. CO_2 sequestration by enzyme immobilized onto bioinspired silica. *Chemical Communications* **49**, 3191–3193 (2013).
24. Glueck, S. M., Gumu, S., Fabian, W. M. F., Faber, K. Biocatalytic carboxylation. *Chemical Society Reviews* **39**, 313–328 (2009).
25. Aresta, M., Dibenedetto, A., Angelini, A. The changing paradigm in CO_2 utilization. *Journal of CO_2 Utilization* **3–4**, 65–73 (2013).
26. Obert, R., Dave, B. C. Enzymatic conversion of carbon dioxide to methanol: Enhanced methanol production in silica sol–gel matrices. *Journal of the American Chemical Society* **121**, 12192–12193 (1999).
27. Reda, T., Plugge, C. M., Abram, N. J., Hirst, J. Reversible interconversion of carbon dioxide and formate by an electroactive enzyme. *Proceedings of the National Academy of Sciences of the United States of America* **105**, 10654–10658 (2008).
28. Sanna, A., Hall, M. R., Maroto-Valer, M. Post-processing pathways in carbon capture and storage by mineral carbonation (CCSM) towards the introduction of carbon neutral materials. *Energy & Environmental Science* **5**, 7781–7796 (2012).
29. Chang, R., Choi, D., Kim, M. H., Park, Y. Tuning crystal polymorphisms and structural investigation of precipitated calcium carbonates for CO_2 mineralization. *ACS Sustainable Chemistry & Engineering* **5**, 1659–1667 (2017).
30. Xie, H. et al. Feedstocks study on CO_2 mineralization technology. *Environmental Earth Sciences* **75** (2016).
31. Bhardwaj, R., van Ommen, J. R., Nugteren, H. W., Geerlings, H. Accelerating natural CO_2 mineralization in a fluidized bed. *Industrial & Engineering Chemistry Research* **55**, 2946–2951 (2016).
32. Seifritz, W. CO_2 disposal by means of silicates. *Nature* **345**, 486 (1990).
33. Chu, D., Vinoba, M., Bhagiyalakshmi, M., Hyunbaek, I., Nam, S. CO_2 mineralization into different polymorphs of $CaCO_3$ using an aqueous-CO_2 system. *RSC Advances* **3**(44), 21722–21729 (2013).
34. Ma, J., Yoon, R. Use of reactive species in water for CO_2 mineralization. *Energy & Fuels* **27**, 4190–4198 (2013).
35. Balucan, R. D., Dlugogorski, B. Z. Thermal activation of antigorite for mineralization of CO_2. *Environmental Science & Technology* **47**, 182–190 (2013).
36. Naraharisetti, P. K., Yeo, T. Y., Bu, J. Factors influencing CO_2 and energy penalties of CO_2 mineralization processes. *ChemPhysChem* **18**, 3189–3202 (2017).
37. Zevenhoven, R., Fagerlund, J. *Developments and Innovation in Carbon Dioxide (CO_2) Capture and Storage Technology*, pp. 433–462 (Woodhead Publishing, Oxford, UK, 2010).
38. Olajire, A. A. A review of mineral carbonation technology in sequestration of CO_2. *Journal of Petroleum Science and Engineering* **109**, 364–392 (2013).
39. Ihsan, B. (ed.) *Thermochemical Data of Pure Substances* (WILEY-VCH Verlag GmbH, New York, 1993).
40. Yuen, Y. T., Sharratt, P. N., Jie, B. Carbon dioxide mineralization process design and evaluation: Concepts, case studies, and considerations. *Environmental Science and Pollution Research* **23**, 22309–22330 (2016).
41. Ji, L. et al. CO_2 sequestration by direct mineralisation using fly ash from Chinese Shenfu coal. *Fuel Processing Technology* **156**, 429–437 (2017).
42. Velts, O., Kindsigo, M., Uibu, M., Kallas, J., Kuusik, R. CO_2 Mineralisation: Production of $CaCO_3$-type material in a continuous flow disintegrator-reactor. *Energy Procedia* **63**, 5904–5911 (2014).

43. Lackner, K. S., Butt, D. P., Wendt, C. H. (eds.) Magnesite disposal of carbon dioxide, *2nd International Technical Conference on Coal Utilization & Fuel Systems* (Los Alamos National Laboratory, Los Alamos, NM, 1997).
44. Koivisto, E., Erlund, R., Fagerholm, M., Zevenhoven, R. Extraction of magnesium from four Finnish magnesium silicate rocks for CO_2 mineralisation-Part 1: Thermal solid/solid extraction. *Hydrometallurgy* **166**, 222–228 (2016).
45. Erlund, R., Koivisto, E., Fagerholm, M., Zevenhoven, R. Extraction of magnesium from four Finnish magnesium silicate rocks for CO_2 mineralisation—part 2: Aqueous solution extraction. *Hydrometallurgy* **166**, 229–236 (2016).
46. Kakizawa, M., Yamasaki, A., Yanagisawa, Y. A new CO_2 disposal process via artificial weathering of calcium silicate accelerated by acetic acid. *Energy* **26**, 341–354 (2001).
47. Xie, H. et al. Using electrochemical process to mineralize CO_2 and separate Ca^{2+}/Mg^{2+} ions from hard water to produce high value-added carbonates. *Environmental Earth Sciences* **73**, 6881–6890 (2015).
48. Sun, Y., Yang, G., Li, K., Zhang, L., Zhang, L. CO_2 mineralization using basic oxygen furnace slag: Process optimization by response surface methodology. *Environmental Earth Sciences* **75**(2016).
49. Stolaroff, J. K., Lowry, G. V., Keith, D. W. Using CaO- and MgO-rich industrial waste streams for carbon sequestration. *Energy Conversion and Management* **46**, 687–699 (2005).
50. Huntzinger, D. N., Gierke, J. S., Kawatra, S. K., Eisele, T. C., Sutter, L. L. Carbon dioxide sequestration in cement kiln dust through mineral carbonation. *Environmental Science & Technology* **43**, 1986–1992 (2009).
51. Bang, J. et al. Leaching of metal ions from blast furnace slag by using aqua regia for CO_2 mineralization. *Energies* **9**, 996 (2016).
52. Kelly, K. E., Silcox, G. D., Sarofim, A. F., Pershing, D. W. An evaluation of ex situ, industrial-scale, aqueous CO_2 mineralization. *International Journal of Greenhouse Gas Control* **5**, 1587–1595 (2011).
53. Verduyn, M., Geerlings, H., Mossel, G. V., Vijayakumari, S. Review of the various CO_2 mineralization product forms. *Energy Procedia* **4**, 2885–2892 (2011).
54. Xie, H. et al. Effect of water on carbonation of mineral aerosol surface models of kaolinite: A density functional theory study. *Environmental Earth Sciences* **73**, 7053–7060 (2015).
55. Lackner, K. S. Carbonate chemistry for sequestring fossil carbon. *Annual Review of Energy and the Environment* **27**, 193–232 (2002).
56. Xie, H. et al. Thermodynamics study on the generation of electricity via CO^- mineralization cell. *Environmental Earth Sciences* **74**, 6481–6488 (2015).
57. Xie, H. et al. Generation of electricity from CO_2 mineralization: Principle and realization. *Science China Technological Sciences* **57**, 2335–2343 (2014).

4 Thermochemical and Photochemical/Photoelectrochemical Conversion Process of CO_2 Conversion

4.1 THERMOCHEMICAL PROCESS OF CO_2 CONVERSION

The thermocatalytic process of CO_2 conversion, with its long history of research and application, is an important alternative way in various method to catalytically convert CO_2 into high calorific fuels.[1–5] The thermocatalytic process is simplified into a schematic diagram, as shown in Figure 4.1. Because of its molecular structure, the CO_2 molecule usually attracts electrons from catalyst sites. Thus, the process of thermocatalytic conversion is accompanied by electron injection to the antibonding orbital of CO_2. Therefore, H_2 (g) is used as a reducing agent for the thermocatalytic conversion of CO_2. In the reaction process, the hydrogen molecule dissociates into atoms and then adsorbs on the surface active sites.[6]

Sabatier and Senderens[7] reported hydrogenation of CO_2 for the first time in the early 20th century, and this is the major method of the thermochemical route. Since then, many researchers have investigated this field.[6,8–10] The thermochemical conversion reaction is usually catalyzed by VIIIB metals on the solid-gas interface, as shown in Figure 4.1. In this section, three kinds of conversion routes producing fuels/chemicals products, including methanol (CH_3OH), formic acid (HCOOH) and carbon monoxide (CO), are introduced.

4.1.1 CO_2 Hydrogenation to CH_3OH

CH_3OH is an important chemical industrial feedstock and an environmentally friendly fuel.[11–17] CH_3OH synthesis via CO_2 hydrogenation has received more and more attention. Wang et al.[18] have summarized catalytic systems of CH_3OH generation, especially for catalytic reactivity and reaction mechanisms. Cu catalytic agent exhibits superior catalytic effect among metal catalytic agents and is usually studied in the lab and widely used in chemical industry. Catalytic Cu usually functions together with other modifiers, like Ti, Zn, Ce, Al, Si, and so on, as summarized in Table 4.1.[11,12,19–21] A typical example is industrial conversion from CO_2 and H_2 to CH_3OH catalyzed by $Cu/ZnO/Al_2O_3$ catalytic system (200°C–300°C, 5–10 MPa).[22,23]

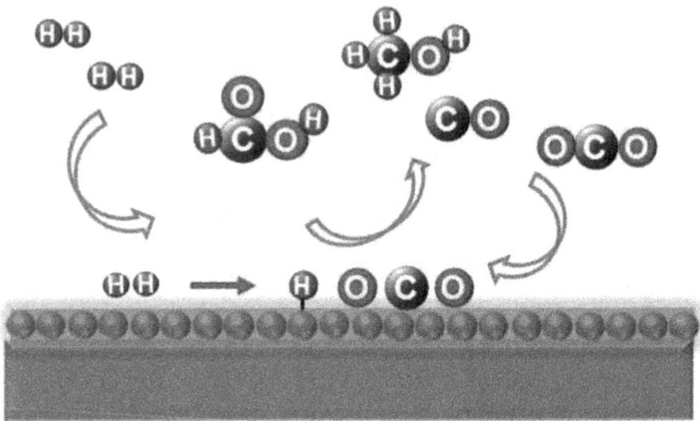

FIGURE 4.1 Simplified schematic diagram of thermochemical conversion reaction catalyzed by VIIIB metals. (Reprinted by permission from Macmillan Publishers Ltd. *Nature*, Ashcroft, A.T. et al., 1991, Copyright 1991.)

TABLE 4.1
Catalyst Systems for CH_3OH Synthesis via CO_2 Hydrogenation

Catalyst Systems	T. (°C)	CO_2 Conversion Rate (%)	CH_3OH Selectivity (%)	Reference
$Cu/Zn/Ga/SiO_2$	270	5.6	99.5	24
Cu/Ga/ZnO	270	6.0	88.0	25
Cu/ZrO_2	240	6.3	48.8	26
$Cu/Ga/ZrO_2$	250	13.7	75.5	27
$Cu/B/ZrO_2$	250	15.8	67.2	27
$Cu/Zn/Ga/ZrO_2$	250	–	75.0	28
$Cu/Zn/ZrO_2$	220	21.0	68.0	29
$Cu/Zn/ZrO_2$	220	12.0	71.1	30
$Cu/Zn/Al/ZrO_2$	240	18.7	47.2	31
$Ag/Zn/ZrO_2$	220	2.0	97.0	29
Au/Zn/ZrO2	220	1.5	100	29
Pd/Zn/CNTs	250	6.3	99.6	32

Source: Wang, W. et al., *Chem. Soc. Rev.*, 40, 3703–3727, 2011.

The HCOO/r-HCOO mechanism, the reverse water gas shift mechanism (RWGS) and the trans-COOH mechanism are three possible functional mechanisms of CH_3OH generation.[6] Here the detailed reaction introduction of the HCOO/r-HCOO routes on catalytic Cu is proposed. Kakumoto et al.[33] proposed the consistency of CO_2 hydrogenation routes and HCOO mechanism: $CO_2 \rightarrow HCOO^* \rightarrow H_2COO^* \rightarrow CH_2O^* \rightarrow CH_3O^* \rightarrow CH_3OH$. However, the activation energy for $HCOO^*$ to H_2COO^* is much

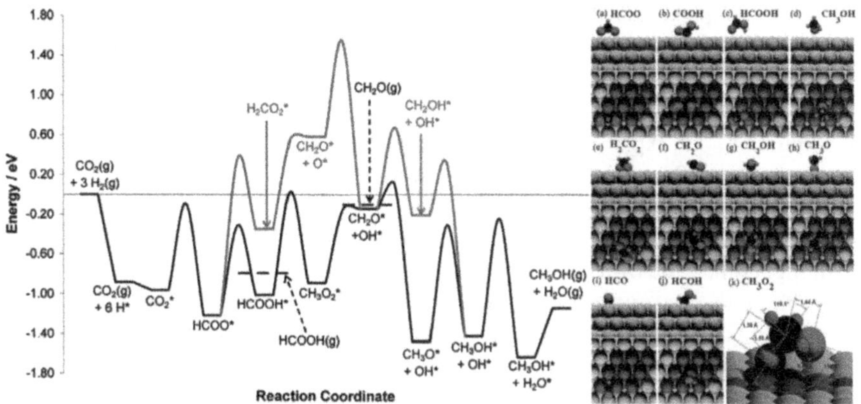

FIGURE 4.2 Potential energy surface (PES) CO_2 hydrogenation with the HCOO/r-HCOO mechanism on Cu(111) surface. (Liu, X. et al., *Appl. Catal. A Gen.*, 279, 241–245, 2005. Reproduced by permission of The Royal Society of Chemistry; Reprinted with permission from Słoczyński, J. et al. *Appl. Catal. A Gen.*, 278, 11–23, 2004. Copyright 2011 American Chemical Society.)

higher than that of HCOO* to HCOOH*, and the conversion from CH_3O_2* to CH_2O* is easier than the conversion from H_2COO* to CH_2O* (Figure 4.2). Thus, Grabow and Mavrikakis[8] have proposed a revised HCOO mechanism of conversion from CH_3O_2* to CH_2O* on Cu (111). In fact, the reaction mechanisms of CH_3OH synthesis via CO_2 hydrogenation is still complicated and lacks thorough understanding. Large numbers of studies at the atomic level devoted to the reaction mechanisms of CO_2 hydrogenation are needed.

4.1.2 CO_2 Hydrogenation to HCOOH

HCOOH is commonly used in the chemical industry in the production of textiles, rubber, food, pharmaceuticals, and so on. Since the 1990s, the synthesis of HCOOH through CO_2 hydrogenation process has received much attention.[34] Farlow and Adkins[35] reported for the first time in 1935 the direct synthesis of HCOOH via CO_2 hydrogenation using a catalytic Raney nickel (80°C–150°C, 20–40 MPa). Until now, transition metals of groups VIII to IIB, such as Ru, Rh, Pd, Ir, Ni, Fe, Co, Cu, Ag, are the main catalytic systems for HCOOH generation action ($CO_2 + H_2 \rightarrow$ HCOOH).[36–41] Here we submit a detailed description of the catalytic system.

In regard to Rh catalytic systems, Musashi and Sakaki[42] presented theoretical research about the reaction mechanisms of HCOOH generation, especially for the differences between the Rh(I), Rh(II), and Rh(III) complexes. The overall conversion routes include two steps. The first step, which mainly influences the reaction rate, is the insertion of CO_2 into Rh-H bonds. The following step has two potential processes: one contains several elementary reactions like the reductive elimination and oxidative addition, while the other includes isomerization of inserted CO_2 and complex metathesis steps.

The Ru complexes have attracted much theoretical and experimental research for their excellent catalytic activity and selectivity in the process of HCOOH generation.[44–46] A typical route of ruthenium dihydride catalyzing HCOOH generation is presented in Figure 4.3.[43,44] There are two probable intermediate sequences proposed by simulation and experiment research which supply more information about the reaction kinetics. The discovery possibly illustrates that the insertion of H_2 into the formate-Ru complex mainly determines the reaction rate, while the insertion of CO_2 is a fast step with a lower energy barrier. The content of the *cis* isomer (**1**, 94.5%) is much higher than the *trans* isomer (**4**, 5.5%) because of the higher activation energy (E_a) in the rate-determining step.[44] Further, a novel catalyst [Ru (Acriphos) (PPh_3) (Cl) ($PhCO_2$)] (Acriphos = 4,5-bis (diphenylphosphino) acridine) for CO_2 conversion to HCOOH has been presented recently for excellent turnover ability.[45] In DMSO or DMSO/H_2O solvent, the turnover numbers and frequencies of HCOOH generation are as high as 4200 and 260 h^{-1}, respectively.

The reaction media also have an important effect on HCOOH generation rate. Various media systems include water, organic solvents, ionic liquids or supercritical CO_2.[38,46] Ionic liquids in particular are used in HCOOH generation for their unique properties, as shown in Figure 4.4.[47,48] It has been reported that HCOOH generation with a Ru-related catalyst happening in a basic ionic liquid exhibits high activity and selectivity. More research about Ru-related catalytic systems in ionic liquid are being conducted, such as Ru/TiO_2,[49,50] Ru nanoparticles,[51] and so on.

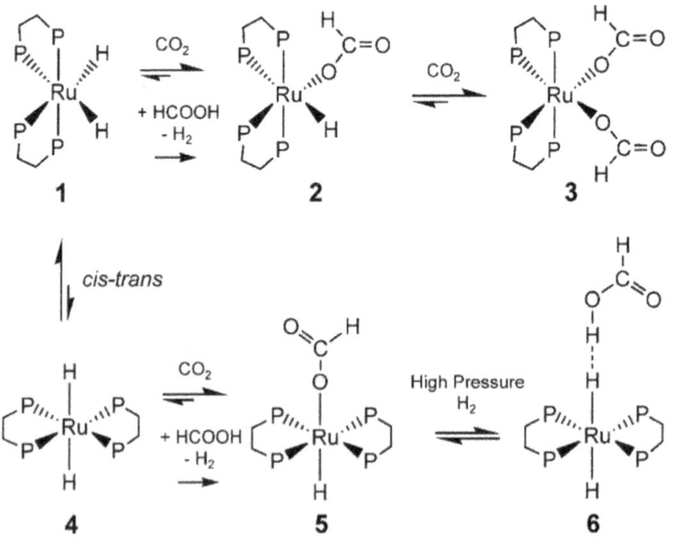

FIGURE 4.3 A typical route of ruthenium dihydride catalyzing HCOOH generation. (From Urakawa, A. et al.: Towards a rational design of ruthenium CO_2 hydrogenation catalysts by Ab Initio metadynamics. *Chemistry—A European Journal*, 2007, 13, 6828–6840. Copyright Wiley-VCH Verlag GmbH & Co. kGaA. Reproduced with permission.)

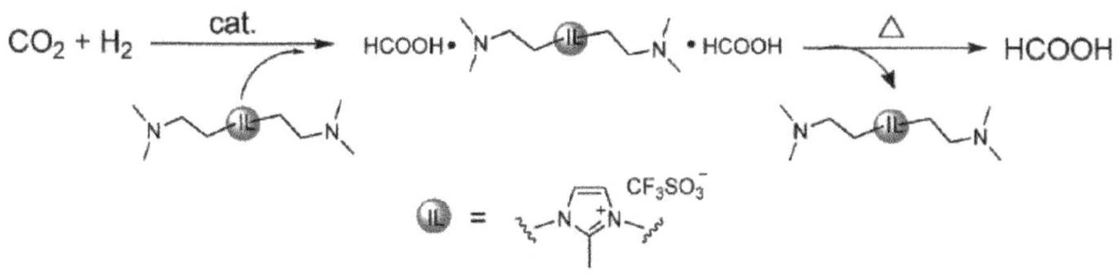

FIGURE 4.4 Reaction process of HCOOH generation with a Ru-related catalyst using an ionic liquid. (From Zhang, Z. et al.: Hydrogenation of CO_2 to formic acid promoted by a diamine-functionalized ionic liquid. *ChemSusChem*, 2009, 2, 234–238. Copyright Wiley-VCH Verlag GmbH & Co. kGaA. Reproduced with permission.)

4.1.3 CO_2 (Dry) Reforming of Methane (DRM)

In addition to the CO_2 hydrogenation routes discussed above, there is another thermochemical conversion way to transform CO_2 into syngas, namely, CO_2 reforming. The generated syngas is applied to produce various fuels/chemicals such as higher alkanes through Fischer-Tropsch synthesis.[52] In 1928, Fischer and Tropsch[53] first made contributions to CO_2 reforming of the methane (DRM) process with Ni and Co catalysts. Unfortunately, the CO_2 reforming process is accompanied by carbon deposition, which leads to severe catalyst activity deactivation. Thus, much research has been conducted to resolve this issue and improve the stability of this DRM process.[9,52,54–56] The key reaction of DRM is[52]:

$$CH_4 + CO_2 \rightleftharpoons 2CO + 2H_2 \ (\Delta H^{\theta} = +247 \text{ kJ mol}^{-1})$$

$$\Delta G^{\theta} = 61770 - 67.32\,T \tag{4.1}$$

where T is Kelvin temperature (K). The high positive ΔH^{θ} illustrates that the DRM process is greatly endothermic and requires reaction temperatures as high as 1173 K to promote the conversion reaction forward to obtain high syngas yields. Kinetic analysis is conducted to get a basic understanding of the elementary reactions of DRM process. The reaction schemes are shown in Figure 4.5; all possible reaction kinetic equations are proposed and summarized by Quiroga and Luna.[56] In the DRM kinetic process, the reactants and the products are adsorbed on the catalyst surface. In detail, CO_2 molecular is adsorbed on the catalyst surface in the form of radical CO_2L, then CO is liberated, and meanwhile an oxygen atom is generated. Notably, the surface chemical reaction is proven to be the rate-determining step by kinetics research.

As to the complex catalyst systems for the DRM process, many factors influence the activity and stability of catalysts, including the intrinsic feature of the metal, the type of support used, the morphological structures such as particle size or surface area of the metal and support, as well as the interaction between the support and metal.[57] Among catalyst systems, Ni-based catalysts are most commonly used

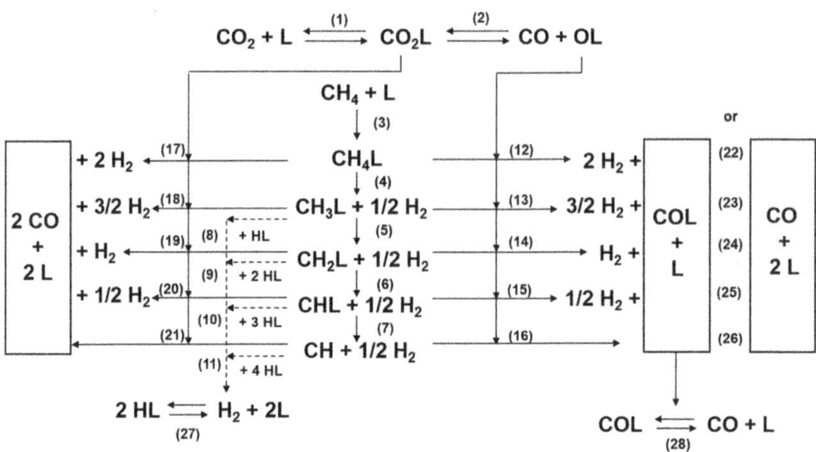

FIGURE 4.5 Kinetic scheme of the DRM mechanism (the species containing "L" are radicals). (Reprinted with permission from Barroso Quiroga, M.M. and Castro Luna, A.E., *Ind. Eng. Chem. Res.*, 46, 5265–5270, 2007. Copyright 2017 American Chemical Society.)

and investigated for their high catalytic performance and relatively low cost, but they are severely limited by the deactivation caused by the carbon deposition.[9,55,58] Therefore, other metal catalysts, such as Ce, Rh, Ru, Pt, Pd, and so on, have been widely researched and developed.[52] Ni-based catalysts modified with other metals such as Pd, Pt, Ru, and so on are viable options for better performance and lower cost. For example, the bimetallic Ni-Pt/ZrO_2 catalyst exhibits better activity and stability compared to that of monometallic Ni/ZrO_2.[59] Buelens et al.[60] proposed a super-dry reforming process using Ni/$MgAl_2O_4$ (see Figure 4.6). In this case, CO_2 is adsorbed on the solid oxygen carrier Fe_2O_3/$MaAl_2O_4$ and CaO/Al_2O_3. During the reforming process, the CH_4 oxidation step is promoted by Ni, and Fe_3O_4 can be reduced into Fe accompanied with CO_2 and H_2O generation. Meanwhile, the CO_2 can be removed by reaction with CaO. The reactions are opposite in the CO_2-reduction stage. Thus, much higher CO production can be achieved and the reactions with H_2O can be blocked by absorbing CO_2.

In addition to CO_2 reforming of methane, other routes for CO_2 reforming are also available to produce organic chemicals. By using CO_2 as the C source, the organic synthesis and products are various, with related reactions shown from Equations (4.2) through (4.6).[61] This chapter mainly focuses on energy-related CO_2 conversion and utilization; therefore, the introduction of conversion to organic compounds will not be detailed.

$$\text{R-epoxide} + CO_2\,(1\,\text{atm}) \xrightarrow[\text{Ionic liquids, RT}]{-2.4\text{ V vs. Ag/AgCl}} \text{R-cyclic carbonate} \tag{4.2}$$

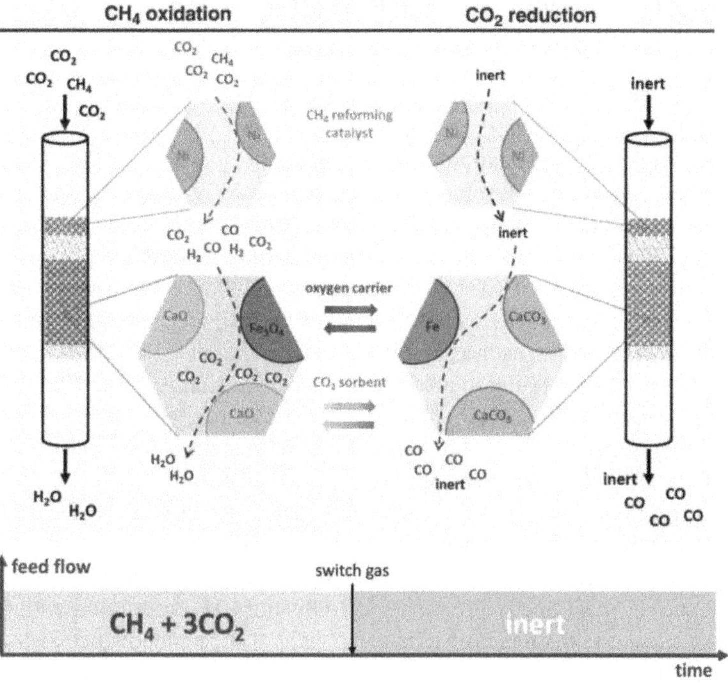

FIGURE 4.6 Schematic of the super-dry reforming process using $Ni/MgAl_2O_4$ catalysts. (From Buelens, L.C. et al., *Science*, 354, 449–452, 2016. Reprinted with permission from AAAS.)

$$\text{R-epoxide} + CO_2 \longrightarrow \text{Cyclic carbonate} \xrightarrow{\text{Polymerization}} \text{Polycarbonates} \tag{4.3}$$

$$\text{propylene oxide} + O{=}C{=}O \xrightarrow{\text{cat.}} \left[\text{polycarbonate}\right]_n \left(+ \text{cyclic carbonate} \right) \tag{4.4}$$

$$\text{oxazole} + CO_2 \xrightarrow[\text{KOH}]{[\text{Au}]} \xrightarrow{\text{aq. HCl}} \text{oxazole-2-COOH} \tag{4.5}$$

$$R{-}C{\equiv}C{-}H + CO_2 \xrightarrow[\text{DMF, R.T, 1 atm}]{\text{Base, 2-5 mol\% CuCl, Ligand}} \xrightarrow{\text{HCl}} R{-}C{\equiv}C{-}COOH \tag{4.6}$$

4.2 PHOTOCATALYTIC AND PHOTOELECTROCHEMICAL PROCESSES OF CO_2 CONVERSION

In the past few decades, many researchers have contributed to the investigation of the photocatalytic and photoelectrochemical processes of CO_2 conversion into carbon-based fuels. Although oxide and non-oxide semiconductors such as TiO_2, ZnO, CdS, and so on, had not yet been developed, Honda et al. in 1979 succeeded in conducting CO_2 photocatalytic reduction to hydrocarbons.[62] Since then, continuous research on photocatalytic CO_2 reduction has been carried out and reported.[63–66]

The basic mechanism of CO_2 photocatalytic reduction is shown in Figures 4.7 and 4.8. When catalysts (such as semiconductor material) are exposed to photon irradiation, an electron can be excited and jump from its balance band (VB) to a higher conduction band (CB), leading to a hole with a positive charge in the CB simultaneously. As soon as these so-called photo-generated electron (e^-) – hole (h^+) pairs are generated, they separate from each other and move to active sites at the catalyst surface. With the full participation of photo-generated electrons at the catalyst surface, CO is converted into various useful carbonous fuels, such as CO, CH_4, CH_3OH, and so on, with or without the presence of H_2O. The specific products are influenced by the number of participative electrons in the CO_2 reduction. For instance, the generation of CO or HCOOH needs $2e^-$, while CH_3OH and CH_4 are formed with $6e^-$ and $8e^-$. Notably, the establishment of an electric field at the semiconductor-electrolyte interface can prolong the effective duration of the excited state of the electron-hole pairs by hindering the recombination of electrons and holes. Through this, CO_2 can be reduced through a photoelectrocatalysis process.[67,68]

The activity of the catalyst mainly determines the kinetics of the photocatalysis process. In this chapter, three common kinds of photocatalyst materials are summarized: semiconductor materials, graphene-based nano-catalyst, and organometallic complexes. Semiconductor materials consist of various types of inorganic binary

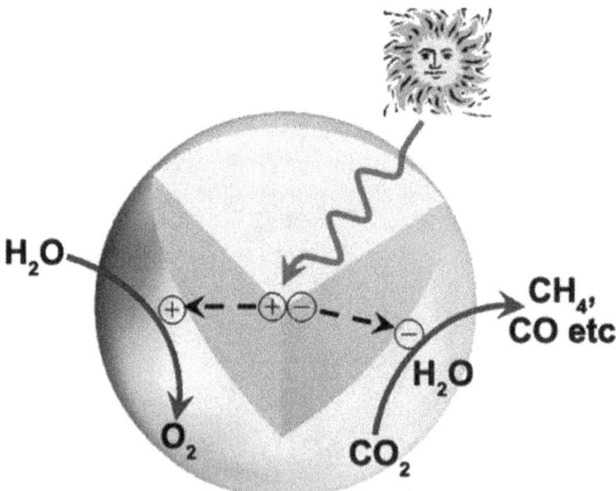

FIGURE 4.7 Basic mechanism of photocatalytic reduction of CO_2. ⊖ represents an electron with a negative charge, while ⊕ represents a hole with a positive charge.

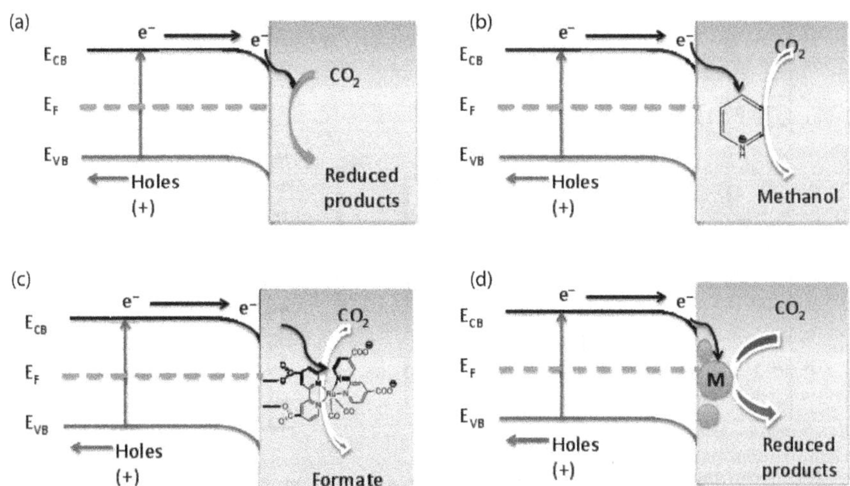

FIGURE 4.8 Illustrative band bending diagrams for p-type semiconductor photoelectrode for photocatalytic reduction of CO_2: (a) semiconductor material itself as CO_2 reduction catalyst; (b) homogeneous unattached catalyst (pyridinium, for example); (c) molecular catalyst $[Ru(bpy)_2(CO)_2]$ attached on photoelectrode surface; (d) electrodeposited or nanoparticle casted metallic catalyst on the surface of semiconductor materials for CO_2 reduction. (Reprinted from Yan, Y. et al., Chapter 12: Photoelectrocatalytic reduction of carbon dioxide, in Styring, P. et al. (Eds.), *Carbon Dioxide Utilization: Closing the Carbon Cycle*, pp. 211–233, Elsevier, Amsterdam, the Netherlands, Copyright 2015, with permission from Elsevier.)

compounds, such as TiO_2, ZnO, CdS, SiC, and so on.[69–71] It has been inferred that at the surface of a semiconductor, the correlation of energy levels between the semiconductor and the redox agents in the solution plays the critical role in the charge transfer kinetics between the photo-generated carriers (e^- and h^+).[62] This hypothesis of energy correlation between various semiconductor catalysts and the redox potentials of different species is illustrated in Figure 4.9. Three factors are widely believed to influence the efficiency of semiconductor photocatalysts. First, the width of the semiconductor band gap is closely related to the absorption scope of photons to generate electron-hole pairs. Second, the effective separation of electron-hole pairs from each other and migration to the surface also have an effect on the reforming performance. Third, the amount and availability of active sites that participate in photocatalytic reactions are meaningful.

Semiconductor materials have been thoroughly investigated for high conversion efficiency and selectivity of the photocatalysis process.[71–75] As a typical, environmentally friendly and low-cost photocatalyst, TiO_2 has been widely studied over the last several decades.[76–79] Low et al.[80] have reviewed numerous approaches of surface modification of TiO_2 photocatalysts, such as impurity doping, metal deposition, heterojunction construction, and so forth. Zhao et al.[79] have reviewed the significant effect of TiO_2 surface point defects, offering inspiration to the design of highly efficient semiconductor photocatalysts for CO_2 conversion. In addition, the research about CO_2 reduction mechanisms and pathways on the surface of TiO_2

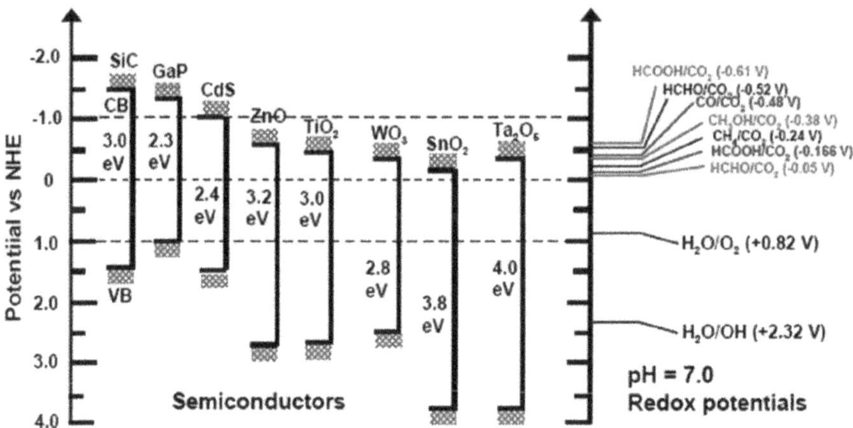

FIGURE 4.9 The energy correlation relationship between semiconductor catalysts and redox couples in water condition. (Reprinted by permission from Macmillan Publishers Ltd. *Nature*, Inoue, T., 1979, Copyright 1979.)

have also been conducted. Lee et al.[81] have reported research about electron-induced dissociation of CO_2 adsorbed at the oxygen vacancy defect on the TiO_2(110) surface by using advanced scanning tunneling microscopy (STM). The application of STM revealed the electron-transfer process at STM tip/CO_2/TiO_2 interface. Furthermore, the catalyst photoreduction initiated by ultrafast electron injection into TiO_2, followed by a rapid (ps-ns) and sequential two-electron oxidation are revealed using another advanced experimental method, time-resolved infrared rectroscopy (TRIR).[82] However, compared to thermodynamic potentials research, experimental evidence of the photocatalytic mechanisms of semiconductor materials still remains uncertain. Some more advanced experimental methods such as in situ time-resolved spectroscopy is used to put forward the studies about the key governing catalytic performances in semiconductors.

Another type of potential catalyst material, graphene-based composite photocatalysts, has attracted increasing attention in recent years.[83–87] The graphene not only plays a synergistic role in composite graphene-semiconductor photocatalysts but also induces significant performance enhancements for CO_2 conversions. For example, Liang et al.[70] have synthesized a series of SEG-P25 (graphene-TiO_2 P25 with the appropriate addition of solvent-exfoliated graphene) composites with significantly higher photoactivities than that of blank P25 for CO_2 reduction under visible light or ultraviolet (UV) irradiation. Ong et al.[69] have prepared nitrogen doped-TiO_2-001/graphene (N-TiO_2-001/GR) composites by in situ photocatalytic reduction. The N-TiO_2-001/GR exhibited superior yield, as high as 3.7 mmol $g_{catalyst}^{-1}$ of CH_4, after reacting for 10 hours, which was about 10 times higher than that of TiO_2-001. In addition, the doping of graphene has proven to notably improve the CO_2-reducing performance of C_3N_4,[84] while the reduced graphene oxide-CdS significantly improved the photocatalytic activities for the reduction of CO_2 to CH_4 in visible-light compared to bare CdS.[86] Low et al.[85] have discussed the advantages of the performance enhancements by uniting graphene with a photocatalyst. In these graphene-semiconductor photocatalysts, the semiconductor plays the role of light harvester, while the graphene plays the role of an efficient co-catalyst and acts as an electron conductive platform. With the assistance of graphene, the energy level in composites is lower than the CB of the semiconductor, which reduces the energy barrier for electronic transition and CO_2 reduction.

Except for the co-catalyst effect, graphene has also been reported for its intrinsic activity and some related derivatives, such as graphene oxide (GO),[88–90] as reported by Hsu et al.[91] have indicated that CO_2 with H_2O vapor could convert into CH_3OH under simulated solar light irradiation with the participation of GO. Particularly, the photocatalytic conversion rates of CO_2 can reach as high as 0.172 μmol g cat^{-1} h^{-1}, which is six times higher than that of state-of-the-art TiO_2 P25. They also demonstrated that Cu particles modification on the surface of GO can amazingly enhance the solar fuel formation rate to 6.4 μmol g cat^{-1} h^{-1}, higher than that of blank GO.[92] In addition, a series of photo-responsive metal-organic complexes based on GO supports is found to significantly improve photocatalytic activities of CO_2 reduction compared to bare GO.[93,94] Although a comprehensive explanation about these performance enhancements in different systems remains incomplete, all the experiment results above have indicated that graphene is a promising photocatalyst for achieving high-efficiency photochemical CO_2 conversions.

Several organometallic complexes have been found to assist the photocatalysis CO_2 reduction, which is analogous to an artificial simulation for photosynthesis catalytic CO_2 reduction in nature. Inspired by this, researchers are searching for a photocatalyst similar to chlorophyll which can facilitate the CO_2 reduction to hydrocarbons in natural light conditions. According to different photoelectric features, organometallic complexes can be classified into three categories: (1) the photo-active complex, which can only play a role of electron transfer medium; (2) the catalytic-active complex, which needs to work together with organic or inorganic photosensitizers; and (3) the complex that is both catalytic-active and photo-active, such as a series of $Ru(bpy)_2(CO)X^{n+}$ complexes (bpy: 2, 2′-bipyridine, X: CO, Cl, H, etc.).[95] Previous studies have demonstrated that the second type of materials exhibits the highest quantum efficiency and product selectivity in general. Therefore, the idea of combining this type of material with semiconductors is considered promising for high-efficient and high-selective CO_2 conversions. For example, Sato et al.[96] have done research about combined p-type $N\text{-}Ta_2O_5$ with $[Ru(dcbpy)(bpy)(CO)_2]^{2+}$ and $[Ru(dcbpy)_2(CO)_2]^{2+}$ to implement CO_2 reduction to CO in a triethanolamine-MeCN solution, and they have reported results about a preferable light quantum photon yield as high as 1.9 with 75% selectivity at a wavelength of 405 nm.

Photoelectrocatalytic CO_2 reduction is expected to possess advantages of both photocatalytic reduction and electrocatalytic reduction approaches to achieve efficient and highly selective conversions of CO_2.[97] For example, Kaneco et al.[98] have outperformed CO_2 photoelectrocatalytic conversions with p-InP catalysts that are modified with different metals (Pb, Ag, Au, Pd, Cu, Ni) in LiOH-CH_3OH electrolyte, while Barton et al.[99] used photoelectrocatalytic reaction devices to achieve CO_2 reduction to CH_3OH. Here, the current efficiency is surprisingly high, as high as 100% at a reduction potential of −0.3 V (versus saturated calomel electrode [SCE]) with narrow band single crystal p-GaP (111) semiconductors used as photocathodes. Yu et al.[66] have successfully prepared a conformal TiO_2 film on the top surface of a black silicon (b-Si) photoanode and enhanced remarkably the efficiency and stability with a photocurrent density of 32.3 mA cm^{-2} at an external potential of 1.48 V versus reversible hydrogen electrode [RHE]. This finding provides a novel perspective about photoelectrochemical water splitting and even CO_2 conversion research. For high conversion efficiency and product selectivity, research about hybrid catalysts have also been conducted; Zhao et al.[100] have summarized the recent studies on semiconductor/metal complex co-catalyst systems. Sahara et al.[101] first reported a molecular and semiconductor photocatalyst hybrid-constructed photoelectrochemical cell for CO_2 reduction under visible light in water. The cell is composed of a photocathode with a Ru(II)–Re(I) complex photocatalyst and a photoanode with CoO_x/TaON complex. Except for a synergistic effect, modification of the nanostructured ocatalysts or electrodes can also dramatically promote CO_2 reduction reaction kinetics. For instance, Song et al.[102] reported an Si photoelectrodes with a nano-porous Au thin film for photoelectrochemical CO_2 conversion with high efficiency and selectivity. Kong et al.[103] reported that a directed assembly of nanoparticle catalysts on vertical nanowire photoelectrodes can achieve similar conversion under illumination.

4.3 SUMMARY

In this chapter, many approaches to the photoelectric catalytic reduction of CO_2 were introduced. The key research objective in this field is to rationally design a photoelectric catalyst with high efficiency and high selectivity by combining photocatalytic and electrocatalytic techniques. To achieve this goal, further investigation about the mechanisms of photoelectrocatalytic CO_2 reduction, the specific collaborative effect mechanisms, as well as the reaction device structures is needed. Moreover, understanding and designing different reaction routes for CO_2 photocatalytic reduction in different reaction media are essential in this field and will provide guidance for highly efficient and highly selective CO_2 conversions.

REFERENCES

1. Sakakura, T., Choi, J., Yasuda, H. Transformation of carbon dioxide. *Chemical Reviews* **107**, 2365–2387 (2007).
2. Valverde, J. M., Medina, S. Limestone calcination under calcium-looping conditions for CO_2 capture and thermochemical energy storage in the presence of H_2O: An in situ XRD analysis. *Physical Chemistry Chemical Physics* **19**, 7587–7596 (2017).
3. Rao, C. N. R., Dey, S. Generation of H_2 and CO by solar thermochemical splitting of H_2O and CO_2 by employing metal oxides. *Journal of Solid State Chemistry* **242**, 107–115 (2016).
4. Zhang, Y. et al. Splitting of CO_2 via the heterogeneous oxidation of zinc powder in thermochemical cycles. *Industrial & Engineering Chemistry Research* **55**, 534–542 (2016).
5. Scheffe, J. R., Steinfeld, A. Oxygen exchange materials for solar thermochemical splitting of H_2O and CO_2: A review. *Materials Today* **17**, 341–348 (2014).
6. Li, Y., Chan, S. H., Sun, Q. Heterogeneous catalytic conversion of CO_2: A comprehensive theoretical review. *Nanoscale* **7**, 8663–8683 (2015).

7. Sabatier, P., Senderens, J. B. New methane synthesis. Hebd. *Seances Academie Science* **134**, 514–516 (1902).
8. Grabow, L. C., Mavrikakis, M. Mechanism of methanol synthesis on Cu through CO_2 and CO hydrogenation. *ACS Catalysis* **1**, 365–384 (2011).
9. Fidalgo, B., Menendez, J. Á. Carbon materials as catalysts for decomposition and CO_2 reforming of methane: A review. *Chinese Journal of Catalysis* **32**, 207–216 (2011).
10. Ashcroft, A. T., Cheetham, A. K., Green, M. Partial oxidation of methane to synthesis gas using carbon dioxide. *Nature* **352**, 225–226 (1991).
11. An, B. et al. Confinement of ultrasmall Cu/ZnO_x nanoparticles in metal–organic frameworks for selective methanol synthesis from catalytic hydrogenation of CO_2. *Journal of the American Chemical Society* **139**, 3834–3840 (2017).
12. Larmier, K. et al. CO_2-to-methanol hydrogenation on zirconia-supported copper nanoparticles: Reaction intermediates and the role of the metal-support interface. *Angewandte Chemie International Edition* **56**, 2318–2323 (2017).
13. Li, M. M., Zeng, Z., Liao, F., Hong, X., Tsang, S. C. E. Enhanced CO_2 hydrogenation to methanol over CuZn nanoalloy in Ga modified Cu/ZnO catalysts. *Journal of Catalysis* **343**, 157–167 (2016).
14. Liao, F., Wu, X. P., Zheng, J., Li, M. J., Kroner, A. A promising low pressure methanol synthesis route from CO_2 hydrogenation over Pd@Zn core–shell catalysts. *Green Chemistry* **19**(1), 270–280 (2016).
15. Martin, O. et al. Indium oxide as a superior catalyst for methanol synthesis by CO_2 hydrogenation. *Angewandte Chemie International Edition* **55**, 6261–6265 (2016).
16. Rungtaweevoranit, B. et al. Copper nanocrystals encapsulated in Zr-based metal–organic frameworks for highly selective CO_2 hydrogenation to methanol. *Nano Letters* **16**, 7645–7649 (2016).
17. Lei, H., Nie, R., Wu, G., Hou, Z. Hydrogenation of CO_2 to CH_3OH over Cu/ZnO catalysts with different ZnO morphology. *Fuel* **154**, 161–166 (2015).
18. Wang, W., Wang, S., Ma, X., Gong, J. Recent advances in catalytic hydrogenation of carbon dioxide. *Chemical Society Reviews* **40**, 3703–3727 (2011).
19. Angelo, L. et al. Study of CuZnMOx oxides (M=Al, Zr, Ce, CeZr) for the catalytic hydrogenation of CO_2 into methanol. *Comptes Rendus Chimie* **18**, 250–260 (2015).
20. Gao, P. et al. Influence of modifier (Mn, La, Ce, Zr and Y) on the performance of Cu/Zn/Al catalysts via hydrotalcite-like precursors for CO_2 hydrogenation to methanol. *Applied Catalysis A: General* **468**, 442–452 (2013).
21. Arena, F. et al. Synthesis, characterization and activity pattern of Cu–ZnO/ZrO_2 catalysts in the hydrogenation of carbon dioxide to methanol. *Journal of Catalysis* **249**, 185–194 (2007).
22. Kunkes, E. L., Studt, F., Abild-Pedersen, F., Schlögl, R., Behrens, M. Hydrogenation of CO_2 to methanol and CO on Cu/ZnO/Al_2O_3: Is there a common intermediate or not? *Journal of Catalysis* **328**, 43–48 (2015).
23. Behrens, M. et al. The active site of methanol synthesis over $CuZnOAl_2O_3$ industrial catalysts. *Science* **336**, 893–897 (2012).
24. Toyir, J., de la Piscina, P. R. R., Fierro, J. L. G., Homs, N. S. Highly effective conversion of CO_2 to methanol over supported and promoted copper-based catalysts: Influence of support and promoter. *Applied Catalysis B: Environmental* **29**, 207–215 (2001).
25. Toyir, J., Ram Rez De La Piscina, P., Fierro, J. L. G., Homs, N. S. Catalytic performance for CO_2 conversion to methanol of gallium-promoted copper-based catalysts: Influence of metallic precursors. *Applied Catalysis B: Environmental* **34**, 255–266 (2001).
26. Liu, J. et al. Surface active structure of ultra-fine Cu/ZrO_2 catalysts used for the CO_2+H_2 to methanol reaction. *Applied Catalysis A: General* **218**, 113–119 (2001).
27. Liu, X., Lu, G. Q., Yan, Z. Nanocrystalline zirconia as catalyst support in methanol synthesis. *Applied Catalysis A: General* **279**, 241–245 (2005).

28. Sloczynski, J. et al. Effect of metal oxide additives on the activity and stability of Cu/ZnO/ZrO_2 catalysts in the synthesis of methanol from CO_2 and H_2. *Applied Catalysis A: General* **310**, 127–137 (2006).
29. Słoczyński, J. et al. Catalytic activity of the M/(3ZnO·ZrO_2) system (M=Cu, Ag, Au) in the hydrogenation of CO_2 to methanol. *Applied Catalysis A: General* **278**, 11–23 (2004).
30. Guo, X., Mao, D., Lu, G., Wang, S., Wu, G. Glycine–nitrate combustion synthesis of CuO–ZnO–ZrO_2 catalysts for methanol synthesis from CO_2 hydrogenation. *Journal of Catalysis* **271**, 178–185 (2010).
31. An, X. et al. A Cu/Zn/Al/Zr fibrous catalyst that is an improved CO_2 hydrogenation to methanol catalyst. *Catalysis Letters* **118**, 264–269 (2007).
32. Liang, X., Dong, X., Lin, G., Zhang, H. Carbon nanotube-supported Pd–ZnO catalyst for hydrogenation of CO_2 to methanol. *Applied Catalysis B: Environmental* **88**, 315–322 (2009).
33. Kakumoto, T. A theoretical study for the CO_2 hydrogenation mechanism on Cu_ZnO catalyst. *Energy Conversion and Management* **36**, 661–664 (1995).
34. Federsel, C., Jackstell, R., Beller, M. State-of-the-Art catalysts for hydrogenation of carbon dioxide. *Angewandte Chemie International Edition* **49**, 6254–6257 (2010).
35. Farlow, M. W., Adkins, H. The hydrogenation of carbon dioxide and a correction of the reported synthesis of urethans. *Journal of the American Chemical Society* **57**, 2222–2223 (1935).
36. Liu, T. et al. Facile Pd-catalyzed chemoselective transfer hydrogenation of olefins using formic acid in water. *Tetrahedron Letters* **57**, 4845–4849 (2016).
37. Peng, G., Sibener, S. J., Schatz, G. C., Mavrikakis, M. CO_2 hydrogenation to formic acid on Ni(110). *Surface Science* **606**, 1050–1055 (2012).
38. Hao, C., Wang, S., Li, M., Kang, L., Ma, X. Hydrogenation of CO_2 to formic acid on supported ruthenium catalysts. *Catalysis Today* **160**, 184–190 (2011).
39. Federsel, C. et al. A well-defined iron catalyst for the reduction of bicarbonates and carbon dioxide to formates, alkyl formates, and formamides. *Angewandte Chemie International Edition* **49**, 9777–9780 (2010).
40. Gunasekar, G. H., Park, K., Jung, K., Yoon, S. Recent developments in the catalytic hydrogenation of CO_2 to formic acid_formate using heterogeneous catalysts. *Inorganic Chemistry* **3**, 882–895 (2016).
41. Peng, G., Sibener, S. J., Schatz, G. C., Ceyer, S. T., Mavrikakis, M. CO_2 hydrogenation to formic acid on Ni(111). *The Journal of Physical Chemistry C* **116**, 3001–3006 (2012).
42. Musashi, Y., Sakaki, S. Theoretical study of rhodium(III)-catalyzed hydrogenation of carbon dioxide into formic acid. Significant differences in reactivity among rhodium(III), rhodium(I), and ruthenium(II) complexes. *Journal of the American Chemical Society* **124**, 7588–7603 (2002).
43. Urakawa, A., Jutz, F., Laurenczy, G., Baiker, A. Carbon dioxide hydrogenation catalyzed by a ruthenium dihydride: A DFT and high-pressure spectroscopic investigation. *Chemistry—A European Journal* **13**, 3886–3899 (2007).
44. Urakawa, A., Iannuzzi, M., Hutter, J., Baiker, A. Towards a rational design of ruthenium CO_2 hydrogenation catalysts by Ab Initio metadynamics. *Chemistry—A European Journal* **13**, 6828–6840 (2007).
45. Rohmann, K. et al. Hydrogenation of CO_2 to formic acid with a highly active ruthenium acriphos complex in DMSO and DMSO/water. *Angewandte Chemie* **128**, 9112–9115 (2016).
46. Tanaka, R., Yamashita, M., Nozaki, K. Catalytic hydrogenation of carbon dioxide using Ir(III)–pincer complexes. *Journal of the American Chemical Society* **131**, 14168–14169 (2009).
47. Zhang, Z. et al. Hydrogenation of CO_2 to formic acid promoted by a diamine-functionalized ionic liquid. *ChemSusChem* **2**, 234–238 (2009).

48. Zhang, Z. et al. Hydrogenation of carbon dioxide is promoted by a task-specific ionic liquid. *Angewandte Chemie International Edition* **47**, 1127–1129 (2008).
49. Upadhyay, P. R., Srivastava, V. Selective hydrogenation of CO_2 gas to formic acid over nanostructured Ru-TiO_2 catalysts. *RSC Advances* **6**, 42297–42306 (2016).
50. Upadhyay, P., Srivastava, V. Synthesis of monometallic Ru/TiO_2 catalysts and selective hydrogenation of CO_2 to formic acid in ionic liquid. *Catalysis Letters* **146**, 12–21 (2016).
51. Srivastava, V. In situ generation of Ru nanoparticles to catalyze CO_2 hydrogenation to formic acid. *Catalysis Letters* **144**, 1745–1750 (2014).
52. Pakhare, D., Spivey, J. A review of dry (CO_2) reforming of methane over noble metal catalysts. *Chemical Society Reviews* **43**, 7813–7837 (2014).
53. Fischer, F., Tropsch, H. Conversion of methane into hydrogen and carbon monoxide. *Brennst.-Chem.* **9**, 39 (1928).
54. Jafarbegloo, M., Tarlani, A., Mesbah, A. W., Sahebdelfar, S. Thermodynamic analysis of carbon dioxide reforming of methane and its practical relevance. *International Journal of Hydrogen Energy* **40**, 2445–2451 (2015).
55. Kathiraser, Y., Oemar, U., Saw, E. T., Li, Z., Kawi, S. Kinetic and mechanistic aspects for CO_2 reforming of methane over Ni based catalysts. *Chemical Engineering Journal* **278**, 62–78 (2015).
56. Barroso Quiroga, M. M., Castro Luna, A. E. Kinetic analysis of rate data for dry reforming of methane. *Industrial & Engineering Chemistry Research* **46**, 5265–5270 (2007).
57. Avetisov, A. K. et al. Steady-state kinetics and mechanism of methane reforming with steam and carbon dioxide over Ni catalyst. *Journal of Molecular Catalysis A: Chemical* **315**, 155–162 (2010).
58. Yuan, K. et al. Dynamic oxygen on surface: Catalytic intermediate and coking barrier in the modeled CO_2 reforming of CH_4 on Ni (111). *ACS Catalysis* **6**, 4330–4339 (2016).
59. Menegazzo, F., Signoretto, M., Pinna, F., Canton, P., Pernicone, N. Optimization of bimetallic dry reforming catalysts by temperature programmed reaction. *Applied Catalysis A: General* **439–440**, 80–87 (2012).
60. Buelens, L. C., Galvita, V. V., Poelman, H., Detavernier, C., Marin, G. B. Super-dry reforming of methane intensifies CO_2 utilization via Le Chatelier's principle. *Science* **354**, 449–452 (2016).
61. Hu, B., Suib, S. L. Chapter 3: Synthesis of useful compounds from CO_2 utilization, synthesis of useful compounds from CO_2. In Centi, G., Perathoner, S. (Eds.), *Green Carbon Dioxide: Advances in CO_2 Utilization*, pp. 51–97 (Wiley & Sons, New York, 2014).
62. Inoue, T., Fujishima, A., Konishi, S., Honda, K. Photoelectrocatalytic reduction of carbon dioxide in aqueous suspensions of semiconductor powders. *Nature* **277**, 637–638 (1979).
63. Fujita, E. Photochemical carbon dioxide reduction with metal complexes. *Coordination Chemistry Reviews* **185–186**, 373–384 (1999).
64. Tu, W., Zhou, Y., Zou, Z. Photocatalytic conversion of CO_2 into renewable hydrocarbon fuels: State-of-the-art accomplishment, challenges, and prospects. *Advanced Materials* **26**, 4607–4626 (2014).
65. Li, M. et al. Morphology and doping engineering of Sn-doped hematite nanowire photoanodes. *Nano Letters* **17**, 2490–2495 (2017).
66. Yu, Y. et al. Enhanced photoelectrochemical efficiency and stability using a conformal TiO_2 film on a black silicon photoanode. *Nature Energy* **2**, 17045 (2017).
67. Aresta, M., Dibenedetto, A., Angelini, A. The changing paradigm in CO_2 utilization. *Journal of CO_2Utilization* **3–4**, 65–73 (2013).

68. Yan, Y., Gu, J., Zeitler, E. L., Bocarsly, A. B. Chapter 12: Photoelectrocatalytic reduction of carbon dioxide. In Styring, P., Quadrelli, E. A., Armstrong, K. (Eds.), *Carbon Dioxide Utilization: Closing the Carbon Cycle*, pp. 211–233 (Elsevier, Amsterdam, the Netherlands, 2015).
69. Ong, W., Tan, L., Chai, S., Yong, S., Mohamed, A. R. Self-assembly of nitrogen-doped TiO_2 with exposed {001} facets on a graphene scaffold as photo-active hybrid nanostructures for reduction of carbon dioxide to methane. *Nano Research* **7**, 1528–1547 (2014).
70. Liang, Y. T., Vijayan, B. K., Gray, K. A., Hersam, M. C. Minimizing graphene defects enhances titania nanocomposite-based photocatalytic reduction of CO_2 for improved solar fuel production. *Nano Letters* **11**, 2865–2870 (2011).
71. Indrakanti, V. P., Kubicki, J. D., Schobert, H. H. Photoinduced activation of CO_2 on Ti-based heterogeneous catalysts: Current state, chemical physics-based insights and outlook. *Energy & Environmental Science* **2**, 745 (2009).
72. Pougin, A., Pougin, A., Dilla, M., Strunk, J. Identification and exclusion of intermediates of photocatalytic CO_2 reduction on TiO_2 under conditions of highest purity. *Physical Chemistry Chemical Physics* **16**, 10809–10817 (2016).
73. Wu, J. et al. Enhanced charge separation of rutile TiO_2 nanorods by trapping holes and transferring electrons for efficient cocatalyst-free photocatalytic conversion of CO_2 to fuels. *Chemical Communications* **28**, 5027–5029 (2016).
74. Sasan, K., Zuo, F., Wang, Y., Feng, P. Self-doped Ti^{3+}-TiO_2 as a photocatalyst for the reduction of CO_2 into a hydrocarbon fuel under visible light irradiation. *Nanoscale* **32**, 13369–13372 (2015).
75. Zhou, H. et al. Biomimetic polymeric semiconductor based hybrid nanosystems for artificial photosynthesis towards solar fuels generation via CO_2 reduction. *Nano Energy* **25**, 128–135 (2016).
76. Wang, T. et al. Photoreduction of CO_2 over the well-crystallized ordered mesoporous TiO_2 with the confined space effect. *Nano Energy* **9**, 50–60 (2014).
77. Woolerton, T. W. et al. Efficient and clean photoreduction of CO_2 to CO by enzyme-modified TiO_2 nanoparticles using visible light. *Journal of the American Chemical Society* **132**, 2132–2133 (2010).
78. Xu, Q., Yu, J., Zhang, J., Zhang, J., Liu, G. Cubic anatase TiO_2 nanocrystals with enhanced photocatalytic CO_2 reduction activity. *Chemical Communication* **51**, 7950–7953 (2015).
79. Zhao, H., Pan, F., Li, Y. A review on the effects of TiO_2 surface point defects on CO_2 photoreduction with H_2O. *Journal of Materiomics* **3**, 17–32 (2017).
80. Low, J., Cheng, B., Yu, J. Surface modification and enhanced photocatalytic CO_2 reduction performance of TiO_2: A review. *Applied Surface Science* **392**, 658–686 (2017).
81. Lee, J., Sorescu, D. C., Deng, X. Electron-induced dissociation of CO_2 on TiO_2 (110). *Journal of the American Chemical Society* **133**, 10066–10069 (2011).
82. Abdellah, M. et al. Time-resolved IR spectroscopy reveals a mechanism with TiO_2 as a reversible electron acceptor in a TiO_2–Re catalyst system for CO_2 photoreduction. *Journal of the American Chemical Society* **139**, 1226–1232 (2017).
83. Yang, M., Xu, Y. Photocatalytic conversion of CO_2 over graphene-based composites: Current status and future perspective. *Nanoscale Horizons* **1**, 185–2 (2016).
84. Ong, W., Tan, L., Chai, S., Yong, S. Graphene oxide as a structure-directing agent for the two-dimensional interface engineering of sandwich-like graphene–g-C_3N_4 hybrid nanostructures with enhanced visible-light photoreduction of CO_2 to methane. *Chemical Communications* **51**, 858–861 (2015).
85. Low, J., Yu, J., Ho, W. Graphene-based photocatalysts for CO_2 reduction to solar fuel. *The Journal of Physical Chemistry Letters* **6**, 4244–4251 (2015).

86. Yu, J., Jin, J., Chenga, B., Jaroniec, M. A noble metal-free reduced graphene oxide–CdS nanorod composite for the enhanced visible-light photocatalytic reduction of CO_2 to solar fuel. *Jounal of Materials Chemistry A* **2**, 3407–3416 (2014).
87. Li, F. et al. Photocatalytic CO_2 conversion to methanol by Cu_2O/graphene/TNA heterostructure catalyst in a visible-light-driven dual-chamber reactor. *NanoEnergy* **27**, 320–329 (2016).
88. Yeh, T., Chan, F., Hsieh, C., Teng, H. Graphite oxide with different oxygenated levels for hydrogen and oxygen production from water under illumination: The band positions of graphite oxide. *The Journal of Physical Chemistry C* **115**, 22587–22597 (2011).
89. Yeh, T., Syu, J., Cheng, C., Chang, T., Teng, H. Graphite oxide as a photocatalyst for hydrogen production from water. *Advanced Functional Materials* **20**, 2255–2262 (2010).
90. Iwase, A. et al. Water splitting and CO_2 reduction under visible light irradiation using Z-scheme systems consisting of metal sulfides, CoO_x-loaded $BiVO_4$, and a reduced graphene oxide electron mediator. *Journal of the American Chemical Society* **138**, 10260–10264 (2016).
91. Hsu, H. et al. Graphene oxide as a promising photocatalyst for CO_2 to methanol conversion. *Nanoscale* **5**, 262–268 (2013).
92. Shown, I. et al. Highly efficient visible light photocatalytic reduction of CO_2 to hydrocarbon fuels by Cu-Nanoparticle decorated graphene oxide. *Nano Letters* **14**, 6097–6103 (2014).
93. Kumar, P., Sain, B., Jain, S. L. Photocatalytic reduction of carbon dioxide to methanol using a ruthenium trinuclear polyazine complex immobilized on graphene oxide under visible light irradiation. *Journal of Materials Chemistry A* **2**, 11246 (2014).
94. Kumar, P., Bansiwal, A., Labhsetwar, N., Jain, S. L. Visible light assisted photocatalytic reduction of CO_2 using a graphene oxide supported heteroleptic ruthenium complex. *Green Chemistry*, **17**, 1605–1609 (2015).
95. Doherty, M. D., Grills, D. C., Muckerman, J. T., Polyansky, D. E., Fujita, E. Toward more efficient photochemical CO_2 reduction: Use of $scCO_2$ or photogenerated hydrides. *Coordination Chemistry Reviews* **254**, 2472–2482 (2010).
96. Sato, S., Morikawa, T., Saeki, S., Kajino, T., Motohiro, T. Visible-light-induced selective CO_2 reduction utilizing a ruthenium complex electrocatalyst linked to a p-Type nitrogen-doped Ta_2O_5 semiconductor. *Angewandte Chemie* **122**, 5227–5231 (2010).
97. Halmann, M. Photoelectrochemical reduction of aqueous carbon dioxide on p-type gallium phosphide in liquid junction solar cells. *Nature* **275**, 115–116 (1978).
98. Kaneco, S., Katsumata, H., Suzuki, T., Ohta, K. Photoelectrocatalytic reduction of CO_2 in LiOH/methanol at metal-modified p-InP electrodes. *Applied Catalysis B: Environmental* **64**, 139–145 (2006).
99. Barton, E. E., Rampulla, D. M., Bocarsly, A. B. Selective solar-driven reduction of CO_2 to methanol using a catalyzed p-GaP based photoelectrochemical cell. *Journal of the American Chemical Society* **130**, 6342–6344 (2008).
100. Zhao, J., Wang, X., Xu, Z., Loo, J. S. C. Hybrid catalysts for photoelectrochemical reduction of carbon dioxide: A prospective review on semiconductor/metal complex co-catalyst systems. *Journal of Materials Chemistry A* **2**, 15228–15233 (2014).
101. Sahara, G. et al. Photoelectrochemical reduction of CO_2 coupled to water oxidation using a photocathode with a Ru(II)–Re(I) complex photocatalyst and a CoO_x/TaON photoanode. *Journal of the American Chemical Society* **138**, 14152–14158 (2016).
102. Song, J. T. et al. Nanoporous Au thin films on Si photoelectrodes for selective and efficient photoelectrochemical CO_2 reduction. *Advanced Energy Materials* **7**, 1601103 (2017).
103. Kong, Q. et al. Directed assembly of nanoparticle catalysts on nanowire photoelectrodes for photoelectrochemical CO_2 reduction. *Nano Letters* **16**, 5675–5680 (2016).

5 Low-Temperature Electrochemical Process of CO_2 Conversion

5.1 BRIEF INTRODUCTION TO THE ELECTROCHEMICAL PROCESS OF CO_2 CONVERSION

Multiple pathways have been investigated and proposed by researchers for the electrochemical conversion of CO_2, as shown in Figure 5.1.[1] These pathways can be divided into two categories: high-temperature (>800°C) conversions using gaseous techniques such as solid oxide cells (SOCs), including solid oxide fuel cell (SOFC) and solid oxide electrolysis cell (SOEC)[2]; and low-temperature (<200°C) conversions using aqueous and non-aqueous techniques such as transition metal electrodes in liquid electrolytes (e.g., methanol, acetonitrile, etc.). The variety of products generated with high-temperature conversions is much less than that of low-temperature conversions, although the selectivity and efficiency of SOCs are generally superior to that of low-temperature conversions. Therefore, catalysts in low-temperature conversions of CO_2 need to have high selectivity and can speed up the reaction rate. The major products of low-temperature conversion methods (CO_2 electroreduction or electrolysis) are low-carbon products such as CH_4, CO, CH_3OH, HCOOH, HCHO, C_2H_5OH, and so on.[3] These electrochemical CO_2 conversion pathways will be discussed in this chapter.

5.2 THERMODYNAMICS OF LOW-TEMPERATURE ELECTROCHEMICAL PROCESS

The electrochemical conversion of CO_2 has attracted great attention because it provides an approach to store renewable electricity in a convenient form with high energy density.[4–6] Although the electrochemical conversion of CO_2 is of great potential, to improve its energy efficiency and reaction rates, significant technological advances are necessary.

The thermodynamics of CO_2 was discussed in Chapter 2. In respect to the electrochemical process, the thermodynamics of CO_2 can be studied using Equations (5.1) and (5.2) in which the enthalpy change between the reactants and products can be provided by both electricity and heat.[7] Table 5.1 presents calculations for common relevant CO_2 products, including syngas (H_2 and CO) production.

$$E^{*} = -\frac{\Delta G}{nF} \tag{5.1}$$

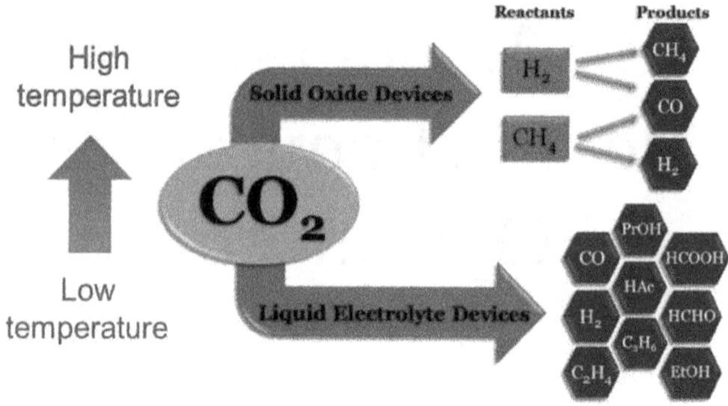

FIGURE 5.1 Pathways for electrochemical conversion of CO_2. (Spinner, N.S. et al., *Catal. Sci. Technol.*, 43, 3–9, 2012, Reproduced by permission of The Royal Society of Chemistry.)

TABLE 5.1
Standard Nernst (E*) and Thermoneutral (E^n) Voltages. Their Difference Is the Overvoltage Required to Account for the Entropic Term of the Global Reaction

Products	Global Reaction	E* (V)	E^n (V)	E*–E^n (mV)
CO	$CO_2 \rightarrow CO + 0.5O_2$	1.33	1.47	134
HCOOH	$CO_2 + H_2O \rightarrow HCOOH + 0.5O_2$	1.34	1.34	0
CH_4	$CO_2 + 2H_2O \rightarrow CH_4 + 2O_2$	1.06	1.15	94
C_2H_4	$CO_2 + 2H_2O \rightarrow C_2H_4 + 3O_2$	1.15	1.22	69
CH_3OH	$CO_2 + 2H_2O \rightarrow CH_3OH + 2O_2$	1.21	1.25	41
C_2H_5OH	$2CO_2 + 3H_2O \rightarrow C_2H_5OH + 3O_2$	1.14	1.18	36
CO, H_2	$CO_2 + 2H_2O \rightarrow CO + H_2 + 0.5O_2$	1.28	1.47	193
H_2	$H_2O \rightarrow H_2 + 0.5O_2$	1.23	1.48	250

Source: Antonio, J. and Martín, G.O.L.J., *Green Chem.*, 17, 5114–5130, 2015.

$$E^n = -\frac{\Delta H}{nF} \tag{5.2}$$

where E^* is the Nernst potential (equilibrium potential) (V), E^n is the thermo-neutral potential (V), n is the number of electrons involved per reaction for the electrolysis reaction, and F is the Faraday constant (A s mol^{-1}).

Multi-electron reduction of CO_2 via Ru-CO_2, _C(O)OH, _CO, _CHO, and _CH_2OH species has been studied by Tanaka and Ooyama.[8] The results demonstrated that the equilibrium potentials (E versus SCE, pH 7.0) progressively shift in a positive direction by increasing the number of electrons participating in the reduction (as shown in Table 5.2). Thermodynamically, the free energy required in the reduction of CO_2 has an inverse association with the number of electrons participating in the reduction of CO_2.

TABLE 5.2
Electrode Reaction and Corresponding Equilibrium Potential with Different Electrons

Electrode Reaction	E°(V) versus RHE(V)
$CO_2 + e^- \rightarrow CO_2^-$	–2.14
$CO_2 + 2e^- + 2H^+ \rightarrow HCOOH$	–0.85
$CO_2 + 2e^- + 2H^+ \rightarrow CO + H_2O$	–0.76
$CO_2 + 4e^- + 4H^+ \rightarrow HCHO + H_2O$	–0.72
$CO_2 + 6e^- + 6H^+ \rightarrow CH_3OH + H_2O$	–0.62
$CO_2 + 8e^- + 8H^+ \rightarrow CH_4 + 2H_2O$	–0.48

Source: Tanaka, K. and Ooyama, D., *Coord. Chem. Rev.*, 226, 211–218, 2002.
Abbreviation: RHE, the Reversible Hydrogen Electrode.

5.3 ELECTROLYZER USED IN LOW-TEMPERATURE ELECTROCHEMICAL PROCESS

Electrolyzers are the most significant devices for CO_2 electroreduction or electrolysis at low-temperatures. There is no standard and well-established configuration for CO_2 electrolyzers, but several reactor concepts have been proposed in the literature. The typical electrolyzer resembles the design of proton exchange membrane fuel cells (PEMFCs), which contain a conducting polymer membrane separating the cathodic and anodic chambers. Delacourt et al.[9] summed up most reported configurations as shown in Figure 5.2. The simplest alternative needs a solid electrolyte to conduct via OH^- or H^+ (Figure 5.2a). Figure 5.2b shows an electrolyzer with an additional liquid buffer layer that functions as a solid electrolyte but with a little more ohmic resistance. An electrolyzer without a liquid buffer layer includes a transport specie (e.g., HCO_3^- and K^+) as shown in Figure 5.2c and d.

5.4 CATALYST/ELECTRODE USED IN LOW-TEMPERATURE ELECTROCHEMICAL PROCESS

In consideration of the stable molecule structure, hybridization form, thermodynamics, and even kinetics of CO_2, the conversion of CO_2 into CO is normally difficult, as shown in Figure 5.3 (left). It can be clearly seen that the intermediate state in CO_2 conversions is $\bullet CO_2^-$ and that the activation energy of this conversion reaction without catalysts is too high. Therefore, during the electrochemical conversion process, the provided overpotential ($\eta_{no\ cat.}$) is relatively high to activate C = O bonds in CO_2 molecules. Using a catalyst, the overpotential will be significantly reduced. Energetic efficiency is an important parameter to determine the recoverable energy contained in the product, which is closely related to the energy cost of production. The Faraday and energy efficiencies can be calculated using the equations as shown in Figure 5.3 (right). It has been recognized that high-energy efficiency is attainable through a combination of low overpotentials and high Faraday efficiency.[10]

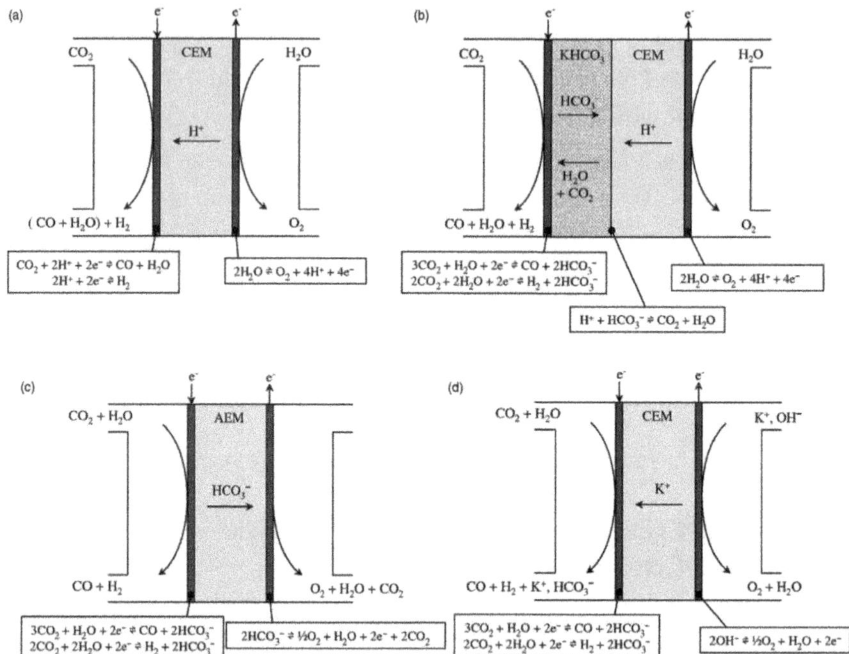

FIGURE 5.2 Schematics of four typical configurations with cation or/and anion exchange membrane for CO_2 reduction via electrochemical process. Electrochemical cells with (a) a cation-exchange membrane (CEM) as electrolyte, (b) a pH-buffer layer of aqueous $KHCO_3$ between the cathode catalyst layer and the CEM, (c) an anion exchange membrane (AEM) as electrolyte, (d) a CEM in the K^+-form. (Reproduced from Delacourt, C. et al., *J. Electrochem. Soc.*, 155, B42–B49, 2008. With permission from The Electrochemical Society, Copyright 2008.)

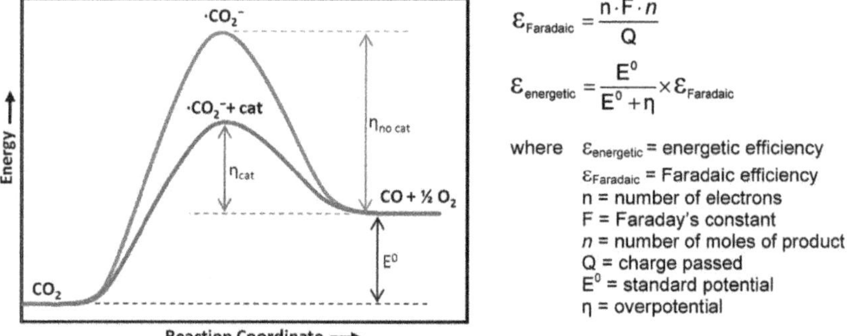

FIGURE 5.3 Reaction pathway energy diagrams for reactions of CO_2 with/without catalyst (left) and the calculation formula for energetic efficiency (right). $\varepsilon_{Faradaic}$ and $\varepsilon_{energetic}$ are the Faradaic and energetic efficiencies, respectively. (Reproduced with permission from Whipple, D.T. and Kenis, P.J.A., *J. Phys. Chem. Lett.*, 1, 3451–3458, 2010. Copyright 2010, American Chemical Society.)

Although there are challenges related to CO_2 electrocatalytic reductions, such as low catalytic activity and insufficient stability, electrochemical catalysts for CO_2 conversion have attracted great attention. Particularly, transition metals and post-transition metals are frequently studied in the literature. Normally, onset overpotentials are all lager than 300 mV even in cases with electrocatalysts.[7] On bulk metal surfaces, some display high selectivity for HCOOH (Sn, In, Pb, Hg)[11,12] or CO (Au, Ag).[13,14] Cu has attracted much interest for its high selectivity for hydrocarbons.[15–17] The comprehensive mechanisms for the reduction of CO_2 on copper was proposed by Kortlever et al.[18] (Figure 5.4). As shown in Figure 5.4, both C1 (e.g., methane) and C2 (e.g., ethylene) pathways are presented. In the C1 pathway, the CO intermediate is first reduced to a formyl species (*CHO) or a *COH species, which

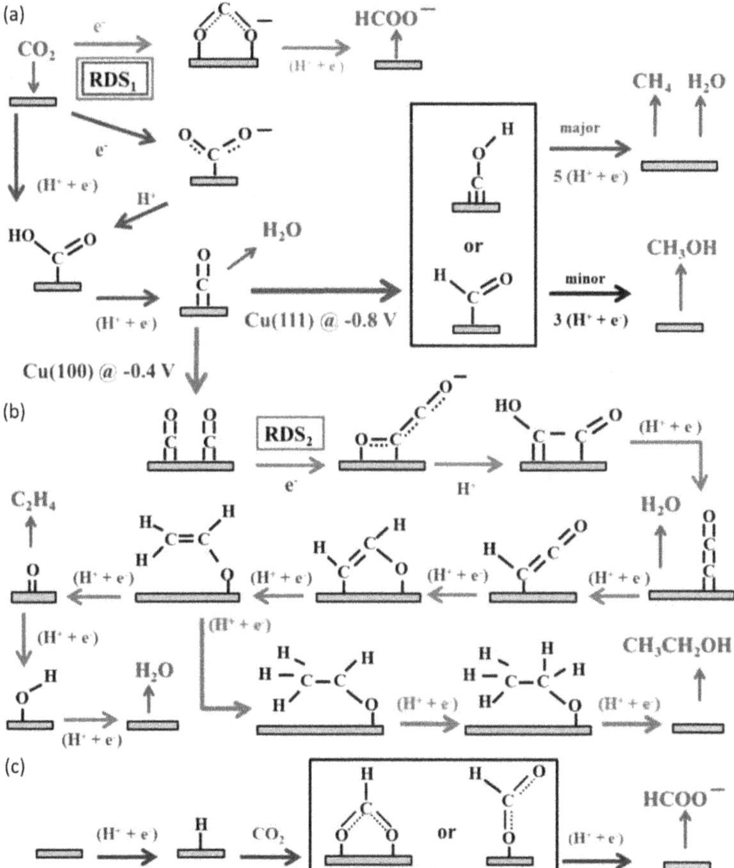

FIGURE 5.4 Possible reaction pathways on molecular catalysts and transition metals for the electrocatalytic reduction of CO_2 to products: (a) pathways of CO_2 to CO, CH_3OH (black arrows), CH_4 (blue arrows), and $HCOO^-$ (orange arrows); (b) pathways of CO_2 to ethanol (green arrows) and ethylene (gray arrows); (c) pathway of CO_2 insertion into a metal-H bond yielding formate (purple arrows). (Reproduced with permission from Kortlever, R. et al., *J. Phys. Chem. Lett.*, 6, 4073–4082, 2015. Copyright 2015, American Chemical Society.)

is then further reduced to methane. In the C2 pathway, the key C–C bond-making step at low overpotentials is a CO dimerization step mediated by electron transfers rendering a $*C_2O_2^-$ intermediate. In general, insufficient knowledge of the reduction mechanisms and insufficient understanding of the active sites are the main barriers for the selection of new selective and active electrode materials. Qiao et al.[3,19–21] conducted a series of studies on catalysts, especially copper oxide/copper electrocatalysts, for the electroreduction of CO_2 to produce useful low-carbon fuels. A summary of CO_2 electroreduction catalysts was also systematically presented by her group.[3] A case in point is where SnO_2-CuO nanocomposites were synthesized and the high catalytic activity for CO_2 electroreduction was investigated. Here, Sn was reported to be the most active electrode towards CO_2 electroreduction in aqueous electrolytes for producing HCOOH.[20,22]

The main research of new electroactive materials for CO_2 electrolytic reduction is exhibited in Figure 5.5, as proposed by Martín et al.[7] Most are based on the catalytic activity of d- and p-metals found in the 1980s. Compared with bulk metals, nanoparticles and alloys can exhibit very different catalytic properties (size/composition effects).[23] In addition, the selectivity of CO varies with the size of Pd nanoparticles, as shown in Figure 5.5. Oxides of precursor materials can produce highly selective and active catalysts (oxide-derived catalysts).[24] Some metal/oxides and oxides are selected as catalysts, which greatly enhances the activity and Faradaic effciencies (oxides and metal/oxides),[25,26] and some complexes and ionic liquids are

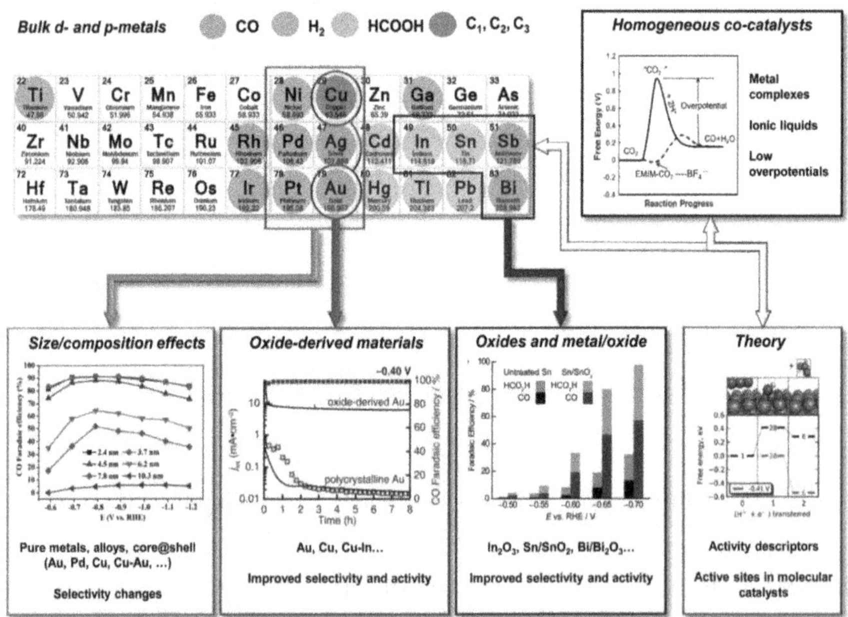

FIGURE 5.5 Main research on new electroactive materials for CO_2 electroreduction. (Reproduced with permission from Antonio, J. and Martín, G.O.L.J., *Green Chem.*, 17, 5114–5130, 2015. Reproduced by permission of The Royal Society of Chemistry.)

chosen to be potential mediators or co-catalysts for CO_2 reduction (homogeneous co-catalysts).[14,27,28] Here, computing methods such as density functional theory (DFT) can be used in parallel with experimental results to elucidate activity descriptors (theory) and reaction mechanisms.[29,30]

5.5 SUMMARY AND OUTLOOK

The thermodynamics, electrolyzer, and catalyst/electrode used in low-temperature electrochemical processes of CO_2 conversion are briefly introduced in this chapter. To overcome the challenges such as low catalytic activity, low selectively of products, and insufficient stability, the innovative electrocatalysts for the CO_2 electroreduction at low temperature to produce low-carbon fuels (e.g., CO, $HCOOH/HCOO^-$, CH_2O, CH_4, $H_2C_2O_4/HC_2O_4^-$, C_2H_4, CH_3OH, CH_3CH_2OH, and others) must be explored. Thus, further fundamental understanding through both experiments and theoretical modeling is also necessary to guide the development of those novel catalysts.

REFERENCES

1. Spinner, N. S., Vega, J. A., Mustain, W. E. Recent progress in the electrochemical conversion and utilization of CO_2. *Catalysis Science & Technology* **43**, 3–9 (2012).
2. Zheng, Y., et al. A review of high temperature co-electrolysis of H_2O and CO_2 to produce sustainable fuels using solid oxide electrolysis cells (SOECs): Advanced materials and technology. *Chemical Society Reviews* **16**, 127–1463 (2017).
3. Qiao, J., Liu, Y., Hong, F., Zhang, J. A review of catalysts for the electroreduction of carbon dioxide to produce low-carbon fuels. *Chemical Society Reviews* **43**, 631–675 (2014).
4. Ni, Y., Eskeland, G. S., Giske, J., Hansen, J. The global potential for carbon capture and storage from forestry. *Carbon Balance and Management* **11**, 3–11 (2016).
5. Zeng, X., Jin, F., Yao, G., Huo, Z. Theoretical study on highly efficient reduction of CO_2 by in-situ produced hydrogen using Al under mild hydrothermal conditions. *International Journal of Hydrogen Energy* **41**, 9140–9144 (2016).
6. Alves, D. C. B., Silva, R., Voiry, D., Asefa, T., Chhowalla, M. Copper nanoparticles stabilized by reduced graphene oxide for CO_2 reduction reaction. *Materials for Renewable and Sustainable Energy* **4**, 2–8 (2015).
7. Antonio, J., Martín, G. O. L. J. Towards sustainable fuels and chemicals through the electrochemical reduction of CO_2: Lessons from water electrolysis. *Green Chemistry* **17**, 5114–5130 (2015).
8. Tanaka, K., Ooyama, D. Multi-electron reduction of CO_2 via Ru CO_2, C(O)OH, CO, CHO, and CH_2OH species. *Coordination Chemistry Reviews* **226**, 211–218 (2002).
9. Delacourt, C., Ridgway, P. L., Kerr, J. B., Newman, J. Design of an electrochemical cell making syngas (CO + H_2) from CO_2 and H_2O reduction at room temperature. *Journal of The Electrochemical Society* **155**, B42–B49 (2008).
10. Whipple, D. T., Kenis, P. J. A. Prospects of CO_2 utilization via direct heterogeneous electrochemical reduction. *The Journal of Physical Chemistry Letters* **1**, 3451–3458 (2010).
11. Kumar, B., et al. Reduced SnO_2 porous nanowires with a high density of grain boundaries as catalysts for efficient electrochemical CO_2-into-HCOOH conversion. *Angewandte Chemie International Edition* **56**, 3645–3649 (2017).
12. Li, F., et al. Towards a better Sn: Efficient electrocatalytic reduction of CO_2 to formate by Sn/SnS_2 derived from SnS_2 nanosheets. *Nano Energy* **31**, 270–277 (2017).

13. Fang, Y., Flake, J. C. Electrochemical reduction of CO_2 at functionalized Au electrodes. *Journal of the American Chemical Society* **139**, 3399–3405 (2017).
14. Singh, M. R., Kwon, Y., Lum, Y., Ager, J. W., Bell, A. T. Hydrolysis of electrolyte cations enhances the electrochemical reduction of CO_2 over Ag and Cu. *Journal of the American Chemical Society* **138**, 13006–13012 (2016).
15. Ma, M., Djanashvili, K., Smith, W. A. Controllable hydrocarbon formation from the electrochemical reduction of CO_2 over Cu nanowire arrays. *Angewandte Chemie International Edition* **55**, 6680–6684 (2016).
16. Min, S., et al. Low overpotential and high current CO_2 reduction with surface reconstructed Cu foam electrodes. *Nano Energy* **27**, 121–129 (2016).
17. Yin, Z., et al. Highly selective palladium-copper bimetallic electrocatalysts for the electrochemical reduction of CO_2 to CO. *Nano Energy* **27**, 35–43 (2016).
18. Kortlever, R., Shen, J., Schouten, K. J. P., Calle-Vallejo, F., Koper, M. T. M. Catalysts and reaction pathways for the electrochemical reduction of carbon dioxide. *The Journal of Physical Chemistry Letters* **6**, 4073–4082 (2015).
19. Qiao, J., Jiang, P., Liu, J., Zhang, J. Formation of Cu nanostructured electrode surfaces by an annealing–electroreduction procedure to achieve high-efficiency CO_2 electroreduction. *Electrochemistry Communications* **38**, 8–11 (2014).
20. Fan, M., Fu, S., Liu, Y., Ma, C., Qiao, J. Synthesis of SnO_2-CuO nanocomposite and its high catalytic activity for CO_2 electroreduction. in *2014 Electrochemical Conference on Energy & the Environment (ECEE)* **3**, 377–377 (2014).
21. Fan, M., Bai, Z., Zhang, Q., Ma, C., Zhou, X. D. Aqueous CO_2 reduction on morphology controlled Cu_xO nanocatalysts at low overpotential. *Rsc Advances* **4**, 44583–44591 (2014).
22. Machunda, R. L., Ju, H., Lee, J. Electrocatalytic reduction of CO_2 gas at Sn based gas diffusion electrode. *Current Applied Physics* **11**, 986–988 (2011).
23. Huang, H., et al. Understanding of strain effects in the electrochemical reduction of CO_2: Using Pd nanostructures as an ideal platform. *Angewandte Chemie International Edition* **56**, 3594–3598 (2017).
24. Chen, Y., Li, C. W., Kanan, M. W. Aqueous CO_2 reduction at very low overpotential on oxide-derived Au nanoparticles. *Journal of the American Chemical Society* **134**, 19969–19972 (2012).
25. Li, Q., et al. Tuning Sn-catalysis for electrochemical reduction of CO_2 to CO via the Core/Shell Cu/SnO_2 structure. *Journal of the American Chemical Society* **139**, 4290–4293 (2017).
26. Wu, J., Sun, S., Zhou, X. Origin of the performance degradation and implementation of stable tin electrodes for the conversion of CO_2 to fuels. *Nano Energy* **27**, 225–229 (2016).
27. Bonin, J., Maurin, A., Robert, M. Molecular catalysis of the electrochemical and photochemical reduction of CO_2 with Fe and Co metal based complexes. Recent advances. *Coordination Chemistry Reviews* **334**, 184–198 (2017).
28. Rosen, B. A., et al. Ionic liquid–mediated selective conversion of CO_2 to CO at low overpotentials. *Science* **334**, 643–644 (2011).
29. Peterson, A. A., Abild-Pedersen, F., Studt, F., Rossmeisl, J., Nørskov, J. K. How copper catalyzes the electroreduction of carbon dioxide into hydrocarbon fuels. *Energy & Environmental Science* **3**, 1311–1316 (2010).
30. Barton Cole, E., et al. Using a one-electron shuttle for the multielectron reduction of CO_2 to methanol: Kinetic, mechanistic, and structural insights. *Journal of the American Chemical Society* **132**, 11539–11551 (2010).

6 High-Temperature Electrochemical Process of CO_2 Conversion with SOCs 1

Introduction and Fundamentals

6.1 INTRODUCTION OF SOLID OXIDE CELLS (SOCs)

A SOC is a device that can convert electrical energy into chemical energy in the solid oxide electrolysis cell (SOEC) mode and convert chemical energy into electrical energy in the solid oxide full cell (SOFC) mode.[1–4] As one of the most effective conversion approaches, CO_2/H_2O co-electrolysis using SOECs (600°C–1000°C) will be the focus in this book. The schematic diagram of a SOEC for CO_2/H_2O co-electrolysis is shown in Figure 6.1.[5] This high temperature (HT) electrolyzer includes a cathode (fuel electrode), an anode (oxygen electrode), and a dense electrolyte (ionic conductor), in which CO_2 (g) and H_2O (g) are fed into the cathode. The oxide ions are then conducted across the dense electrolyte from the cathode to the anode. Finally, oxygen is evolved at the anode, and the syngas is produced at the cathode. It is worth mentioning that the CO_2 can be captured from the emissions of power plants or industries and the syngas produced can be transformed into synfuel, thus turning waste into fuel. In addition, this process can utilize clean energy sources such as nuclear or other renewable energies.

As identified, high temperature CO_2/H_2O co-electrolysis (HTCE) using SOECs has greater potential for future cost reductions and efficiency improvements[6] as it is more energy efficient than those of the two separate electrolysis processes in terms of both lower energy consumption and fewer electrolysis steps, as well as the fact that it requires only one necessary reactor to operate.[7–11] The syngas (H_2 + CO) that is produced is an effective energy carrier beyond electricity that can be used for large-scale energy storage. It can also be further processed to generate chemicals or liquid fuels via Fischere-Tropsch (F-T) synthesis.[12] Additionally, the fuels produced can be used to generate power within the same SOEC device in the reversible mode.[13] Considering the above, SOEC technology can provide an attractive route to reduce CO_2 emissions, enable scalable energy storage capabilities, and facilitate the integration of renewable energies into the electric grid. Minh and Mogensen[13]

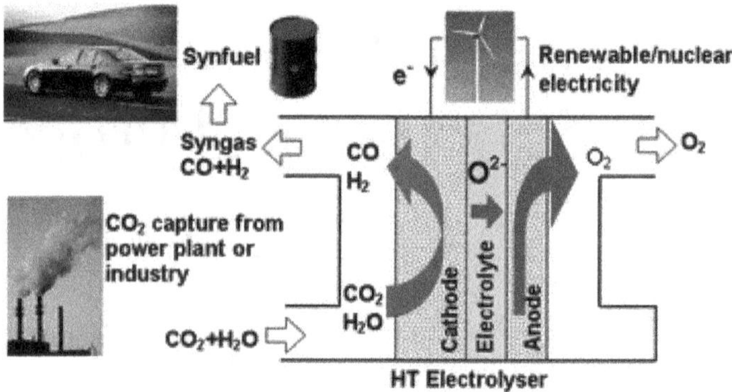

FIGURE 6.1 Schematic diagram of CO_2/H_2O co-electrolysis using a solid oxide electrolysis cell (SOEC) for syngas and O_2 production. (Ebbesen, S.D. et al., *Chem. Rev.*, 114, 10697–10734, 2014. Reproduced by permission of The Royal Society of Chemistry.)

believed that SOECs should have the following attractive features: (1) compatibility (environmentally compatible with reduced CO_2 emissions), (2) flexibility (fuel flexible and suitable for integration with various energy sources, especially sustainable energies), (3) capability (can be used for different functions), (4) adaptability (suitable for a variety of applications or different local energy needs), and (5) affordability (competitiveness in cost).

On the other hand, it also appears that SOECs for CO_2 conversion and utilization are not mature enough to meet the requirements for practical implementation and commercialization. This can be attributed to major challenges such as the low activation/conversion rate of CO_2 as well as the degradation of SOEC stacks. To facilitate research and development in overcoming these challenges, it is necessary to specifically focus on high-temperature CO_2/H_2O co-electrolysis using SOECs. Aside from the fundamentals presented in this chapter, other key topics discussed in following several chapters include the key materials/technologies, characterizations, analysis methods, performance validation/optimizations, product analysis methods, and economic analysis. In regard to theoretical approaches, the oxygen ionic transport mechanisms of oxygen electrodes in SOECs are also reviewed. Finally, future research directions are proposed to address challenges for technology application and commercialization. In particular, insights into CO_2 electrochemical conversion, SOEC material behaviors and degradation mechanisms are highlighted to provide a better understanding into the co-electrolysis process. We believe that high temperature CO_2/H_2O co-electrolysis using SOECs can be widely used for energy storage and conversion with further research and development in the future.

6.2 FUNDAMENTALS OF HIGH-TEMPERATURE CO_2 CONVERSION THROUGH ELECTROCHEMICAL APPROACHES

6.2.1 Thermodynamics of High-Temperature CO_2/H_2O Co-electrolysis

The high temperature electrolysis reductions of H_2O and CO_2 are both endothermic with their corresponding electrochemical reactions, as shown in Equations (6.1) and (6.2), respectively. As the thermodynamic parameters [the enthalpy change (ΔH), the Gibbs

TABLE 6.1
Thermodynamic Parameters for High-Temperature CO_2/H_2O Co-electrolysis at 298.15 K and 101.325 kPa

	$\Delta_f G$ (kJ·mol⁻¹)	$\Delta_f H$ (kJ·mol⁻¹)	S (kJ·mol⁻¹)	ΔH_{Vap} (kJ·mol⁻¹)	Heat Capacity (C_p) (J·mol⁻¹·K⁻¹)
$H_2(g)$	0	0	0.131	—	$27.28 + 0.00326T + 50200/T^2$
$O_2(g)$	0	0	0.205	—	$29.96 + 0.00418T - 167400/T^2$
$H_2O(l)$	−237.2	−285.8	0.070	40.7	75.44
$H_2O(g)$	−228.6	−241.8	0.189	—	$30.00 + 0.01071T+33500/T^2$
$CO(g)$	−137.2	−110.5	0.198	—	$28.409 + 0.00410T - 46000/T^2$
$CO_2(g)$	−394.4	−393.5	0.214	—	$44.141 + 0.00904T - 853500/T^2$

Source: Ihsan, B., ed., *Thermochemical Data of Pure Substances*, Weinheim, Federal Republic of Germany, WILEY-VCH Verlag GmbH, New York, 1993.

Note: δh_{vap} is the enthalpy of vaporization, the definitions of other parameter symbols are presented in Section 6.2.1.

(l) and (g) refer to liquid phase and gas phase, respectively.

free energy change (ΔG), and the entropy change (ΔS)] are a function of temperature, the ΔH, ΔS, and ΔG of the reactions at 298.15 K can be calculated with reference to Equations (6.3) to (6.5) and the data in Table 6.1 Based on this, ΔH, ΔS, and ΔG at various temperatures can be calculated according to Equations (6.5) to (6.8). Both ΔH and ΔG are related to cell voltage according to Equations (6.9) and (6.10). The calculated results are shown in Figure 6.1, which indicates the relationships between ΔG, ΔS, and ΔH of the electrolysis reactions for $H_2O(g)$, $CO_2(g)$ and $H_2O/CO_2(g)$, respectively. Meanwhile, the relationship between ΔG and ΔH of the water gas shift reaction is also added.

$$H_2O + 2e^- \rightarrow H_2 + O^{2-} \tag{6.1}$$

$$CO_2 + 2e^- \rightarrow CO + O^{2-} \tag{6.2}$$

$$\Delta H\ (298.15\ K) = \sum_B \nu_B \Delta_f H_B (298.15\ K) \tag{6.3}$$

$$\Delta S\ (298.15\ K) = \sum_B \nu_B S_B (298.15\ K) \tag{6.4}$$

$$\Delta G = \Delta H - T\Delta S \tag{6.5}$$

$$\Delta H\ (T) = \Delta H\ (298.15\ K) + \int_{298.15K}^{T} \Delta C_p dT \tag{6.6}$$

$$\Delta S\ (T) = \Delta S\ (298.15\ K) + \int_{298.15K}^{T} \Delta C_p dT \tag{6.7}$$

$$\Delta C_p = \sum_{B} \nu_B C_p(B) \tag{6.8}$$

$$E^* = -\frac{\Delta G}{nF} \tag{6.9}$$

$$E^n = -\frac{\Delta H}{nF} \tag{6.10}$$

where ΔH is the enthalpy change (kJ mol^{-1}), B is one of the reactants or products in the reaction, ν_B is the stoichiometric number of reactants ($\nu_B > 0$) or products ($\nu_B < 0$) in the reaction, $\Delta_f H_B$ is the enthalpy of formation for B, S is the entropy (kJ mol^{-1} K^{-1}), ΔS is the entropy change (kJ mol^{-1} K^{-1}), T is the co-electrolysis temperature (K), ΔG is the Gibbs free energy change (kJ mol^{-1}), C_p is the heat capacity(J mol^{-1} K^{-1}), ΔC_p is the heat capacity change (J mol^{-1} K^{-1}), E^* is the Nernst potential (equilibrium potential) (V), E^n is the thermo-neutral potential (V), n is the number of electrons involved per reaction for the electrolysis reaction (n is equal to 2 for CO_2/H_2O co-electrolysis), and F is the Faraday constant (A s mol^{-1}).

Figure 6.2 illustrates the energy demands of the two electrolysis reactions for H_2O or CO_2.[15,17,18] It is well known that enthalpy is made up of a Gibbs free energy and an entropy term in accordance with Equation (6.5). As indicated in Figure 6.2, ΔH, ΔG, and TΔS are the total energy demand, electrical energy demand, and heat demand, respectively. With the increase in temperature, the total energy demand is almost invariant, but the increasing heat demand and the decreasing electrical energy demand are shown for both H_2O splitting and CO_2 splitting, which is due to positive entropy change (ΔS > 0). Compared to low- or intermediate-temperature electrolysis, high-temperature electrolysis can use both electricity and heat effectively (the efficiency of electricity-to-syngas is up to about 100%), and can achieve high reaction rates, which in turn results in reduced cell internal resistance and greater productivity at the same voltage.[15] In other words, as for H_2O or CO_2 electrolysis, from a thermodynamic viewpoint, high-temperature operation is the better choice for both efficiency and cost.[16] With respect to the relationship between ΔH, ΔG, and ΔS of H_2O/CO_2 co-electrolysis, similar trends can be seen in Figure 6.2b. The ΔH is also insensitive to the increasing temperature, which includes ΔG and TΔS. Thus, as a complement, the decrease of electrical energy demand is almost equal to the increase of heat demand, with temperature increasing. The ΔG is still more than 300 kJ/mol at 1000°C, which definitely restricts the competitiveness of co-electrolysis. However, the competitiveness can be increased using the industrial waste heat or the electricity from renewable energy.[17]

Compared to the separate electrolysis processes of H_2O and CO_2, co-electrolysis is complicated due to the water gas shift (WGS) reaction [Equation (6.11)], where H_2O is reduced in the forward reaction, and CO_2 is reduced in the reverse process (reverse water gas shift [RWGS]). The ΔH and ΔG of the WGS reaction with increasing temperatures are shown in Figure 6.2c. The curves decrease sharply at 100°C due to a phase change from a liquid state to a gaseous state. In addition, it can be seen clearly that the ΔG of the WGS is 0 at 816°C, suggesting that the forward reaction

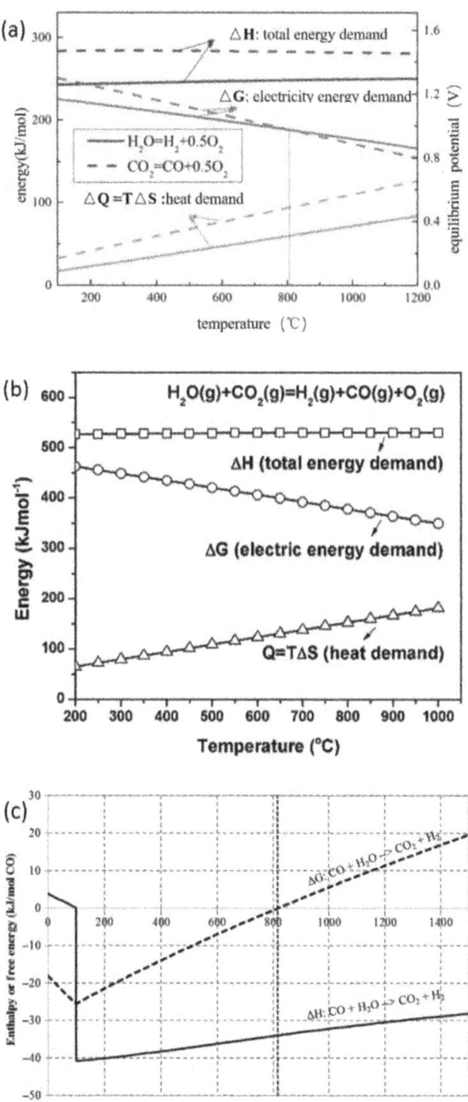

FIGURE 6.2 (a) Relationships between enthalpy change (ΔH), Gibbs free energy change (ΔG), and entropy change (ΔS) of electrolysis reactions for CO_2 (g, dotted lines), H_2O (g, solid lines) and (b) H_2O/CO_2 (g), as well as (c) the relationship between ΔG and ΔH of the water gas shift reaction. (From Li, W.Y., Mechanism and characteristics research on CO_2/H_2O co-electrolysis by solid oxide electrolysis cells, Doctoral dissertation, Tsinghua University, April 2015; Reprinted from *J. Power Sources*, 277, Wang, Y. et al., A novel clean and effective syngas production system based on partial oxidation of methane assisted solid oxide co-electrolysis process, 261–267, Copyright 2015, with permission from Elsevier; Reprinted from Elder, R. et al., High temperature electrolysis, in Styring, P. et al. (Eds.), *Carbon Dioxide Utilisation: Closing the Carbon Cycle*, pp. 183–209, Elsevier, Amsterdam, the Netherlands, 2015, Copyright 2014, with permission from Elsevier.)

is more favorable at lower temperatures (ΔG < 0), but the reverse process is opposite. When the temperature is higher than 3000°C, the reaction of CO_2 electrolysis becomes spontaneous, but the stability of the material becomes affected. Thus, it is clear that operations at 600°C–1000°C (high-temperature electrolysis) with a gas feed stream is much easier than operations at room temperature using liquid/gas feeds.[15] It can be seen from this that the cell configurations and material requirements in high-temperature electrolysis are obviously different from those of low-temperature electrolysis systems.

$$H_2O + CO \underset{RWGS}{\overset{WGS}{\rightleftarrows}} H_2 + CO_2 \tag{6.11}$$

6.2.2 Kinetics of High-Temperature CO_2/H_2O Co-electrolysis

At the fuel electrode, the electrochemical reductions of H_2O or CO_2 occurs at the triple phase boundary (TPB), and both the RWGS and WGS reactions occur on the surfaces of the electrode catalyst (e.g., nickel).[19] This can be expressed by Equations (6.1), (6.2), and (6.11). As mentioned above, compared to separate H_2O or CO_2 electrolysis, co-electrolysis is complicated due to the WGS reaction. In the literature, based on electrochemical analysis, different researchers have seen different contributions of the WGS reaction to the overall process.[7,20] This is probably due to the fact that the reaction mechanisms depend on many variables, such as temperature, pressure, gas flow rates, compositions, materials, morphology, and experimental setups and conditions. Therefore, it is important to consider the exact conditions employed when discussing experimental results.

In fact, disputes still exist concerning co-electrolysis reaction pathways. As shown in Table 6.2, in 2011, Kim-Lohsoontorn and Bae[21] investigated the electrodes of solid oxide electrolysis cells (SOECs) during H_2O electrolysis, CO_2 electrolysis, and CO_2/H_2O co-electrolysis by using Ni-YSZ, Ni-GDC, and Ni/Ru-GDC as fuel electrodes. The results showed that the performances of H_2O electrolysis and CO_2/H_2O co-electrolysis at various electrodes are comparable. A similar result was obtained by Stoots et al.,[22] from the Idaho National Laboratory (INL), in 2009. The electrolysis and co-electrolysis polarization curves were overlapping during SOEC stack operation, and the area specific resistances (ASRs) of co-electrolysis and H_2O electrolysis were very close. They believed that, at these electrolysis conditions, the reduction of CO_2 in co-electrolysis is mainly due to RWGS but not due to CO_2 electrolysis. However, when Graves et al.[20] from the Risø National Laboratory (RNL) and Li et al.[23] from Tsinghua University (THU) examined the co-electrolysis performance of a SOEC, they observed that the polarization curves and the electrochemical impedance spectroscopy (EIS) of co-electrolysis were both located between those of CO_2 electrolysis and H_2O electrolysis. Therefore, they believed that H_2O electrolysis and CO_2 electrolysis both occurred in co-electrolysis, and that the CO was produced from RWGS and CO_2 electrolysis.

As shown in Figure 6.3, a reaction pathway for CH_4 production via CO_2/H_2O co-electrolysis was proposed by Li et al.[23] In their proposal, at the Ni-YSZ cathode (650°C), CO can be generated by CO_2 electrolysis through both the R1 pathway and the RWGS reaction (R2). CH_4 is then generated from the reactions of CO and H_2

TABLE 6.2
Some Experiments of CO_2/H_2O Co-electrolysis and Main Results about the Reaction Mechanisms

Units[a]	Material	Cell Type	Gas Composition	Main Conclusions
KAIST[21]	Ni-YSZ YSZ Pt	Button cell, electrolyte support	$38H_2O/38H_2/24N_2$ $25CO_2/25CO/50N_2$ $22CO_2/22CO/22H_2O/22H_2/12N_2$	Almost no CO_2 electrolysis CO is mainly generated from RWGS
INL[22]	Ni-YSZ YSZ LSM	Cell stacks, electrolyte support	$54.8H_2O/22.5H_2/22.7N_2$ $100CO_2$ $54.9H_2O/22.5H_2/22.6CO_2$	
RNL[20]	Ni-YSZ YSZ LSM-YSZ	Planar cell, fuel electrode support	$50H_2O/50H_2$ $50CO_2/50CO$ $25CO_2/25CO/25H_2O/25H_2$	With CO_2 electrolysis CO is produced from RWGS and CO_2 electrolysis
THU[23]	Ni-YSZ ScSZ LSM-ScSZ	Button cell, fuel electrode supported	$28.6H_2O/14.3H_2/57.1Ar$ $28.6CO_2/14.3CO/57.1Ar$ $28.6CO_2/28.6H_2O/14.3H_2/28.5Ar$	

Source: Li, W. et al., *J. Power Sources*, 243, 118–130, 2013.

[a] KAIST: Korea Advanced Institute of Science and Technology, INL: Idaho National Laboratory, RNL: Risø National Laboratory, THU: Tsinghua University.

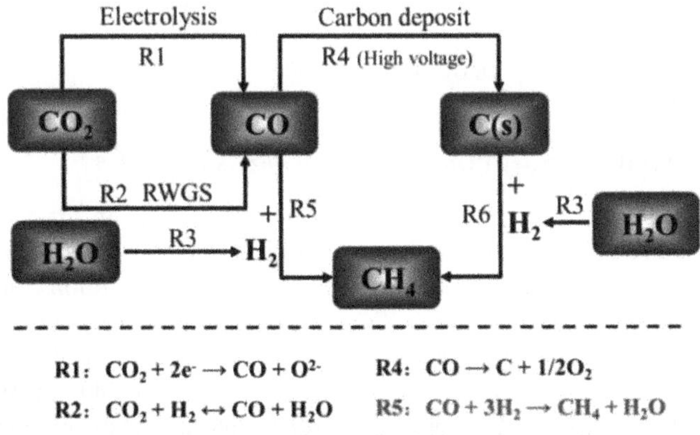

FIGURE 6.3 Proposed reaction pathways for methane production by CO_2/H_2O co-electrolysis. (Reprinted from *Int. J. Hydrogen Energ.*, 38, Li, W. et al., Performance and methane production characteristics of H_2O-CO_2 co-electrolysis in solid oxide electrolysis cells, 11104–11109, Copyright 2013, with permission from Elsevier.)

produced by H_2O electrolysis (R3) on the Ni catalyst (R5). Furthermore, by increasing the operating voltage, CO can be reduced to C(s) by the R4 pathway. It is suspected that CH_4 can also be generated from elemental C hydrogenation reactions through the R6 pathway. When higher operating voltages are applied, the carbon-deposits will be more significant in the presence of Ni catalysts, resulting in more CH_4 through the reaction of elemental C and H_2 (R6). Obviously, adding H_2 into the reactants can also facilitate R5 and R6 pathways to generate more CH_4.

6.2.3 Comparisons between Water Electrolysis, CO_2 Electrolysis, and H_2O/CO_2 Co-electrolysis

The comparable performances for steam or CO_2 electrolysis and co-electrolysis are shown in Figure 6.4. For an individual electrolysis cell, Figure 6.4a shows that the performance of CO_2 electrolysis using SOEC at 850°C is obviously lower than H_2O/CO_2 co-electrolysis, and the lowest ASR can be measured for H_2O electrolysis. Moreover, pure CO_2 electrolysis may also cause nickel oxidation in the Ni-YSZ fuel electrode, leading to degradation of cell performance.[3] Similar results can also be found in Figure 6.4b, which shows a 10-cell SOEC stack in which the ASR of pure CO_2 electrolysis is approximately 1.8 times higher than that of CO_2/H_2O electrolysis.[22] Although the ASR of H_2O electrolysis is similar to CO_2/H_2O electrolysis, the conversion of CO_2 into useful sustainable fuels or chemicals can't be achieved by individual H_2O electrolysis. In addition, the co-electrolysis has many fewer electrolysis steps, and it requires only one necessary reactor to operate as mentioned above. On the other hand, even in regard to H_2O electrolysis only, the performances using alkaline cells, proton exchange membrane cells, or solid oxide electrolysis cells has been investigated and the result indicates that the electrolysis efficiency could be decreased with increasing cell voltage, and that the capital cost was decreased with increasing current density, as shown in Figure 6.4c.[6] Obviously, when taking all these factors into account, the H_2O/CO_2 co-electrolysis using SOEC for CO_2 conversion to useful fuels or chemicals is superior to others.

6.2.4 Key Material Selection and Components of SOCs

A SOEC is a device that mainly consists of a fuel electrode, an oxygen electrode, and a dense electrolyte, as discussed earlier. Similarly, the microstructure of a SOEC also includes the same three compartments. One of the most common configurations of a SOEC to be investigated is Ni-YSZ|YSZ|LSM-YSZ, as shown in Figure 6.5.

During SOEC operation, the electrode materials are used to offer active sites for significant electrochemical reactions, including oxygen evolution reaction and reduction of H_2O and CO_2. Definitely, electrode materials also provide pathways for transport of electrons, ions, reactants, and products. As a good conductor, the electrolyte material provides a pathway for oxide ions from the fuel electrode to the oxygen electrode, which can mainly determine the cell ohmic resistance,[26] whereas the electron transport should be suppressed in the electrolyte to prevent the "short circuit." The properties of the electrolyte have a major impact on electrolysis cell

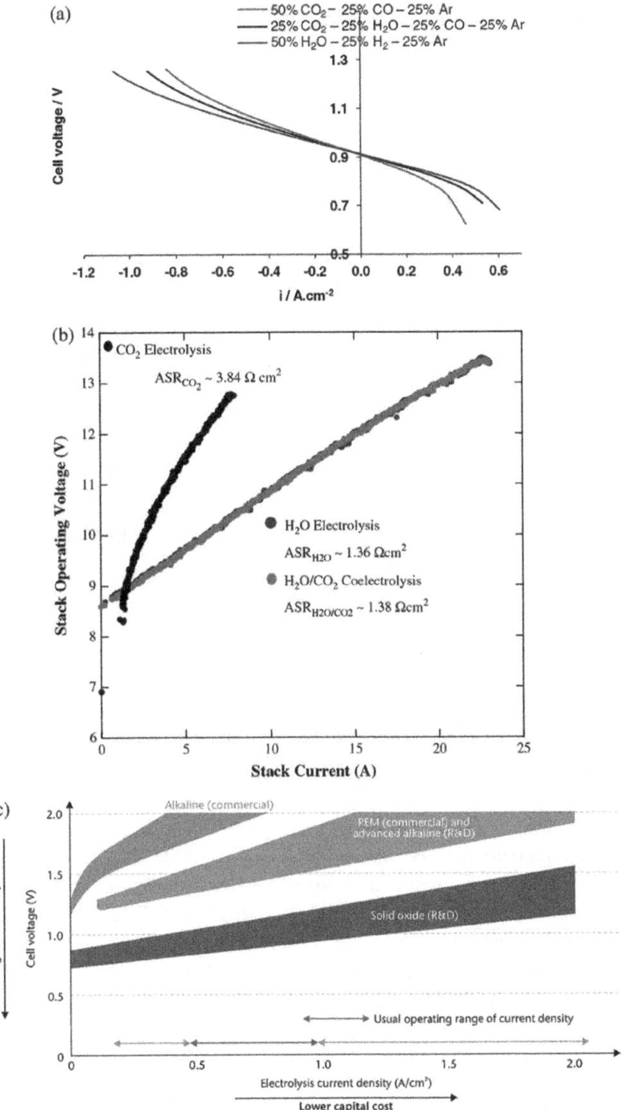

FIGURE 6.4 Polarization characterization of electrolysis cell: (a) a planar Ni-YSZ|YSZ|LSM-YSZ SOEC, H_2O electrolysis (50% H_2O-25% H_2-25% Ar), H_2O/CO_2 co-electrolysis (25% H_2O-25% CO_2-25% CO-25% Ar) and CO_2 electrolysis (50% CO_2-25% CO-25% Ar) and stack: (b) 10 cell stack, H_2O electrolysis (54.8% H_2O-22.5% H_2-22.7% N_2), H_2O/CO_2 co-electrolysis (54.9% H_2O-22.6% CO_2-22.5% H_2) and CO_2 electrolysis (100% CO_2) and (c) H_2O electrolysis in alkaline cells, proton exchange membrane cells, and solid oxide cells. (Reprinted with permission from Ebbesen, S.D. et al., *Chem. Rev.*, 114, 10697–10734, 2014. Copyright 2014 American Chemical Society; Reprinted from *Int. J. Hydrogen Energ.*, 34, Stoots, C. et al., Results of recent high temperature coelectrolysis studies at the Idaho National Laboratory, 4208–4215, Copyright 2009, with permission from Elsevier; From International Energy Agency, Technology Roadmap Hydrogen and Fuel Cells, Copyright 2015, IEA.)

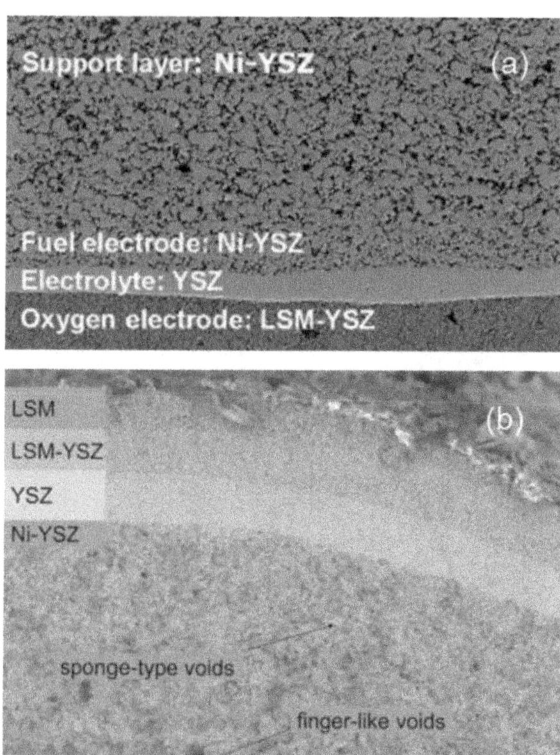

FIGURE 6.5 Microstructures of a reversible solid oxide cell with the configuration Ni-YSZ|YSZ|LSM-YSZ: (a) a planar and (b) a tubular. (Reprinted with permission from Ebbesen, S.D. et al., *Chem. Rev.*, 114, 10697–10734, Copyright 2014, American Chemical Society; Reprinted from *Electrochim. Acta*, 179, Kleiminger, L. et al., Syngas (CO-H_2) production using high temperature micro-tubular solid oxide electrolysers, 565–577, Copyright 2015, with permission from Elsevier.)

performance through its contribution to the ohmic internal resistance. In addition to those materials, the interconnect offers an electrical connection between the oxygen electrode of one individual cell to the fuel electrode of an adjacent one in SOEC mode. It is also considered a physical barrier to protect the oxygen electrode material from the reducing atmosphere of the fuel electrode side.[27] Similarly, the fuel electrode materials can equally be protected from the oxidizing environment. The H_2 and CO are very useful products for H_2O/CO_2 co-electrolysis to generate fuels/chemicals; however, they leak out easily in high temperature due to their small molecule size. Therefore, another material for cell sealing used in SOEC to avoid gas leakage is indispensable; this material must endure both reducing and oxidizing conditions during solid oxide cell operating.[28] Note that the interconnect and cell sealing are of great importance, especially for multiple cell stack.

Since normal SOECs operate at high temperatures, there are many important limitations on electrode, electrolyte, and interconnect materials.[3] Therefore the material requirements for SOEC, such as high conductivities, compatible thermal expansion coefficients, dimensional/mechanical/chemical/thermal stabilities, and cell integration, are strict.[29–34] Ni et al.[35] emphasized the following requirements for SOEC components:

1. In order to achieve high energy conversion efficiency, the dense electrolyte should have good ionic conductivity, low electronic conduction, and chemical stability.
2. The dense electrolyte should be as thin as possible to lower the ohmic overpotential, whereas the syngas and O_2 should be separated completely by a gastight structure of electrolyte.
3. Good electronic conductivity and chemical stability are necessary for fuel and oxygen electrodes, especially in reducing or oxidizing atmosphere.
4. The suitable porosity and pore size should be designed not only to support gas transportation but also to provide sufficient triple phase boundary (3PB), the boundary of electrolyte, electrode, and gas.
5. For the purpose of avoiding material failure, the compatible thermal expansion coefficients of both electrolyte and electrodes should be achieved.
6. With regard to interconnect materials, chemical stability in both the reducing and the oxidizing atmosphere is particularly important because the interconnect materials usually contact with CO (g), H_2O (g), CO_2 (g), and O_2 (g) simultaneously.
7. Last but not least, the cost of raw materials and manufacture should be as low as possible.

6.3 SUMMARY

The typical electrochemical method for CO_2 conversion under high temperature is introduced in this chapter. The operating principle and features, as well as the fundamentals, including thermodynamics and kinetics of CO_2/H_2O co-electrolysis using a solid oxide electrolysis cell, are summarised in this chapter. To facilitate research and development in overcoming some challenges such as the degradation issues at long-time operation, one of the most important topics is materials research; thus, some selection requirements are introduced here. More detail about the materials used in each component of SOECs will be discussed in subsequent chapters.

REFERENCES

1. Zheng, Y., et al. A review of high temperature co-electrolysis of H_2O and CO_2 to produce sustainable fuels using solid oxide electrolysis cells (SOECs): Advanced materials and technology. *Chemical Society Reviews* **46**, 1427–1463 (2017).
2. Irvine, J. T., et al. Evolution of the electrochemical interface in high-temperature fuel cells and electrolysers. *Nature Energy* **1**, 15014 (2016).
3. Ebbesen, S. D., Jensen, S. H., Hauch, A., Mogensen, M. B. High temperature electrolysis in alkaline cells, solid proton conducting cells, and solid oxide cells. *Chemical Reviews* **114**, 10697–10734 (2014).

4. Gao, Z., Mogni, L. V., Miller, E. C., Railsback, J. G., Barnett, S. A. A perspective on low-temperature solid oxide fuel cells. *Energy & Environmental Science* **9**(5), 1602–1644 (2016).
5. Fu, Q., Mabilat, C., Zahid, M., Brisse, A., Gautier, L. Syngas production via high-temperature steam/CO_2 co-electrolysis: An economic assessment. *Energy & Environmental Science* **3**, 1365–1608 (2010).
6. International Energy Agency, Technology Roadmap Hydrogen and Fuel Cells, June, 2015.
7. Ebbesen, S. D., Graves, C., Mogensen, M. Production of synthetic fuels by co-electrolysis of steam and carbon dioxide. *International Journal of Green Energy* **6**, 646–660 (2009).
8. Stoots, C. M., Herring, J. S., Hartvigsen, J. J. Syngas production via high-temperature coelectrolysis of steam and carbon dioxide. *Journal of Fuel Cell Science and Technology* **6**, 0110141–01101412 (2009).
9. Ebbesen, S. D., Graves, C., Hauch, A., Jensen, S. H., Mogensen, M. Poisoning of solid oxide electrolysis cells by impurities. *Journal of the Electrochemical Society* **157**, B1419–B1429 (2010).
10. Isenberg, A. O. Energy conversion via solid oxide electrolyte electrochemical cells at high temperatures. *Solid State Ionics* **3/4**, 431–437 (1981).
11. Ebbesen, S. D., Knibbe, R., Mogensen, M. Co-electrolysis of steam and carbon dioxide in solid oxide cells. *Journal of the Electrochemical Society* **159**, F482–F489 (2012).
12. Menon, V., Fu, Q., Janardhanan, V. M., Deutschmann, O. A model-based understanding of solid-oxide electrolysis cells (SOECs) for syngas production by H_2O/CO_2 co-electrolysis. *Journal of Power Sources* **274**, 768–781 (2015).
13. Minh, N. Q., Mogensen, M. B. Reversible solid oxide fuel cell technology for green fuel and power production. *The Electrochemical Society Interface Winter* **22**, 55–62 (2013).
14. Ihsan, B. (ed.) *Thermochemical Data of Pure Substances*, (Weinheim, Germany, Federal Republic of Germany; WILEY-VCH Verlag GmbH, New York, 1993).
15. Elder, R., Cumming, D., Mogensen, M. B. High temperature electrolysis. In Styring, P., Quadrelli, A., Armstrong, K. (Eds.), *Carbon Dioxide Utilisation: Closing the Carbon Cycle*, pp. 183–209 (Elsevier, Amsterdam, the Netherlands, 2015).
16. Graves, C. R. Recycling CO_2 into sustainable hydrocarbon fuels: Electrolysis of CO_2 and H_2O, PhD dissertation, Columbia University (2010).
17. Wang, Y., et al. A novel clean and effective syngas production system based on partial oxidation of methane assisted solid oxide co-electrolysis process. *Journal of Power Sources* **277**, 261–267 (2015).
18. Li, W. Y. Mechanism and characteristics research on CO_2/H_2O co-electrolysis by solid oxide electrolysis cells, Doctoral dissertation from Tsinghua University, April 2015.
19. Kim, S., et al. Reactions and mass transport in high temperature co-electrolysis of steam/CO_2 mixtures for syngas production. *Journal of Power Sources* **280**, 630–639 (2015).
20. Graves, C., Ebbesen, S. D., Mogensen, M. Co-electrolysis of CO_2 and H_2O in solid oxide cells: Performance and durability. *Solid State Ionics* **192**, 398–403 (2011).
21. Kim-Lohsoontorn, P., Bae, J. Electrochemical performance of solid oxide electrolysis cell electrodes under high-temperature coelectrolysis of steam and carbon dioxide. *Journal of Power Sources* **196**, 7161–7168 (2011).
22. Stoots, C., O'Brien, J., Hartvigsen, J. Results of recent high temperature coelectrolysis studies at the Idaho National Laboratory. *International Journal of Hydrogen Energy* **34**, 4208–4215 (2009).
23. Li, W., Wang, H., Shi, Y., Cai, N. Performance and methane production characteristics of H_2O-CO_2 co-electrolysis in solid oxide electrolysis cells. *International Journal of Hydrogen Energy* **38**, 11104–11109 (2013).

24. Li, W., Shi, Y., Luo, Y., Cai, N. Elementary reaction modeling of CO_2/H_2O co-electrolysis cell considering effects of cathode thickness. *Journal of Power Sources* **243**, 118–130 (2013).
25. Kleiminger, L., Li, T., Li, K., Kelsall, G.H. Syngas (CO-H_2) production using high temperature micro-tubular solid oxide electrolysers. *Electrochimica Acta* **179**, 565–577 (2015).
26. Jiang, S. P. Challenges in the development of reversible solid oxide cell technologies: A mini review. *Asia-Pacific Journal of Chemical Engineering* **11**, 386–391 (2016).
27. Haile, S. M. Fuel cell materials and components. *Acta Materialia* **51**, 5981–6000 (2003).
28. Lessing, P. A. A review of sealing technologies applicable to solid oxide electrolysis cells. *Journal of Materials Science* **42**, 3465–3476 (2007).
29. Kan, W. H., Thangadurai, V. Challenges and prospects of anodes for solid oxide fuel cells (SOFCs). *Ionics* **21**, 301–318 (2015).
30. Tao, G., et al. Study of carbon dioxide electrolysis at electrode/electrolyte interface: Part I. Pt/YSZ interface. *Solid State Ionics* **175**, 615–619 (2004).
31. Yue, X., Irvine, J. T. Alternative cathode material for CO_2 reduction by high temperature solid oxide electrolysis cells. *Journal of the Electrochemical Society* **159**, F442–F448 (2012).
32. Yang, C., Yang, Z., Jin, C., Liu, M., Chen, F. High performance solid oxide electrolysis cells using Pr0.8Sr1.2(Co, Fe)0.8Nb0.2O4+δ–Co–Fe alloy hydrogen electrodes. *International Journal of Hydrogen Energy* **38**, 11202–11208 (2013).
33. Hauch, A., Ebbesen, S. D., Jensen, S. H., Mogensen, M. Solid oxide electrolysis cells: Microstructure and degradation of the Ni/Yttria-Stabilized zirconia electrode. *Journal of the Electrochemical Society* **155**, B1184–B1193 (2008).
34. Rossmeisl, J., Bessler, W.G. Trends in catalytic activity for SOFC anode materials. *Solid State Ionics* **178**, 1694–1700 (2008).
35. Ni, M., Lenng, M., Leung, D. Technological development of hydrogen production by solid oxide electrolyzer cell (SOEC). *International Journal of Hydrogen Energy* **33**, 2337–2354 (2008).

7 High-Temperature Electrochemical Process of CO_2 Conversion with SOCs 2

Research Status

7.1 BRIEF HISTORY

In the mid-1960s, in the history of electrochemical CO_2 conversion and utilization, solid oxide electrolysis of CO_2 was originally presented as a technique for oxygen regeneration in spacecraft habitats in the US manned space-flight program. Westinghouse Research Laboratories in the United States made pioneer contributions to high temperature co-electrolysis (HTCE) for aerospace applications. Chandler et al.[1] succeeded in building and operating a number of small-scale systems with oxygen production rates of 150 cc/min. Elikan et al.[2] first proposed the concept of utilizing solid electrolyte electrolysis batteries to generate oxygen from decomposing respiratory CO_2 and/or H_2O instead of carrying a heavily stored oxygen supply, which appeared to be an important means for reducing launch weight on manned space flights of long durations. This research was carried out at Westinghouse Research Laboratories under the support of National Aeronautics and Space Administration (NASA).

In the 1970s, Ash et al.[3] first proposed the use of solid oxide electrolysis of CO_2 as an in situ resource utilization (ISRU) technology to generate O_2 from the predominately CO_2 atmosphere of Mars. Stancati et al.[4] then proposed the use of solid oxide electrolysis cells (SOECs) for remote automated in situ jet propellant production for an unmanned Mars sample return mission. Richter[5] performed experimental research into the technological investigations using yttria-stabilized zirconia (YSZ) as the solid electrolyte. In the mid-1990s, Sridhar et al.[6,7] considered utilizing solid oxide cells as regenerative cells, utilizing CO and O_2 produced by solid oxide cells during the daylight hours as fuel under fuel cell mode and producing electricity during the night under electrolysis mode.

Since the beginning of the twenty-first century, research groups from various countries and regions have renewed studies to focus on the development of new materials, the enhancement of the durability and performance of stacks, and the reduction of the cost of syngas production. In 2004, NASA began to plan exploratory missions to the moon again and further on to Mars, with research on HTCE for aerospace. Since 2005, there has been renewed international interest in related terrestrial applications for the co-electrolysis process, namely, the feasibility of high-temperature CO_2/H_2O co-electrolysis to produce syngas ($CO+H_2$).[8,9] Based on high-temperature steam electrolysis technology, there has been rapid development of HTCE using SOECs.

In the United States, the Idaho National Laboratory (INL), in cooperation with Ceramatec Inc. and accompanied by Northwestern University, the University of California, and the University of Florida, has played an important role in promoting the feasibility of HTCE using SOECs. With emphasis on renewable energy sources, many research groups from Europe and China have also been involved in the development of HTCE to produce sustainable fuels using SOECs. The research status in terms of technical features and some research groups in different countries and regions are shown in Table 7.1.

TABLE 7.1
Some Research Groups with Various Research Fields in Different Countries and Regions

Countries/ Regions	Unit/Groups	Fields	Ref.
USA	NASA	(1) Solid oxide electrolysis of CO_2 for O_2 regeneration	1,10,11
		(2) Tubular YSZ cell stack with Pt-ZrO_2 composite electrodes	12,13
		(3) CO_2 In Situ Resource Utilization (ISRU) technology	3,14
	INL	(1) The feasibility of HTCE using SOEC	9,15
		(2) "Next Generation Hybrid System: Nuclear/Coal to Liquid"	16
		(3) A novel process called biosyntrolysis	17
		(4) Co-electrolysis and Cabin Air Revitalization Deployment Architecture	18
	Univ. of Florida	(1) Bilayer-electrolyte (erbia-stabilized bismuth oxide/ samaria-doped ceria)	19
	Univ. of Pennsylvania	(1) Electrolyze CO_2 in SOEC using a infiltrated ceramic electrode	20
	Northwestern Univ.	(1) The process of "renewable-to-liquids" (RTL)	21
		(2) Electrolysis through SOEC with H_2O-CO_2-H_2 mixtures	22,23
	Columbia Univ.	(1) Transforming CO_2 into sustainable hydrocarbon fuels with CO_2/H_2O co-electrolysis	24
		(2) Thermodynamic performance of fuel synthesis system with heat from renewable energy	25
	MIT	(1) Degradation mechanism in the contact layer of SOEC stacks	26
		(2) Surface electronic structure transitions	27
		(3) Microscopic level understanding and control of a faster ORR kinetics	28
		(4) Materials for batteries/cells and the thermodynamics and kinetics of electrochemical reactions	29,30

(Continued)

TABLE 7.1 (*Continued*)
Some Research Groups with Various Research Fields in Different Countries and Regions

Countries/ Regions	Unit/Groups	Fields	Ref.
	Univ. of Texas at Austin	(1) Discussed fluorite structure of solid oxides and various solid materials that allow oxide-ion conduction, as well as double perovskite as anode materials for SOFCs	31
	Univ. of South Carolina	(1) A superior solid oxide-ion electrolyte $Sr_{3-3x}Na_{3x}Si_3O_{9-1.5x}$ (x=0.45)	32
		(2) A novel dual-layer Ni–YSZ cathode-supported and various hydrogen electrodes in SOEC	33–36
		(3) Electrode materials $Sr_2Fe_{1.5}Mo_{0.5}O_{6-\delta}$ in HTSE and HTCE	37,38
		(4) Combining HTCE and partial oxidation of methane (POM)	39
Europe	RNL	(1) SOEC and SOFC operation modes under different conditions	40
		(2) A synthetic fuel production system with integrated high-pressure HTCE	41
		(3) The degradation of SOEC during the process of HTCE	42,43
		(4) Large-scale electricity storage utilizing SOEC combined with underground storage of CH_4 and CO_2	44
	EIFER	(1) Technical and economic analysis of HTCE and FT synthesis for syngas and synfuel production	45,46
	FZJ	(2) Materials development covering cell manufacturing, stack design, characterization, and even the subsequent systems design and demonstration	47,48
		(3) A 9000 h operation of a solid oxide cell	49,50
		(4) Tape-casting and sequential tape-casting	51
		(5) 1- and 5-layer planar-type steam pre-reformers	52
	SunFire GmbH	(1) A reversible solid oxide cell electrolyzer with high-efficiency and low-cost hydrogen production	53
	CEA	(1) HTSE for large-scale demonstration	54
		(2) A low-weight stack, degradation rate below 3%/kh with 3-cell stacks	55,56
		(3) Micro models for HTSE and HTCE	57
	Imperial College London	(1) Thickness of SOFC electrodes using a 3D microstructure model	58,59
		(2) H_2 production when coupling SOEC with intermittent renewable energies.	60
		(3) Electrode ($La_{2-x}Sr_xCo_{0.5}Ni_{0.5}O_{4\pm\delta}$) and electrolyte materials (scandia-and ceria-stabilized zirconia 10Sc1CeSZ)	61

(Continued)

TABLE 7.1 (*Continued*)
Some Research Groups with Various Research Fields in Different Countries and Regions

Countries/ Regions	Unit/Groups	Fields	Ref.
	ICMA	(1) Steam electrolysis using a microtubular SOFC, electrode ($Nd_2NiO_{4+\delta}$) and electrolyte materials ($LaNb_{0.84}W_{0.16}O_{4.08}$)	61–64
China	THU	(1) HTR-PM project, Multiple-cells SOEC stacks, design and construction of kilowatt scale HTE facilities	65–74
		(2) Patterned, porous, and tubular SOEC in different scales; theoretical modeling of HTCE	75–80
		(3) Electrochemical performance and stability of electrode or electrolyte materials including LSM–YSZ, LSCF-GDC, LCCZ-GDC, BCFN9721, LSFG, and LSCFN for solid oxide cells	81
	USTC	(1) Manufacturing CH_4 from HTCE in tubular SOECs directly	82
	CAS	(1) $Nd_2NiO_{4+\delta}$ (NNO)-SSZ oxygen electrode	83
		(2) YSZ-SFM composite oxygen electrode	84

Abbreviations: National Aeronautics and Space Administration (NASA), Idaho National Laboratory (INL), Massachusetts Institute of Technology (MIT), Risø National Laboratory (RNL), European Institute for Energy Research (EIFER), Forschungszentrum Jülich (FZJ), French Atomic Energy Commission (CEA), Materials Science Institute of Aragon (ICMA), University of Science and Technology of China (USTC), Tsinghua University (THU), Chinese Academy of Sciences (CAS). High-temperature CO_2/H_2O co-electrolysis (HTCE), High-temperature steam electrolysis (HTSE), High-temperature electrolysis (THE, including HTCE and HTSE).

7.2 RESEARCH STATUS OF HTCE WITH SOC IN THE UNITED STATES

Westinghouse Research Laboratories in the United States made a pioneer contribution to HTCE for aerospace application. Chandler and his co-workers[1] succeed in building and operating a number of small-scale systems with an oxygen production rate of 150 cc/min. Weissbart and Smart[85] studied alternate electrolyte materials and began to pursue the topic using alternate cell geometry of flat disc. Elikan et al.[12,13] also built a breadboard regeneration system (shown in Figure 7.1) to reduce a mixture of CO_2 and H_2O to recover oxygen for spacecraft life-support applications. The cells consisted of a tubular YSZ cell stack with Pt-ZrO_2 composite electrodes. The feasibility of constructing a multi-person system for manned testing had been demonstrated based on the successful operation for 180 days of an integrated, closed-loop breadboard system that produced 125 cc/min of oxygen. The system could supply pure oxygen to a 4-person, 100-day mission with a weight of about 121 pounds that would require

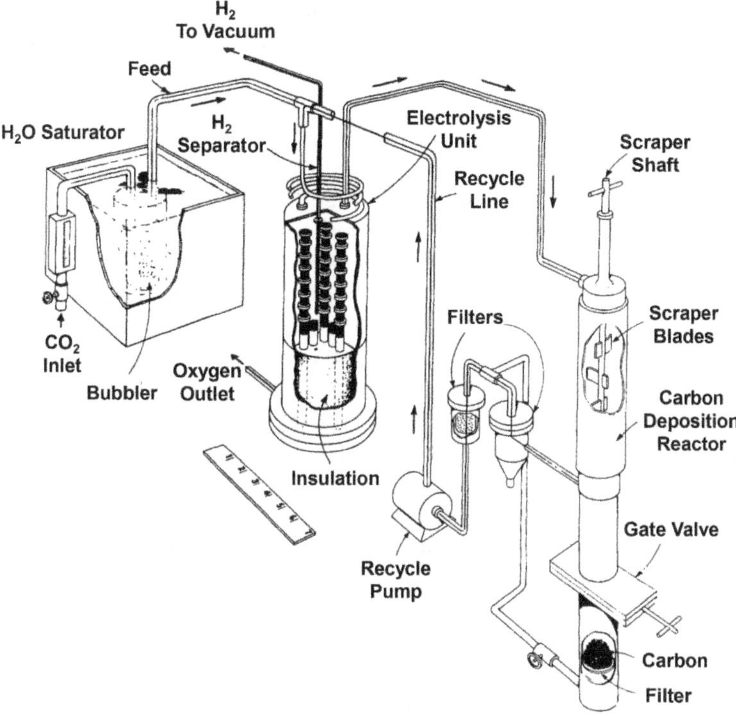

FIGURE 7.1 Schematic diagram of breadboard regeneration system.

approximately 1160 watts of power. Aside from the works performed by Ash et al., Stancati et al., Richter, and Sridhar et al. mentioned above, Tao Gege[86,87] carried on the regenerative experiments using Pt/YSZ/Pt and Pt-YSZ/YSZ/Pt-YSZ, in which the current density increased three times due to the improvement of the electrodes.

In 2004, NASA began to plan manned and unmanned exploratory missions to return to the moon and eventually to Mars, as part of the Exploration Vision directed by the US president. For a mission to Mars, the natural resource most readily available is the atmosphere, which consists of roughly 95% carbon dioxide, a potential feedstock for the production of oxygen, methane, and other consumables. Singhal[88] discussed the advantage of CO_2 solid oxide electrolysis over two other potential competing technologies (Sabatier/water electrolysis and reverse water-gas shift reaction processes). On the one hand, solid oxide electrolysis of carbon dioxide can produce a nearly pure, dry oxygen gas stream without any additional chemical separation required. On the other hand, the device could be operated in reverse to be used as a fuel cell that can generate power when solar power is not available, or for portable devices such as rovers. The solid oxide system is the most promising selection for Mar's mission because of its flexibility in being able to handle mixtures of CO_2 and water in any ratio, ease of control, and fewer interfaces. Research on using solid oxide cells for the purpose of aerospace continues today.

From 2006, INL (the United States), in conjunction with Ceramatec Inc. (Salt Lake City, United States) carried out investigations on the feasibility of HTCE using SOC.[8] Three-dimensional computational fluid dynamics (CFD) models had been created to model high-temperature co-electrolysis process utilizing a planar solid oxide electrolyzer.[9,15] A process model was developed to evaluate the potential performance of a large-scale high-temperature co-electrolysis plant for the production of syngas from steam and carbon dioxide in 2007.[89] After that, a one-dimensional chemical equilibrium model was developed for mechanism analysis of simultaneous high-temperature co-electrolysis process, as shown in Figure 7.2.[90]

In 2008, multi-stack testing was conducted at INL. The stacks were fabricated by Ceramatec Inc., with a size of 10×10 cm^2 and an active area of 8×8 cm^2. The stacks tested ranged from 10 cell short stacks to 240 cell modules. Tests were conducted either in the bench-scale test apparatus or in the newly developed 5 kW Integrated Laboratory Scale (ILS) test facility.[91,92] In 2009, INL studied system simulation and economics evaluation of high-temperature co-electrolysis for synthetic fuels production driven by nuclear energy.[93]

In 2010, Department of Energy (DOE) in the United States, underwent rapid and dramatic expansion with an unprecedented influx of funds to execute the Energy Independence mission. Federal laboratories, including INL, were focused on developing "transformational" energy technology to assist the federal government in implementing a clean, secure, efficient energy policy. "Next-Generation Hybrid System: Nuclear/Coal to Liquid" was put forward by INL, as shown in Figure 7.3.[16] On the basis of the 2010 annual report, J. D. Smith pointed out that the co-electrolysis process, when coupled with a high-temperature nuclear reactor in the hybrid systems, can save up to 50% of the carbon from the direct use of liquid fuels, and enable 99% less CO_2 emissions than the "standard" coal-to-liquid fuel conversion. Also, the system can integrate with renewable energy sources (i.e., wind, solar, hydro, geothermal, etc.) to achieve flexible energy storage, which can be on a smaller scale than a dedicated nuclear plant. In addition, a novel process called biosyntrolysis was proposed first by INL in 2010 to investigate the feasibility of syngas production from renewable biomass.[17]

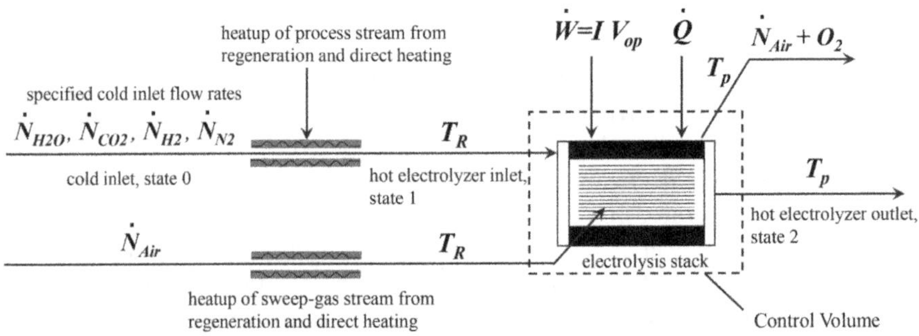

FIGURE 7.2 Schematic of the chemical equilibrium co-electrolysis (CEC) model.

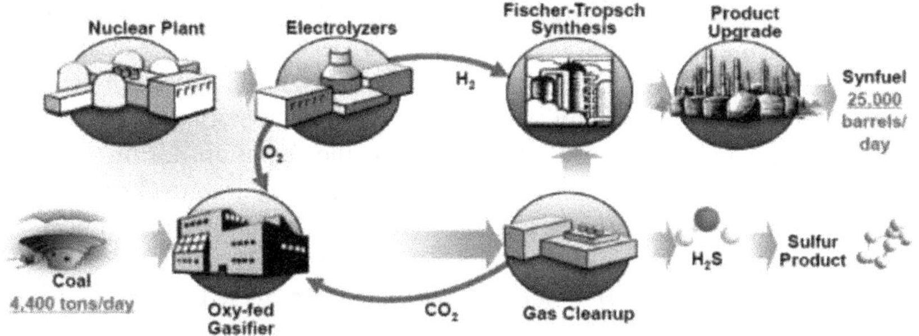

FIGURE 7.3 Next-generation hybrid system: nuclear/coal to liquid. (From Smith, J.D. and Reed, M., Innovative management of carbon emission from fossil plants, *24th Annual ACERC Conference*, Prove, UT, 2010.)

In 2012, the concept of co-electrolysis and cabin air revitalization deployment architecture, in support of the co-electrolysis in situ processes examples, was provided by INL.[18] Figure 7.4 shows the concept deployment schematic architecture in the context of a manned Mars surface exploration mission. In this concept, it was assumed that a bimodal nuclear thermal propulsion system would be used for surface descent and ascent operations, in addition to electrical power production and the provision of supplemental process heat during the course of the surface mission. Also

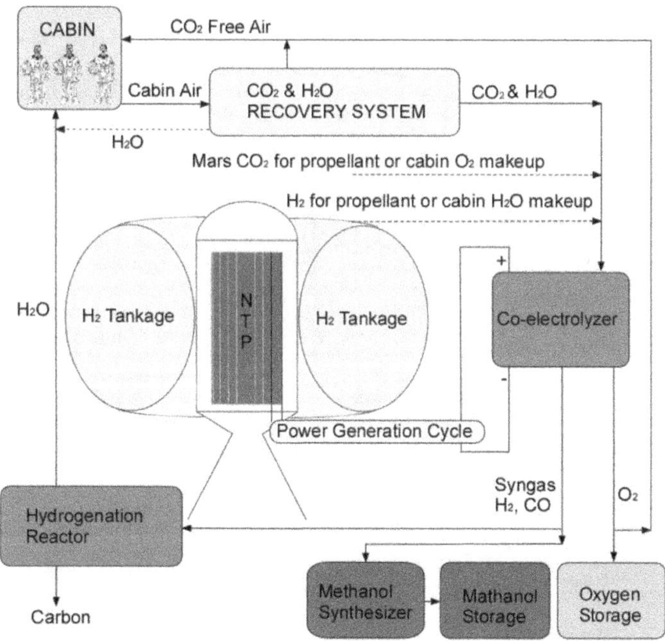

FIGURE 7.4 Schematic of a concept coupled system for cabin air revitalization and fuel production. (From O'Brien, R.C. et al., *Nucl. Emerg. Technol. Space*, 2012, 1–2, 2012.)

in 2012, Sohal[94] summarized various ongoing INL and INL-sponsored activities aimed at addressing SOEC degradation, including stack testing, post-test examination, degradation modeling, and a list of issues that must be addressed in the future.

From 2013 to 2015, the durability of SOC for high-temperature steam electrolysis was evaluated and improved by INL, and it is the important technical foundation of HTCE. In 2013, the durability of SOC for high-temperature electrolysis was evaluated and improved by Zhang at INL.[95] An experimental study has been conducted to demonstrate recent improvements in long-term durability of SOEC and stacks. The electrolyte-supported SOEC 10-cell stacks (provided by Ceramatec Inc.) and electrode-supported SOEC 5-cell stacks (provided by Materials and Systems Research Inc. [MSRI]) were presented. The results showed that long-term degradation rates of 3.2%/khr and 4.6%/khr were observed for MSRI and Ceramatec stacks, respectively. In order to improve the durability of SOEC stacks, optimization of electrode and electrolyte materials, interconnect coatings, and electrolyte electrode interface microstructures were of great importance. In addition, the single solid oxide cells with different configurations from several manufacturers were also evaluated for initial performance and long-term durability by Zhang[96] in 2013.

On the basis of these previous works, in 2015, the experimental design, operation, and results of a 4 kW high-temperature steam electrolysis (HTSE) long-term test was presented by Zhang[97] (Figure 7.5). A demonstration of 830 hours of stable operation was achieved with a degradation rate of 3.1% per 1000 hours. This work will be helpful to promote the technology towards near-term commercialization.

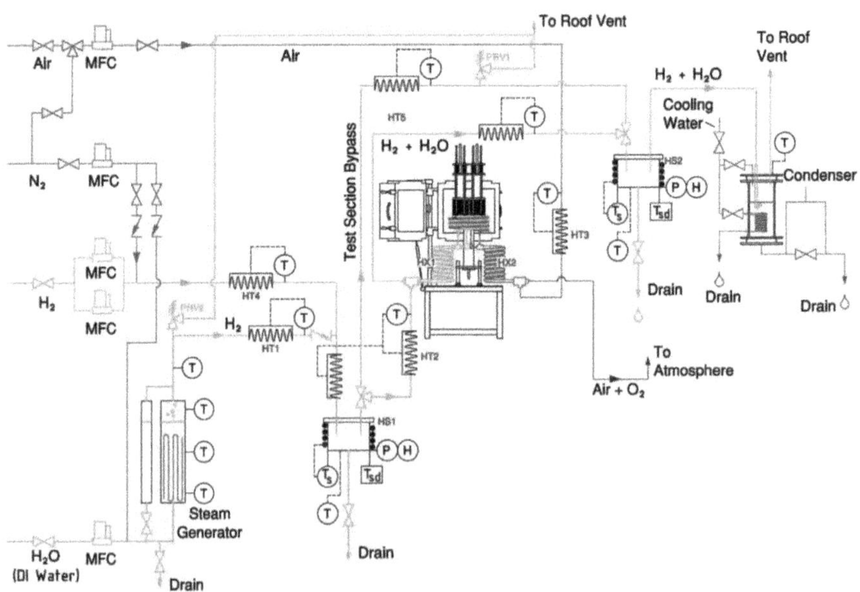

FIGURE 7.5 Flow chart of 4 KW high temperature steam electrolysis P&ID. (From Zhang, X. et al., *J. Power Sources*, 297, 90–97, 2015.)

At the University of Florida, Park[19] developed a bilayer-electrolyte (erbia-stabilized bismuth oxide/samaria-doped ceria [ESB/SDC]) architecture to produce pure oxygen from CO_2 at temperatures below 700°C for potential use in NASA's Mars exploration mission. He found the major factors that influenced oxygen generation included the oxygen-ion conductivity of the solid-oxide electrolyte, applied current, operating temperature, and fuel utilization. Higher temperatures resulted in higher oxygen generation rates due to reduced cell resistance. However, lowering the ceramic oxygen generator (COG) operating temperature was very important, and the bilayer COGs showed promise for operation below 700°C. More work should be carried out in order to optimize the relative thickness of the ESB/SDC electrolytes to decrease electronic conduction. Also, reduction in the total thickness of the bilayer solid oxide electrolysis to several micrometers should decrease area specific resistance (ASR) by orders of magnitude, resulting in a dramatic increase in the oxygen generation rate at lower temperatures with minimization of input power.

With the cooperation of the University of Pennsylvania and the University of St. Andrews (United Kingdom), Bidrawn et al.[20] examined electrolyze CO_2 in SOEC using a ceramic electrode of $La_{0.8}Sr_{0.2}Cr_{0.5}Mn_{0.5}O_3$ (LSCM), infiltrated into a yttria-stabilized zirconia scaffold together with 0.5 wt % Pd supported on 5 wt % $Ce_{0.48}Zr_{0.48}Y_{0.04}O_2$. An SOE with this electrode exhibited a total cell impedance of 0.36 $\Omega{\cdot}cm^2$ at 1073 K for operation in CO/CO_2 mixtures. Also, LSCM exhibited good conductivity in both oxidizing and reducing environments.

At Northwestern University, Barnett proposed the process of renewable-to-liquids (RTL). Note that production of the liquid fuel from syngas is currently practiced in gas-to-liquids (GTL) processes, where natural gas is the feedstock used to produce syngas, and coal-to-liquids (CTL), where coal is the feedstock. These processes are expected to become increasingly important as petroleum resources dwindle. The RTL process could be more frequently adopted in the future as natural gas and coal also become increasingly depleted. It should be a significant advantage that the RTL technology and its infrastructure are quite similar to the GTL and CTL technologies.

Still at Northwestern University, in 2009, Zhan[21] carried out electrolysis through SOEC with H_2O-CO_2-H_2 mixtures at the Ni-YSZ cathode and air at the LSCF-GDC anode under temperatures of 700°C–800°C. In 2011, Bierschenk[22,23] proposed using reversible solid oxide (shown in Figure 7.6) to improve round-trip efficiency. In 2013, Hughes[98] presented a durability testing of solid oxide cell electrodes in reversing-current and constant-current operation modes. $(La_{0.8}Sr_{0.2})_{0.98}MnO_{3-\delta}$–$Zr_{0.84}Y_{0.16}O_{2-\delta}$ (LSM–YSZ) symmetric cells were tested at 800°C with current cycle periods of 1 and 12 hours. The results indicated that lower current densities and reversing-current operation were desirable to maximize the lifetime of SOCs. In 2015, oxygen electrode characteristics of $Pr_2Ni\ O_{4+\delta}$-infiltrated porous $(La_{0.9}Sr_{0.1})(Ga_{0.8}Mg_{0.2})O_{3-\delta}$ were studied by Railsback.[99] In the same year, large-scale electricity storage utilizing SOC combined with underground storage of CH_4 and CO_2 was proposed by S. H. Jensen,[44] enabling large-scale electricity storage with a higher round-trip effciency (>70%) and cheaper storage cost (around 3¢ $kW^{-1}\ h^{-1}$).

At Columbia University, recycling CO_2 into sustainable hydrocarbon fuels with co-electrolysis of CO_2 and H_2O was investigated systematically by Graves for his

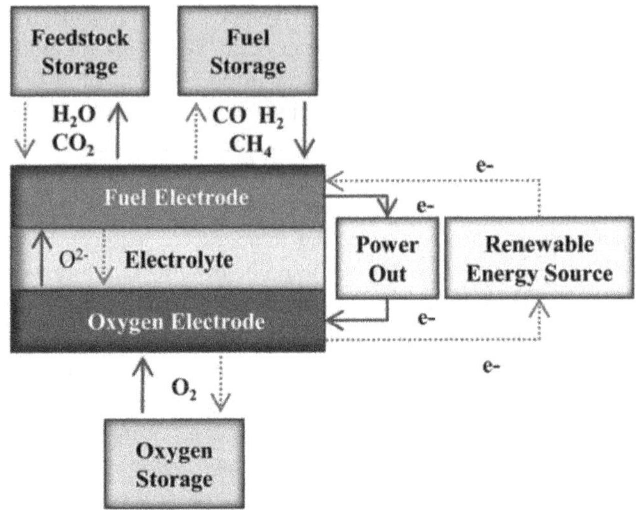

FIGURE 7.6 Schematic diagram of a simplified SOC system. (From Bierschenk, D.M. et al., *Energ. Environ. Sci.*, 4, 944–951, 2011; *ECS Trans.*, 35, 2969–2978, 2011.)

PhD dissertation.[24] Aside from the performance and durability of HTCE in SOC, the molybdate-based ceramic electrode materials and the aspects of metal-YSZ electrode kinetics were also studied in detail. The outline of Graves' dissertation is shown in Figure 7.7. Wang[25] conducted energy and exergy analyses to study the thermodynamic performance of a fuel synthesis system with heat from renewable energy, and CO_2 captured from ambient air or flue gas (shown in Figure 7.8). The research showed that the total energy requirement of fuel synthesis was not sensitive to CO_2 capture technologies, and exergy loss was dominated by thermal processes and chemical reactions. The principal exergy losses occurred in the electricity generation from the high-temperature heat. The thermodynamic efficiency of fuel synthesis could be increased by developing fuel generation technologies with lower electricity requirements, especially when using solar and nuclear energy.

At the Massachusetts Institute of Technology (MIT), Sharma and Yildiz[26] investigated in 2010 the degradation mechanism in the contact layer of SOEC stacks, which were provided by Ceramatec Inc., after 2000 continuous hours of operation at 830°C. In 2011, surface electronic structure transitions at high temperature on perovskite oxides were investigated by Cai[27] as an example of the strained $La_{0.8}Sr_{0.2}CoO_3$ thin films. In the same year, Jalili[100] raised new insights into the strain coupling to surface chemistry, electronic structure, and reactivity of $La_{0.7}Sr_{0.3}MnO_3$. Meanwhile, oxygen diffusion in solid oxide fuel cell cathode and electrolyte materials was studied by Chroneos with insights from

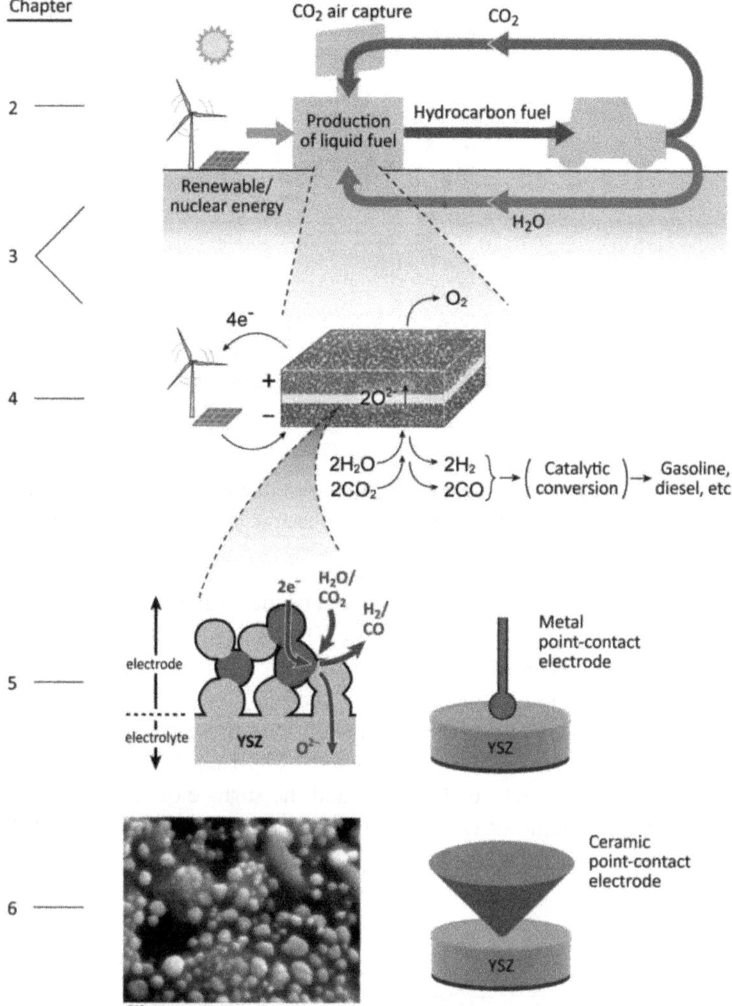

FIGURE 7.7 Illustration of the progression of the studies in the PhD dissertation. (From Graves, C.R., Recycling CO_2 into sustainable hydrocarbon fuels: Electrolysis of CO_2 and H_2O, PhD dissertation, Columbia University, 2010.)

atomistic simulations.[101] In 2012, Cai[102] investigated the chemical heterogeneities on $La_{0.6}Sr_{0.4}CoO_{3-\delta}$ thin films, and the impact of Sr segregation on the electronic structure was studied by Chen.[103]

From 2013 to 2015, to obtain a microscopic level understanding and control of faster oxygen reduction reaction (ORR) kinetics, which was exhibited by $La_{0.8}Sr_{0.2}CoO_3/(La_{0.5}Sr_{0.5})_2CoO_4$ ($LSC_{113/214}$) hetero-interfaces, Y. Chen[28] implemented a novel combination of in situ scanning electron microscopy-focused ion

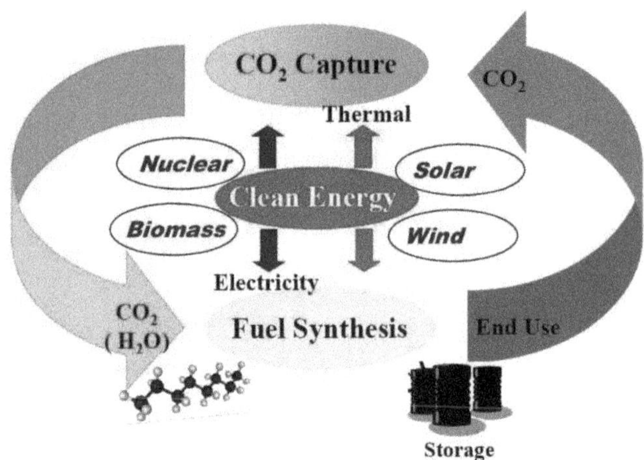

FIGURE 7.8 Closed carbon cycle with air capture and fuel synthesis. (From Wang, T., *ECS Trans.*, 41, 13–24, 2012. With permission from The Electrochemical Society, Copyright 2012.)

beam (SEM-FIB) milling to probe the local electronic structure at nanometer resolution in model multilayer, as shown in Figure 7.9. $LSC_{113/214}$ hetero-interface including reducibility of Co, electronic structure, surface chemistry and oxygen reduction kinetics were also investigated by Tsvetkov[104] and Ma[105] in the same group at MIT. In addition, cation size mismatch and charge interactions drive dopant segregation at the surfaces of manganite perovskites was researched by Lee.[106] The understanding of the relationship between surface chemistry and the surface oxygen exchange kinetics on epitaxial films made of $(La_{1-x}Sr_x)_2CoO_4$ was recently expanded by Chen,[107]

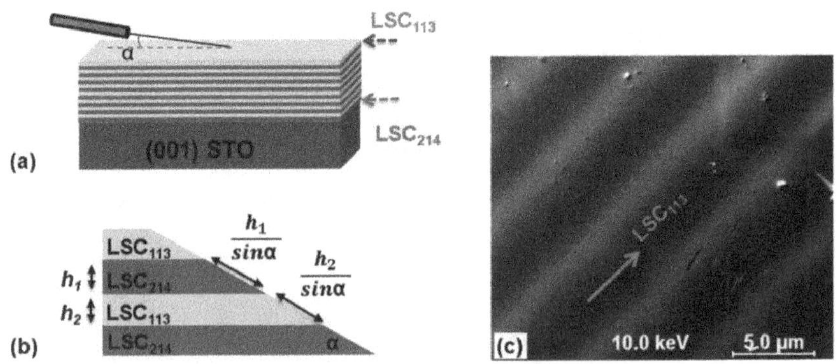

FIGURE 7.9 FIB milling process on the $LSC_{113/214}$ multilayer (ML) structure and characterization of exposed interface: (a) FIB milling process, (b) the cross-sectional view, (c) SEM image of the $LSC_{113/214}$ ML. (From Chen, Y. et al.: Electronic activation of cathode superlattices at elevated temperatures – source of markedly accelerated oxygen reduction kinetics. *Advanced Energy Materials*, 2013, 3, 1221–1229. Copyright Wiley-VCH Verlag GmbH & Co. KGaA. Reproduced with permission.)

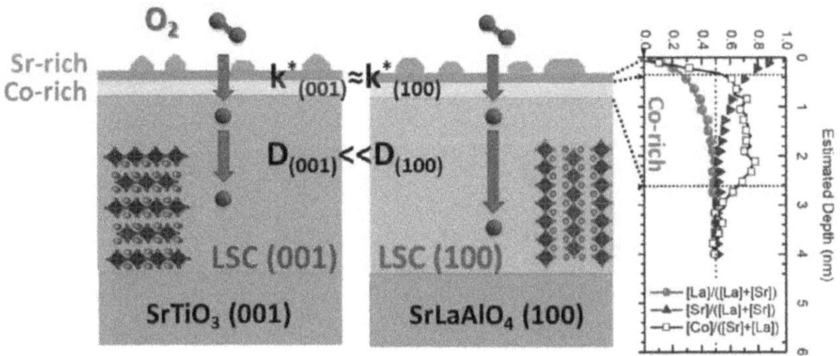

FIGURE 7.10 Two different crystallographic orientations of (001) and (100). (Reprinted with permission from Chen, Y. et al., *Chem. Mater.*, 27, 5436–5450, 2015. Copyright 2015 American Chemical Society.)

as shown in Figure 7.10. LSC thin films with different crystallographic orientations (referred to as LSC (001) and LSC (100), respectively) were fabricated by pulse laser deposition (PLD). The effects of crystal orientation on the surface composition, morphology, oxygen diffusion, and surface exchange kinetics were studied and assessed using molecular dynamics (MD), auger electron spectroscopy (AES), scanning transmission electron microscope (STEM), atomic force microscope (AFM), secondary ion mass spectroscopy (SIMS), low-energy ion scattering (LEIS), X-ray photoelectron spectroscopy (XPS) and X-ray diffraction (XRD). It was found that the surface chemistry played an important role in the oxygen exchange kinetics on perovskite-related oxides; a generalizable approach to probe the surface chemistry on other catalytic complex oxides was also provided.

Yang Shao-Horn at MIT focused on materials for batteries or fuel cells and the thermodynamics and kinetics of electrochemical reactions. A lot of work related to electronic and ionic transport properties of perovskites, oxygen reduction/evolution reactions,[29,30] heterostructured oxide surface,[108–110] surface strontium segregation[111–113] has been done in recent years. In cooperation with Goodenough, Yang Shao-Horn proposed a perovskite oxide optimized for oxygen evolution catalysis from molecular orbital principles[114] and design principles for oxygen-reduction activity on perovskite oxide catalysts for fuel cells and metal-air batteries.[29] On the one hand, more than 10 transition metal oxides were examined systematically to establish a design principle to predict the high activity of $Ba_{0.5}Sr_{0.5}Co_{0.8}Fe_{0.2}O_{3-\delta}$ (BSCF); the result showed that the intrinsic oxygen evolution reaction (OER) activity exhibits a volcano shape with reference to the occupancy of the 3d electron, and the e_g of surface transition metal cations in an oxide was symmetrical.[114] On the other hand, the ORR activity of various perovskite transition-metal-oxide catalysts was compared, as shown in Figure 7.11. The author demonstrated that the σ^*-orbital (e_g) and the extent of B-site transition-metal–oxygen covalency played the primary roles on the ORR activity for oxide catalysts. It was shown that the design principle to improve the ORR activity of transition-metal-oxide perovskites was an e_g-filling having a

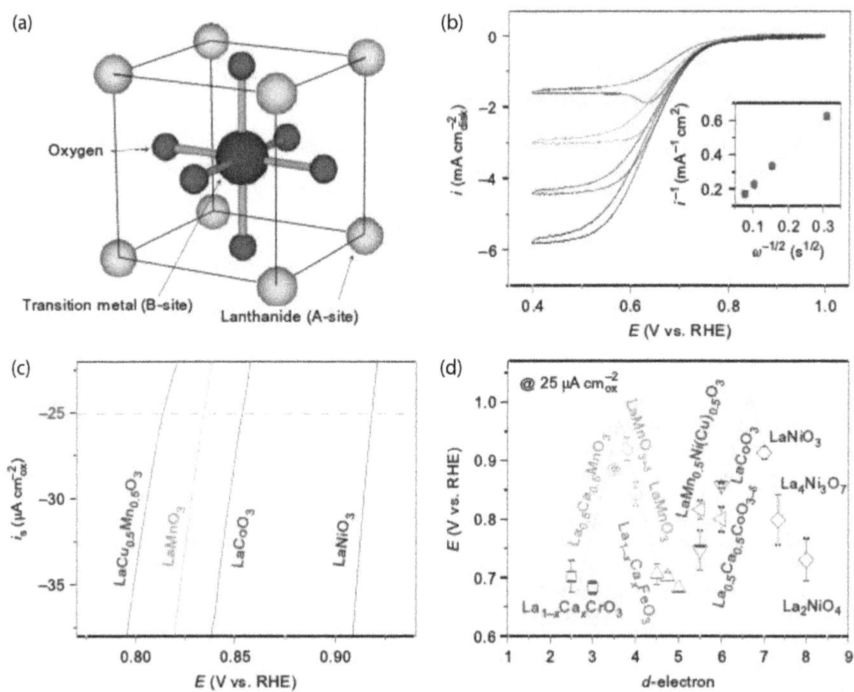

FIGURE 7.11 (a–d) ORR activity of perovskite transition-metal-oxide catalysts. (Reprinted by permission from Macmillan Publishers Ltd. *Nat. Chem.*, Suntivich, J. et al., 2011, Copyright 2011.)

value of ~1 for maximum activity. In addition, the further improvement of the activity could be achieved by increasing the covalency between the metal 3d and oxygen 2p orbitals.[29]

Another interesting phenomenon reported and compared by Yang Shao-Horn was heterostructure, which was sometimes related with strontium segregation.[110,111] A case in point is the research about $(La_{1-y}Sr_y)_2CoO_{4\pm\delta}$ (LSC_{214})/$La_{1-x}Sr_xCoO_{3-\delta}$ (LSC_{113}) heterostructures. The first atomic-scale structure and composition of LSC_{214}/LSC_{113} grown on $SrTiO_3$ was reported, and the anomalous strontium segregation from the perovskite to the interface and the Ruddlesden-Popper phase was observed using direct X-ray methods and density functional theory (DFT) calculations. The results showed that Sr segregation occurred during the film growth, and no obvious changes had been found upon subsequent annealing in O_2. The research findings could give insights into the design of highly active catalysts to enhance the oxygen electrocatalysis.

Goodenough from the University of Texas at Austin (since 1986) is world famous for his pioneering work leading to the invention of the lithium-ion rechargeable battery. He identified and developed the critical materials that provided the high-energy density needed to power portable electronics, initiating the wireless revolution. Today, batteries incorporating Goodenough's cathode materials are used worldwide for mobile phones, power tools, laptops, tablets, and other wireless devices, as well as electric and hybrid vehicles. In 2014, he received the Charles Stark Draper Prize for his contributions to the lithium-ion battery.

Although Goodenough is famous for the research achievements related to lithium-ion rechargeable battery, he also propsed some key points during the development of SOC. First, in 2000, his paper on oxide-ion conductors by design[115] theoretically discussed fluorite structure of solid oxides (shown in Figure 7.12a) and various solid materials that allowed oxide-ion conduction such as YSZ, LSGM ($La_{1-x}Sr_xGa_{1-y}Mg_yO_{3-0.5(x+y)}$), $Bi_4V_{2-x}M_xO_{11-y}$, and so on. Then, in 2004, he also systematically summarized the electronic and ionic transport properties and other physical aspects of perovskites in detail with a rather long article.[116] These important works can help us to get a better understanding for oxide-ion conductors and lay a foundation for later research of SOC.

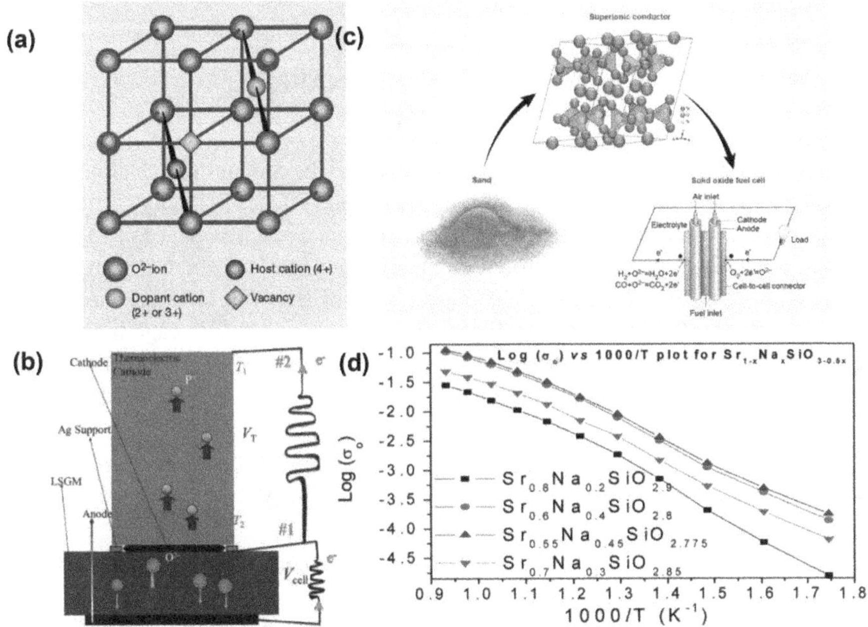

FIGURE 7.12 Diagrams from some published presented by Goodenough: (a) the fluorite structure of solid oxides. (Reprinted by permission from Macmillan Magazines Ltd. *Nature*, Goodenough, J.B., 404, 821–823, 2000. Copyright 1999.); (b) schematic diagram of elongated thermoelectric fuel cell (Wei, T. et al., *RSC Adv.*, 3, 2336–2340, 2013. Reproduced by permission of The Royal Society of Chemistry.); (c) the research from sand to SOFC (in cooperation with Kevin Huang from University of South Carolina and J. B. Goodenough) (Reprinted with permission from Wei, T. et al., *Energ. Environ. Sci.*, 7, 1680–1684, 2014. The Royal Society of Chemistry.); (d) the changes of oxide ion conductivities (σ^o, logarithm) with temperature (T, K^{-1}) for different oxide ion electrolytes. (Reprinted with permission from Singh, P., *J. Am. Chem. Soc.* 135, 10149–10154, Copyright 2013 American Chemical Society.)

After 2 years, a significant paper about double perovskites as anode materials for SOFCs was presented by Goodenough.[31] He reported the identification of double perovskites $Sr_2Mg_{1-x}Mn_xMoO_{6-\delta}$, which faced the requirements for long-term stability with tolerance to sulfur. As a result, double perovskites showed a superior single-cell performance in H_2 and CH_4. In 2007, he reviewed alternative anode materials for SOFCs. In this article, five basic requirements that an anode must satisfy, including, catalytic activity, electronic conductivity, thermal compatibility, chemical stability, and porosity, were presented. It covered (1) Ni/YSZ cermets for H_2 oxidation, (2) Ni/YSZ cermets CH_4 and syngas oxidation, (3) Ni/RDC, and (4) MIECs. This work can help us better understand the choice of anode materials for SOC.

In 2011, as mentioned above, in cooperation with Yang Shao-Horn and Goodenough, the catalysis optimization for OER[114] and design principles for ORR were proposed and published in *Science* and *Nature Chemistry*, respectively.[29] From 2013 to 2015, several materials, including cathode ($Ca_2Co_2O_5$[32] [Figure 7.12b]) and electrolytes ($Sr_{3-3x}Na_{3x}Si_3O_{9-1.5x}$ [x = 0.45] [Figure 7.12c]),[117] monoclinic ($Sr_{1-x}Na_xSiO_{3-0.5x}$ [Figure 7.12d],[118] and monoclinic $SrMO_3$ [M = Si/Ge][119]), for SOC were investigated.

7.3 HTCE RESEARCH TOWARDS SUSTAINABLE HYDROCARBON FUELS IN EUROPE

In Denmark, there is an increasing wish to increase the amount of sustainable CO_2 neutral energy production because of the greenhouse effect (GHE) and an increase in fuel prices. The Danish government aims to be independent of fossil fuels such as coal, oil, and gas in 2050.[120] Production of H_2 or syngas (H_2 and CO) through the use of renewable electricity for SOEC electrolysis of H_2O and CO_2 may benefit the overall energy system, as shown in Figure 7.13. H_2 can act as a storage medium for renewable energy and can be utilized for industrial gas supply, fuel for transport, and a supply of hydrogen to combined heat and power units (CHPs). Syngas can be used

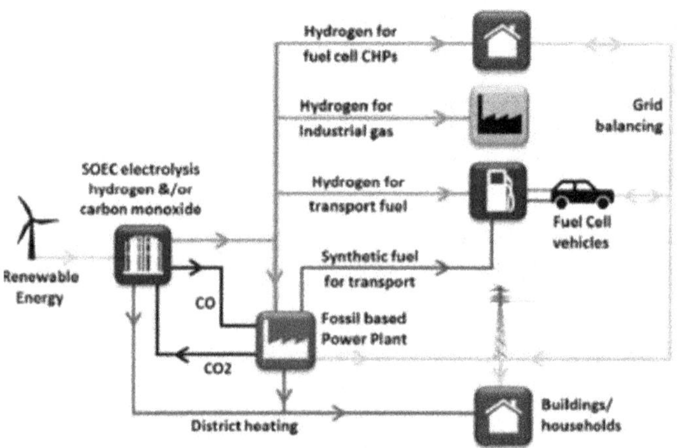

FIGURE 7.13 SOEC electrolysis for fuel production and energy storage in Denmark.

as feedstock for the production of various CO_2 neutral synthetic hydrocarbon fuels. The SOEC production plants, the CHP units, and fuel cell electric vehicles (hydrogen and/or synthetic fuel) may also provide various balancing services to the power grid.

A 6-year strategic electrochemistry research center (SERC) in fundamental and applied aspects of electrochemical cells with a main emphasis on solid oxide cells was started in Risø National Laboratory (RNL) for Sustainable Energy on January 1, 2007, in cooperation with other Danish and Swedish universities and eight industrial partners. The relations between the projects and the industrial partners is shown in Figure 7.14. The center focused on fundamental study of electrochemical cells to significantly extend the understanding of the materials' limitations, which currently impede a widespread commercialization of the technology. HTCE into synthesis gas was one of the most important research directions in the center.[40] The project aims to address different aspects of three crucial challenges to eventually improve electrochemical cells' performance and durability: (1) development of new stable electrode materials with high electrocatalytic properties, (2) improvement of the electrical conduction of segregations to the interfaces and surfaces of an electrochemical cell, and (3) identification and development of an optimized nano-structured region of the electrodes.

In 2007, Jensen[121] reported an electrolysis experiment with a new SOEC that resulted in a record-breaking current density of −3.6 A/cm² at a cell voltage of 1.48 V. with 70% CO_2 + 30% CO at a rate of 33 l/h to the Ni/YSZ electrode and pure O_2 to the LSM/YSZ

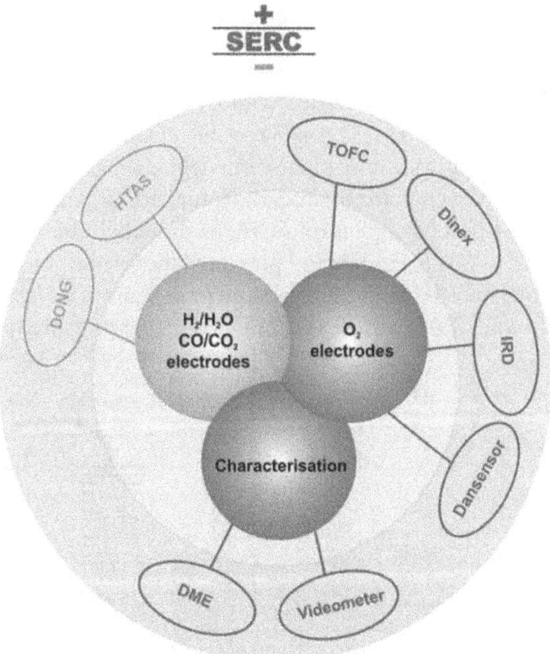

FIGURE 7.14 Schematic diagram of SERC showing the relations between the projects and the industrial partners. (Reproduced from Mogens Mogensen, K.H., *ECS Trans.*, 35, 43–52, 2011. With permission from The Electrochemical Society, Copyright 2011.)

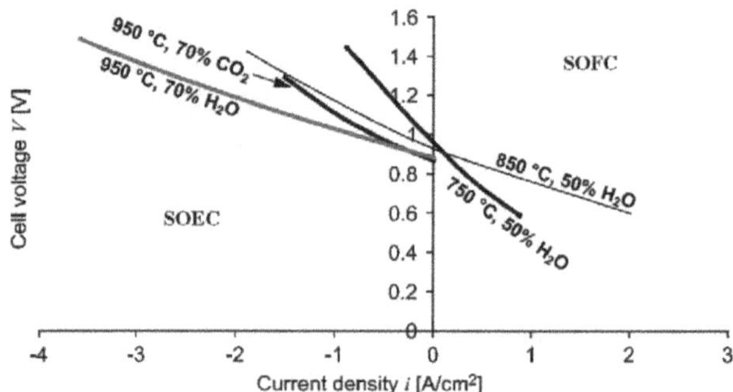

FIGURE 7.15 I–V curves achieved in SOEC and SOFC operation modes under different conditions. (With kind permission from Taylor & Francis: *Int. J. Hydrog. Energ.*, Hydrogen and synthetic fuel production from renewable energy sources 32, 2007, 3253–3257, Jensen, S.H. et al.)

electrode, as shown in Figure 7.15. At −1.5A/cm^2 the cell voltage was 1.29 V, and CO_2 utilization was 21%. In 2009, Ebbesen[122] performed CO_2 electrolysis experiments in SOECs for several thousand hours at 850°C at RNL. The cell passivation corresponds to 0.22–0.44 mV h^{-1} during the first 500 h of operation. After operation for 500 h, the passivation reached a plateau with vibration of only 0.05–0.09 mVh^{-1}. Ebbesen[123,124] also carried out the co-electrolysis experiments of H_2O and CO_2 in SOEC supported by Ni/YSZ electrode. An inexpensive and very efficient method for cleaning the inlet gases was introduced to enable operation without degradation at current densities of up to at least −0.75 A/cm^2. The results showed that cleaning the inlet gases may be a solution for operating SOECs without long-term degradation.[124,125]

In order to produce synthetic hydrocarbon fuel from syngas produced by co-electrolysis followed by conversion into CH_4 or dimethyl ether (DME), the gases must be pressurized to 2–8 MPa. Jensen[41] presented the results of a cell test with pressures ranging from 0.4 to 10 bar in 2010, as shown in Figure 7.16. It was found that the SOEC performance at 700°C was weakly affected by the pressure range.

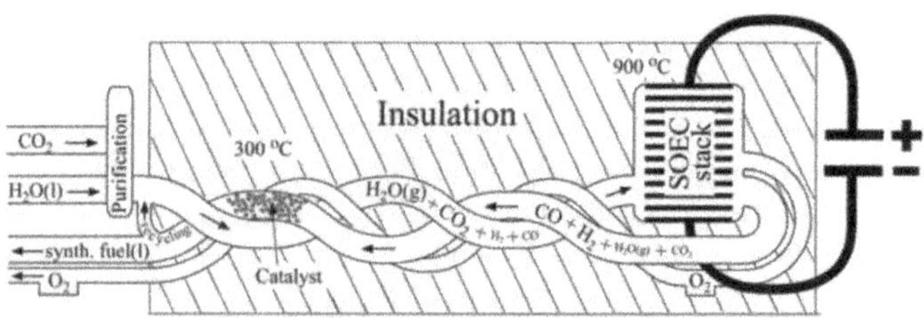

FIGURE 7.16 Schematic of a synthetic fuel production system with integrated high-pressure HTCE. (Reprinted from *Int. J. Hydrog. Energ.*, 35, Jensen, S.H., Hydrogen and synthetic fuel production using pressurized solid oxide electrolysis cells, 9544–9549, Copyright 2010, with permission from Elsevier.)

The performance and durability of SOC in HTCE is one of the most important research fields according to the research status at RNL. In order to improve electrode performance and durability, the degradation of SOC during the process of HTCE was investigated from 2011 to 2015 by researchers including Knibbe,[126] Ebbesen,[42,43] Graves,[127,128] Sun,[129] and Tao.[130] The degradation behaviors and mechanisms were discussed, and the influences of current density operation, operating temperature, inlet gases composition, operating pattern on performance, and the durability of SOC in HTCE were investigated.[42,43,127–131] For example, performance and durability of SOC in HTCE was studied by comparing different operating pattern, a constant-current electrolysis test, and a reversible cycling test.[128] The voltage profiles and impedance spectra of two different operating tests were compared, as shown in Figure 7.17. It was demonstrated that severe electrolysis-induced degradation, which was previously believed to be irreversible, can be completely eliminated by reversibly cycling between electrolysis and fuel-cell modes, similar to a rechargeable battery.

In addition, Graves[132] reviewed many possible technological pathways for recycling CO_2 into fuels using renewable or nuclear energy by considering three stages:

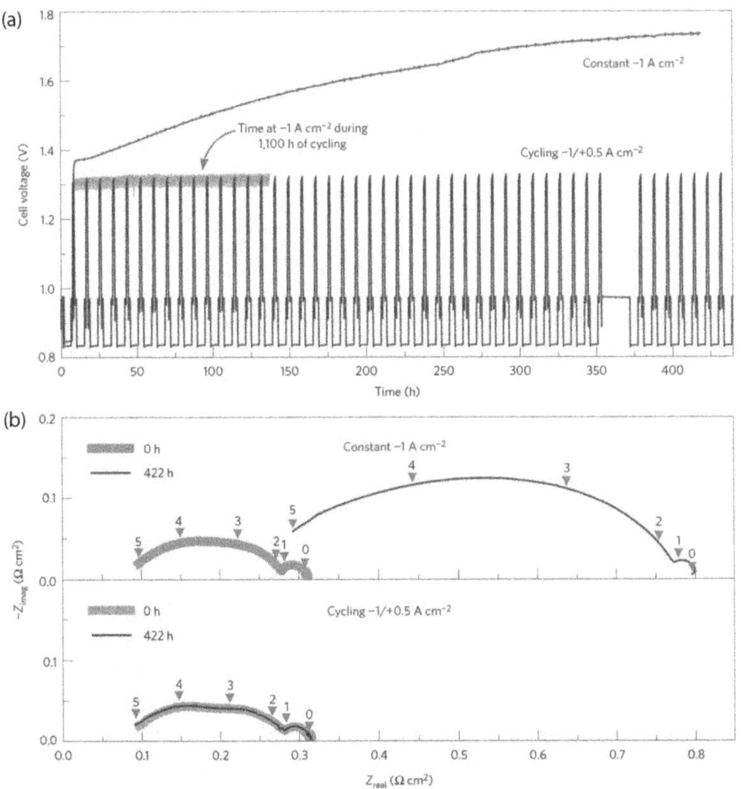

FIGURE 7.17 Comparison of the SOC stability during (a) a constant-current electrolysis test and (b) a reversible cycling test. (Reproduced from Graves, C. et al., *Nat. Mater.*, 14, 239–244, 2014. With permission from The Electrochemical Society, Copyright 2014.)

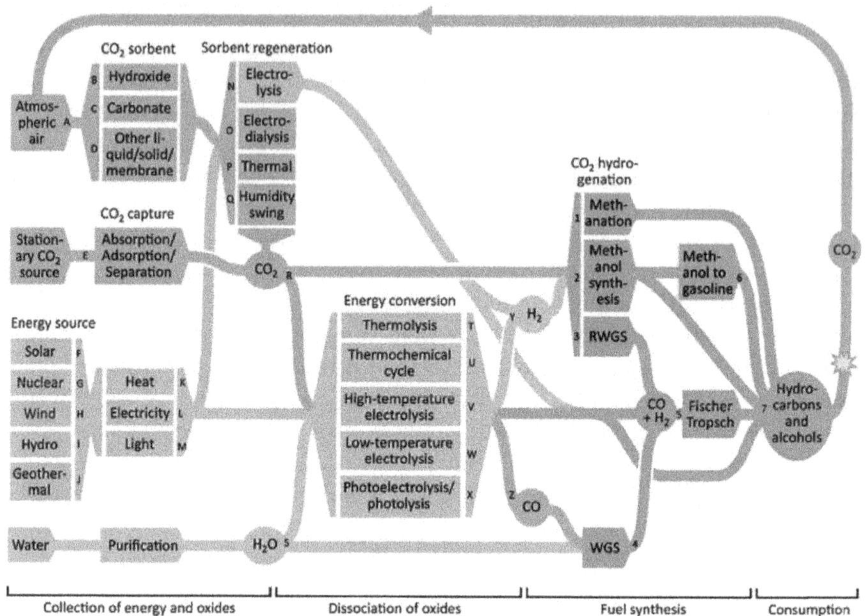

FIGURE 7.18 Map of the possible pathways from H_2O and CO_2 to hydrocarbon fuels. (Reprinted from *Renew. Sust. Energ. Rev.*, 15, Graves, C. et al., Sustainable hydrocarbon fuels by recycling CO_2 and H_2O with renewable or nuclear energy, 1–23, Copyright 2011, with permission from Elsevier.)

CO_2 capture, H_2O and CO_2 dissociation, and fuel synthesis, as shown in Figure 7.18. Dissociation methods include thermolysis, thermochemical cycles, electrolysis, and photoelectrolysis of CO_2 and/or H_2O. It was estimated that the full system could be feasibly operated at 70% electricity-to-liquid fuel efficiency (higher heating value basis).

The quest to develop power-to-gas (P2G) storage is part of Germany's ambitious *Energiewende* ("energy transition"), a long-term plan to clean up the country's energy systems. Enshrined in law, the scheme aims to slash CO_2 emissions by replacing fossil fuels with renewable sources of energy. Germany hopes to generate at least 35% of its electricity from green sources by 2020; by 2050, the share is expected to surpass 80%, as shown in Figure 7.19. *Energiewende* is the world's most extensive embrace of wind and solar power as well as other forms of renewable energy. The plan enjoys the support of all Germany's political parties and most of the population. To reach its goal, Germany is currently investing more than €1.5 billion per year in energy research. One of its chief aims is to improve and build more storage systems, such as the Stuttgart P2G plant. Another is extending and strengthening the electricity grid to connect remote wind turbines and countless small photovoltaic installations. The research program also seeks to improve the efficiency of energy production from sunlight, wind, and biomass, and

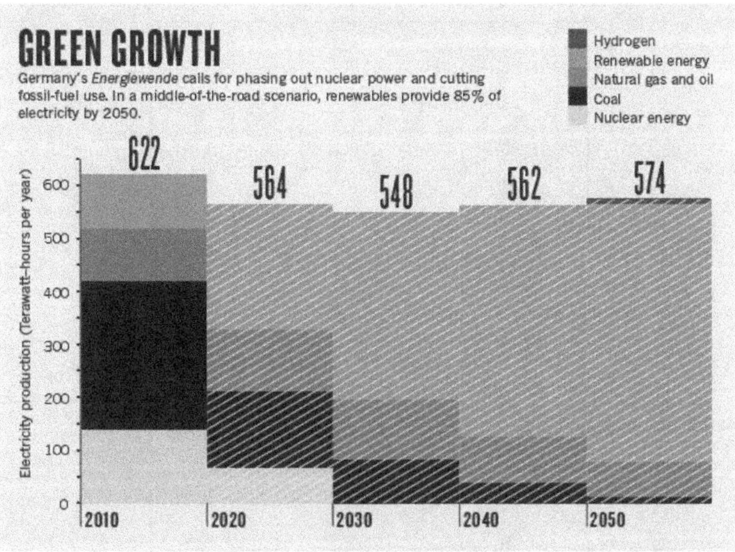

FIGURE 7.19 Schematic of green growth for Germany. (Reproduced from Schiermeier, Q., *Nature*, 496, 156–158, 2013. With permission from The Electrochemical Society, Copyright 2013.)

to encourage people to reduce energy consumption. It is an ambitious plan to slash CO_2 emissions, although it faces some serious problems to be solved.

Fu,[45,46] from the European Institute for Energy Research (EIFER), carried out the technical and economic analysis of HTCE and FT synthesis for syngas and synfuel production in EIFER. Starting from electrolysis, various routes could be followed to produce liquid fuels from water, CO_2, and renewable electricity. Syngas is considered the key intermediate energy carrier that can be produced either by water electrolysis followed by reverse water-gas shift (RWGS) reaction or directly by co-electrolysis. In the investigation, it was found that the product gas composition could be accurately predicted by assuming that the RWGS reaction reached a thermodynamic equilibrium, as shown in Figure 7.20. In comparison to wet CO_2 electrolysis (CO_2 with H_2O or H_2), the cell ASR was only slightly increased under dry CO_2 electrolysis conditions, indicating that, besides the RWGS reaction, electrochemical reduction of CO_2 could also play an important role for CO formation under CO_2/H_2O co-electrolysis conditions.

The research of Forschungszentrum Jülich covers many areas, including materials development over manufacturing of cells, stack design, mechanical and electrochemical characterization, and even the subsequent systems design and demonstration, always supported by feedback from post-test characterization[47,48] For more than 20 years, Forschungszentrum Jülich has been working on developing and improving anode-supported SOFCs.[48] More than 450 SOFC stacks power outputs range from 15 to 100 kW have been manufactured and operated.

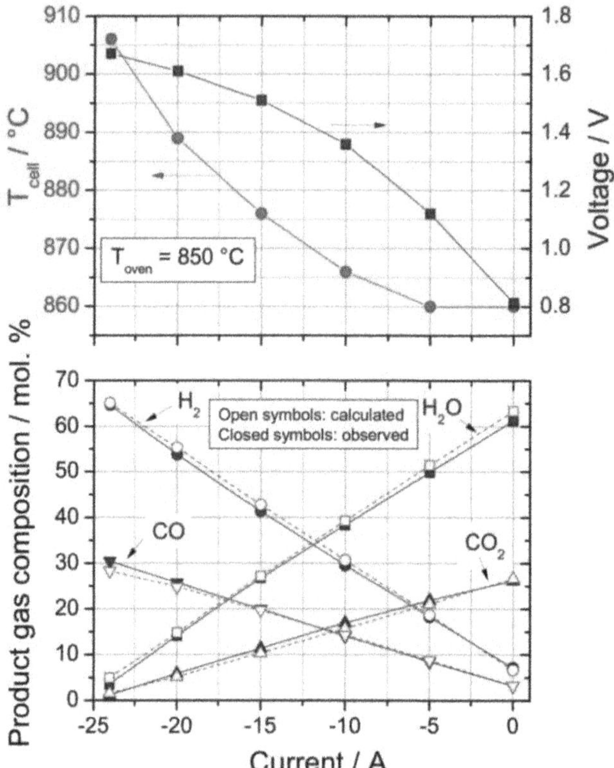

FIGURE 7.20 Product gas composition, cell temperature, and cell voltage as a function of current at an oven temperature of 850°C. (Reproduced from Fu, Q. et al., *ECS Trans.*, 35, 2949–2956, 2011. With permission from The Electrochemical Society, Copyright 2011.)

Blum summarized the recent status of SOFC, stack, and system development at Forschungszentrum Jülich.[47,134] In Figure 7.21, cell development from 1995 to 2009 is shown as a diagram, with the main eight R&D steps of anode-supported SOFCs at this institution from the development of the base materials and material combinations to the research of wet chemical thin film techniques.[47] The cell manufacturing route for 10 × 10 cm^2 cells is shown in Figure 7.22. From 2010 to 2012, cell development was focused on cost reduction by reducing layer thicknesses, improving the co-firing process of the half-cell, and reducing the number of manufacturing steps.[135]

With cooperation between Forschungszentrum Jülich and the European Institute for Energy Research, Tietz and Schefold performed a 9000-hour operation of a solid oxide cell in HTSE at a current density of −1 A cm^{-2} (about 780°C). The cell consisted of YSZ, Ni-YSZ, and LSCF, and the voltage degradation was about 3.8%/1000 h.[49,50] Based on research results of SOC, a two-cell planar stack with solid oxide cells had been demonstrated in both fuel cell and electrolysis cell modes at FZJ in 2013.[136] The long-term aging tests of the reversible solid oxide planar short stack included fuel cell

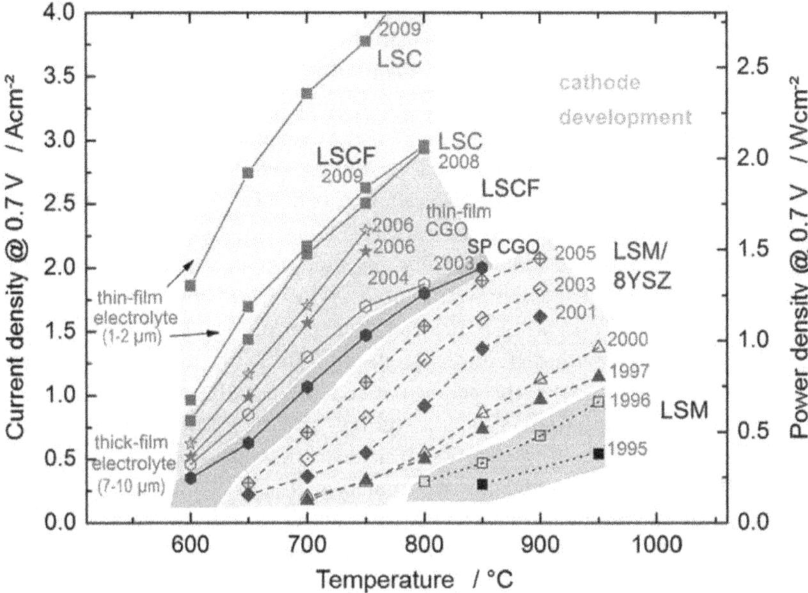

FIGURE 7.21 The development of single cell tests from 1995 to 2009 (current density versus operating temperature). (Reprinted from *J. Power Sources*, 241, Blum, L. et al., Recent results in Jülich solid oxide fuel cell technology development, 477–485, Copyright 2013, with permission from Elsevier.)

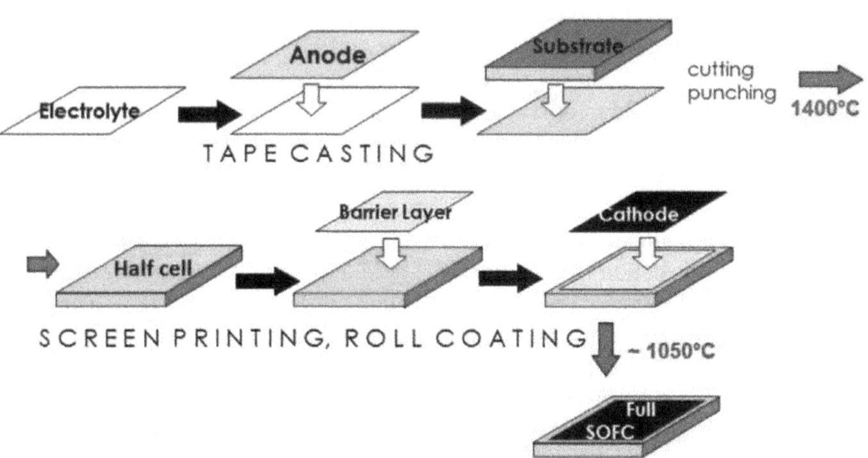

FIGURE 7.22 Manufacturing scheme of anode-supported SOFCs in 2013 at Forschungszentrum Jülich. (Reprinted from *J. Power Sources*, 241, Blum, L. et al., Recent results in Jülich solid oxide fuel cell technology development, 477–485, Copyright 2013, with permission from Elsevier.)

operation for 4000 h at a current density of 0.5 A cm^{-2} (750°C), HTSE for 3450 h, and HTCE for 640 h under various current density from −0.3 to −0.875 A cm^{-2} (800°C). The results showed that the voltage degradation in SOFC mode is 0.6%/1000 h for steam electrolysis; nearly no degradation was observed after 2000 h for HTSE, and there is an increased degradation 1.5%/1000 h at an increased current density of −0.875 A cm^{-2}. By contrast, the voltage degradations for HTCE were slightly higher. In the same year, durability test and degradation behavior of a 2.5 kW SPFC stack with internal reforming of liquefied natural gas (LNG) was studied by Fang.[135]

Cell developments focused recentely on tape-casting and sequential tape-casting.[51] A 20 kW SOFC system for about 7000 hours was performed.[137] Also in 2015, the influence of operating parameters on overall system efficiencies using SOC was investigated with different system configurations and operating conditions by Peters.[138] In addition, the performance and degradation of SOEC in stack consisting of anode-supported cells was studied, and the stack showed an average voltage degradation rate of 0.7%/1000 h after 2300 h of operation.[139] In 2016, two specially designed 1- and 5-layer planar-type steam pre-reformers using Ni/YSZ as a catalyst were performed to investigate the kinetics of steam reforming with methane.[52] The results confirmed the proposed kinetic expression of the Arrhenius type for different steam-to-carbon ratios and anode off-gas recycling.

SunFire GmbH was established in Bremen in June 2010. The company is developing a competitive process to produce synthetic renewable fuels (e.g., petrol, diesel, kerosene, or methane) from carbon dioxide (CO_2) and water (H_2O) using renewable electrical energy. The process developed by SunFire begins with the decomposition of water into hydrogen and oxygen using electrolysis driven by renewable electrical energy (derived from sunlight, wind, or water). A subsequent step is the reaction of hydrogen and the greenhouse gas CO_2 to form renewable, synthetic petrol, diesel, and kerosene. An important precondition for the economic viability of the process is high efficiency of the electrolysis.

Since 2008, the process has been extensively tested from both an economic and a technical point of view during the course of a detailed viability study. In 2010, an initial prototype on a laboratory scale was commissioned. The next milestone will be the establishment of a test plant with a production capacity of one barrel of fuel per day. The first pre-industrial prototypes have been constructed in 2016. SunFire presented a SOC electrolyzer with high-efficiency and low-cost hydrogen production.

In France, researchers have focused on HTSE for large-scale demonstration projects of H_2 production (see Figure 7.23).[140] A low-weight stack to increase performance, durability, and reliability and to reduce its cost has been designed and proposed by the Commissariat à l'Énergie Atomique (CEA).[55,56] This low-weight stack has been proved to be scalable from 3 to 25 cells. Particularly, a 10-cell stack and a 25-cell stack have been tested in electrolysis mode by Reytier.[141] Their H_2 productions were 0.7 Nm^3/h and 1.9 Nm^3/h, respectively, with a steam conversion around 50% at 800°C below 1.3 V. Gas tightness of the stacks has been evaluated, and a cost analysis has been performed that showed all the economical potentialities of this technology (Figure 7.24).

The composition neodymium nickelate $Nd_2NiO_{4+\delta}$ was integrated as an oxygen electrode for SOC; it was made of TZ3Y electrolyte and a Ni-CGO hydrogen

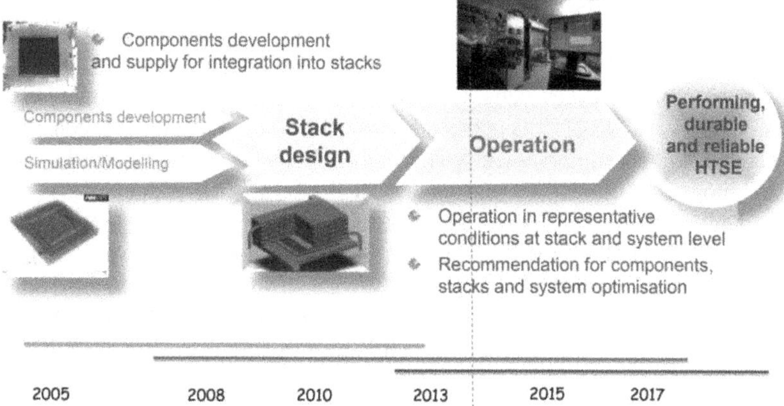

FIGURE 7.23 Project plan for developing high-temperature electrolysis in France. (Reprinted from *Int. J. Hydrog. Energ.*, 41, Odukoya, A. et al., Progress of the IAHE nuclear hydrogen division on international hydrogen production programs, 7878–7891, Copyright 2015, with permission from Elsevier.)

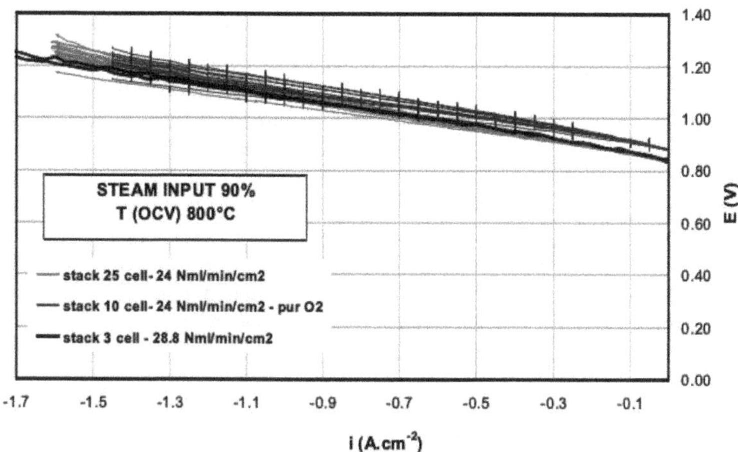

FIGURE 7.24 Performance of 3-, 10-, and 25-cell stack cell on the hydrogen side, air on the oxygen side (800°C, 90% H_2O/10%H_2). (Reprinted from *Int. J. Hydrog. Energ.*, 41, Odukoya, A. et al., Progress of the IAHE nuclear hydrogen division on international hydrogen production programs, 7878–7891, Copyright 2015, with permission from Elsevier.)

electrode, and the cell was performed in both SOEC and SOFC modes. Compared with LSM, it was proved that $Nd_2NiO_{4+\delta}$ can be considered a good candidate as an oxygen electrode for HTSE below 800°C.[142] On the basis of $Nd_2NiO_{4+\delta}$, Chauveau[143] continued to optimize the oxygen electrode by investigating and evaluating two nickelates (with compositions $La_2NiO_{4+\delta}$ and $Nd_2NiO_{4+\delta}$) in HTSE operation at the button cell level. In 2013, the MEIC materials $Ln_2NiO_{4+\delta}$ (Ln=La or Pr) were also investigated as oxygen electrodes for HTSE by Ogier.[55] Two interfacial ceria-based layers (YDC and GDC) have been studied and both yield significant improvements.

The enhancement of performance and durability was studied as a key point using protective coatings, advanced cells, and thin interconnects.[55] A promising degradation rate of 1.9–3.6%/1000 h was achieved on 3-cell stacks. The rate will continue to lower with subsequent improvements. Thermal cycling was also performed on those stacks. As shown in Figure 7.25, very fast transient was applied with no degradation. Lay-Grindler[57] and Aicart[144] built micro models for the HTSE and HTCE, respectively, to develop an understanding of the electrochemical mechanisms.

High-temperature electrolysis can electrolyze steam and also CO_2, in particular to perform H_2O/CO_2 co-electrolysis. Based on years of research of SOC in HTSE, the investigation about HTCE played a more and more important role at CEA.[141,144–146] Aicart[146] presented an experimental modeling approach devoted to a better understanding for the mechanisms of HTCE at 800°C. Experiments at stack level in both electrolysis and co-electrolysis modes have been carried out by Reytier.[141] A 10-cell stack and a 25-cell stack have been tested in electrolysis mode, and a current density of 0.8 A/cm^2 was achieved at 1.15 V with a conversion rate of 52% and with 100% gas recovery at the outlet. Gas tightness of the stacks has been evaluated, and the cost analysis shows all the economical potentialities of this technology.

In 2009, Adjiman et al.[147] summarized the progress in the UK Supergen Fuel Cell Programme, which was the United Kingdom's flagship fuel cell research consortium. It brought together four academic partners (Imperial College, Newcastle University, Nottingham University, and St Andrews University), three industry partners (Johnson Matthey, Rolls-Royce Fuel Cell Systems, and Ceres Power), and the

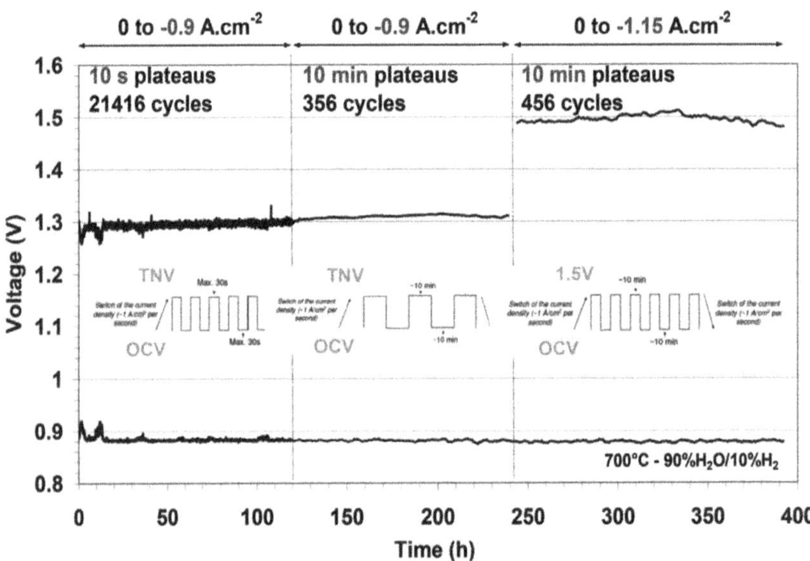

FIGURE 7.25 Different types of cycling (1-cell stack).

United Kingdom's Defense Science and Technology Laboratory (DSTL), to work together to bring fundamental science to address key industry challenges in the fuel cell sector. The research of the consortium is structured within 5 work packages: (1) zero leakage SOFC; (2) significantly improved fuel cell durability; (3) significantly improved fuel cell performance; (4) enhanced fuel flexibility; and (5) outreach, dissemination, and training. Some recent cases follow.

From 2011 to 2014, Qiong Cai,[58,59] at the Imperial College in London, investigated the active thickness of SOFC electrodes using a 3D microstructure model. The computational parameters for a 3D model for solid oxide fuel cell (SOFC) electrodes developed to link the microstructure of the electrode to its performance were also studied. The results showed that, to adequately represent the electrode microstructure, the characterized volume of the electrode should be equivalent to a cube having a minimum length of 7.5 times the particle diameter. Based on previous research, Qiong Cai[60] proposed the optimal control strategies for hydrogen production when coupling SOEC with intermittent renewable energies. The schematic illustration of the SOEC system, including an SOEC stack and an air compressor, is shown in Figure 7.26. Hydrogen production is examined in relation to energy consumption. Several scenarios and optimal control strategies were considered: in Scenario A, in which the current density is a control variable and was allowed to vary over the range 1000–7000 A m^2, the impact of maximizing hydrogen production or minimizing SOEC energy consumption was considered. In Scenario B, in which the input power is treated as a disturbance, with a step-wise profile in the range of 20–100 W, the minimization of the air compressor energy consumption was investigated. It was shown that it is important to impose a limit on the air compressor power in order to constrain the system energy consumption, when this is not considered in the objective function.

From 2010 to 2015, Laguna-Bercero, at the Instituto de Ciència de Materiales de Aragón (ICMA), has done a lot of work on SOC, including steam electrolysis using a microtubular SOFC[61] electrode (such as $La_{2-x}Sr_xCo_{0.5}Ni_{0.5}O_{4\pm\delta}$,[148] $Nd_2Ni\ O_{4+\delta}$[62], and LSM[63]) or electrolyte materials (such as scandia and ceria stabilized zirconia 10Sc1Ce SZ[149] and La $Nb_{0.84}W_{0.16}O_{4.08}$[64]). Microtubular solid oxide fuel cells (m T-SOFCs) with lanthanum strontium manganite (LSM) infiltrated cathodes was

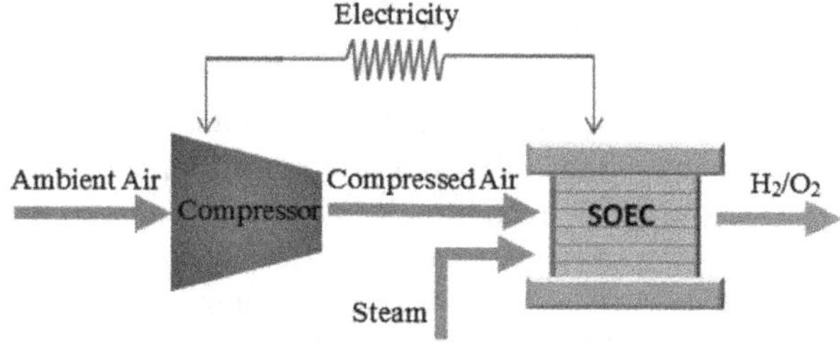

FIGURE 7.26 Schematic illustration of the SOEC system, which includes an SOEC stack for hydrogen production and an air compressor to provide air for temperature control of the SOEC stack. (Reprinted from *J. Power Sources*, 268, Cai, Q. et al., Optimal control strategies for hydrogen production when coupling solid oxide electrolysers with intermittent renewable energies, 212–224, Copyright 2014, with permission from Elsevier.)

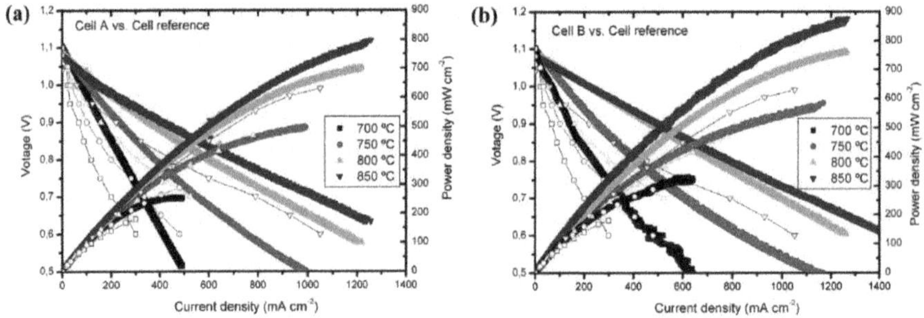

FIGURE 7.27 Electrochemical performance (current and power density versus. voltage) at 700° to 850°C for (a) cell A (a) and (b) cell B. (Reprinted from Laguna-Bercero, M. A. et al., *Int. J. Hydrog. Energ.*, 40, 5469–5474, Copyright 2015, with permission from Elsevier.)

studied by M. A. Laguna-Bercero in 2015.[63] In this research, Ni O-YSZ microtubular supports were fabricated by cold isostatic pressing (CIP) of NiO, YSZ, and pore former powders, followed by spray coating of the YSZ electrolyte and co-sintering at 1400°C. The effect of the infiltrated amount in cell performance was studied. Two cells with identical anode support and thin layer electrolyte and 22 vol% (cell A) and 35 vol% (cell B) infiltrated LSM were prepared. The electrochemical performance is shown in Figure 7.27. The results indicated that the infiltrated cathode with fine distributed LSM particles improved the fuel cell performance using a lower LSM content compared with standard LSM-YSZ composite cathodes.

7.4 RESEARCH STATUS OF HTCE WITH SOC IN CHINA

Researchers from the Institute of Nuclear and New Energy Technology (INNET) at Tsinghua University have made significant advances on the development of HTE and HTCE processes for hydrogen as well as syngas production. The design of the systems relies on high temperature heat and electricity input simultaneously from a generation IV nuclear reactor, such as a high-temperature gas-cooled reactor (HTR).[150] China aims to in around 2017. The project for building a 200 MW modular HTR commercial demonstration power plant (HTR-PM project) is supported by the government in its framework of the "national mid- to long-term science/technology plan" for the period from 2006 to 2020. As an important application of HTR, INET initiated HTE research in 2005. In the past 13 years, the research group mainly focused on the development of novel materials, effective microstructure control of electrode,[65–74] R&D of multiple-cell SOEC stacks, design and construction of kilowatt scale HTE facilities, and theoretically quantitative analysis of gas production efficiency (hydrogen or syngas) through HTE coupled with HTR. The hydrogen production rate of 10-cell stacks is 105 NL/h, and the system is stable. The next phase of the project will investigate a 30-cell stack. Furthermore, the overall efficiency of the HTCE system with HTR was calculated through electrochemical and

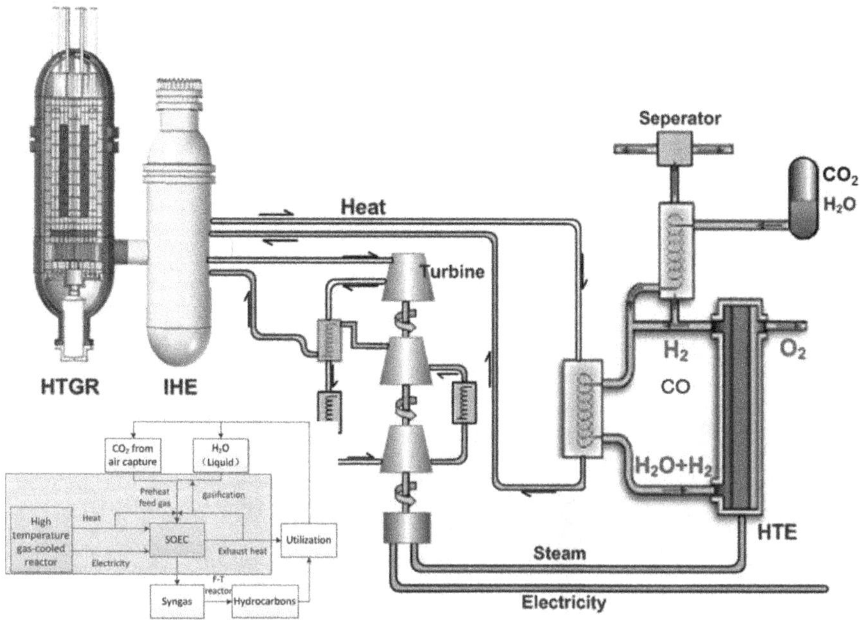

FIGURE 7.28 Schematic of HTCE system powered by HTGR.

thermodynamic analysis (shown in Figure 7.28). The maximum overall efficiency could be up to 60.39% at the temperature of 850°C. What merits attention is that the transformation efficiency of CO_2 is 32.1%.

From 2013 to 2015, Wenying Li and Yixiang Shi at Tsinghua University investigated patterned, porous, and tubular SOECs in different scales.[75–80] In order to understand the performance and methane production characteristics of HTCE in SOC, solid oxide electrolysis button cells at various operating temperature ranging from 550°C to 750°C was performed.[79] The results indicated that the performance of HTCE in Ni-YSZ/Sc SZ/LSM-Sc SZ electrolysis cell increases significantly with temperature, and the reaction pathway for methane production by HTCE was proposed, as shown in Figure 7.29a.

Elementary reaction modeling of the HTCE cell was researched by considering the effects of cathode thickness. The model structures, calculation domains, and boundaries of HTCE are shown in Figure 7.29b. The model proved to be a useful tool for understanding the intricate reaction and transport processes within the SOEC electrode. Subsequently, theoretical modeling of the fuel electrode both in the SOFC and SOEC modes was developed by considering the effects of microstructure and thickness (Figure 7.29c). Based on this previous research, the theoretical modeling of SOEC was applied by considering heterogeneous chemical/electrochemical reactions (Figure 7.29d). In addition to modeling of HTCE, carbon deposition on patterned Ni/YSZ electrodes for SOC was also studied by Yixiang Shi.[76]

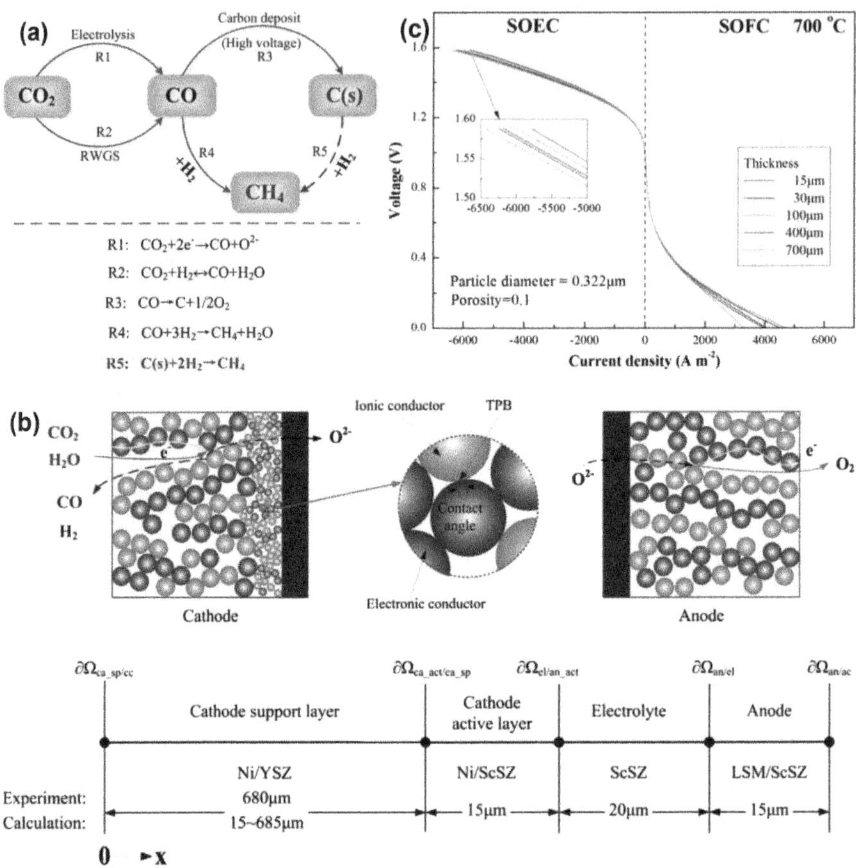

FIGURE 7.29 Diagrams about modeling of H_2O/CO_2 co-electrolysis, proposed by Wenying Li: (a) a proposed reaction pathway for methane production by HTCE (From Li, W. et al., *Int. J. Hydrogen Energy*, 38, 11104–11109, 2013.); (b) model structures, calculation domains, and boundaries of HTCE (From Li, W. et al., *J Power Sources*, 243, 118–130, 2013.); (c) I–V curves for different electrode thickness. (From Li, W. et al., *Int. J. Hydrogen Energy*, 39, 13738–13750, 2014.)

Minfang Han from Tsinghua University and China University of Mining & Technology published many papers investigating SOFC,[81,151–159] most of them focused on the electrochemical performance and stability of electrode or electrolyte materials, including LSM–YSZ,[81] LSCF-GDC,[158] LCCZ ($La_{0.7}Ca_{0.3}Cr_{0.95}Zn_{0.05}O_{3-\delta}$)-GDC,[157] $Ba_{0.9}Co_{0.7}Fe_{0.2}Nb_{0.1}$ $O_{3-\delta}$ (BCFN9721),[153] $La_{0.7}Sr_{0.3}Fe_{0.7}Ga_{0.3}O_{3-\delta}$ (LSFG),[152] and $La_{0.4}Sr_{0.6}Co_{0.2}Fe_{0.7}$ $Nb_{0.1}O_{3-\delta}$ (LSCFN)[151] for SOC. In addition, an innovative design for enhancing the coking resistance of the conventional SOFC anode was presented by Yu Chen and Minfang Han.[159] A thin nano SDC catalyst layer has been deposited efficiently via infiltration on the wall surface of the Ni-YSZ anode internal gas diffusion channel. The microstructure of Ni-YSZ/YSZ half cells with and without the SDC catalyst layer was characterized using a 3D X-ray microscope, as shown

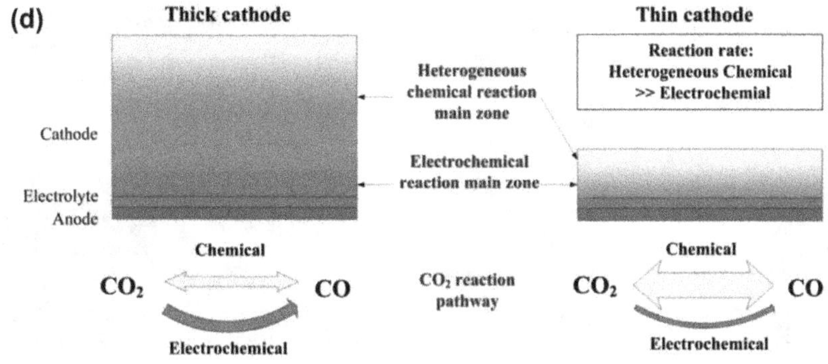

	Governing equation	Boundary of main zones*	Determinant of zone size
Heterogeneous chemical reaction main zone	Mass balance $\nabla\left(-D_k^{eff}\nabla c_{k,g}\right) = R_{k,g}$	90% of $\Delta c_{k,g}$ at OCV	Mass transfer flux $D_k^{eff}\nabla c_{k,g}$
Electrochemical reaction main zone	Charge balance $\nabla\left(-\sigma_{ion}^{eff}\nabla V_{ion}\right) = Q_{ion}$	90% of ΔV_{ion} at 1.4V	Charge transfer flux $\sigma_{ion}^{eff}\nabla V_{ion}$

* Artificial definition in the simulation

FIGURE 7.29 (Continued) Diagrams about modeling of H_2O/CO_2 co-electrolysis, proposed by Wenying Li: (d) reaction pathway and mathematical analysis of heterogeneous chemical/electrochemical reactions. (Reprinted from *J. Power Sources*, 273, Li, W. et al., Elementary reaction modeling of solid oxide electrolysis cells: Main zones for heterogeneous chemical/electrochemical reactions, 1–13, Copyright 2015, with permission from Elsevier; *Int. J. Hydrog. Energ.*, 39, Li, W. et al., Theoretical modeling of air electrode operating in SOFC mode and SOEC mode: The effects of microstructure and thickness, 13738–13750, Copyright 2014, with permission from Elsevier; *J. Power Sources*, 243, Li, W. et al., Elementary reaction modeling of CO_2/H_2O co-electrolysis cell considering effects of cathode thickness, 118–130, Copyright 2013, with permission from Elsevier; *Int. J. Hydrog. Energ.*, 38, Li, W. et al., Performance and methane production characteristics of H_2O-CO_2 co-electrolysis in solid oxide electrolysis cells, 11104–11109, Copyright 2013, with permission from Elsevier.)

in Figure 7.30. The efficiency for catalyst infiltration has been significantly improved using hierarchically porous anode structure with open and straight channels. Single cells with nano SDC layers showed very stable cell performance and a peak power density of 0.65 W cm^{-2} at 800°C using methane as the fuel.

It is well known that the efficient management and utilization of CO_2 is a paramount challenge, as discussed above. In order to face this challenge, Long Chen and Changrong Xia[82] from the University of Science and Technology of China proposed a novel approach for direct synthesis of methane from CO_2/H_2O co-electrolysis in tubular SOECs. Figure 7.31 shows the process from HTCE to Fischer-Tropsch (F-T) synthesis. The 24-hour short-term test showed that the co-electrolysis process presented a generally stable current density at 0.42 A cm^{-2} during the electrolysis operation, while the average CH_4 yield can reach 11.40% (0.84 m L min^{-1}) with an overall CO_2 conversion ratio of 64.1%. In this way, the combining process can be used not only as a scalable method for energy storage but also as an environmentally friendly solution for CO_2 emission.

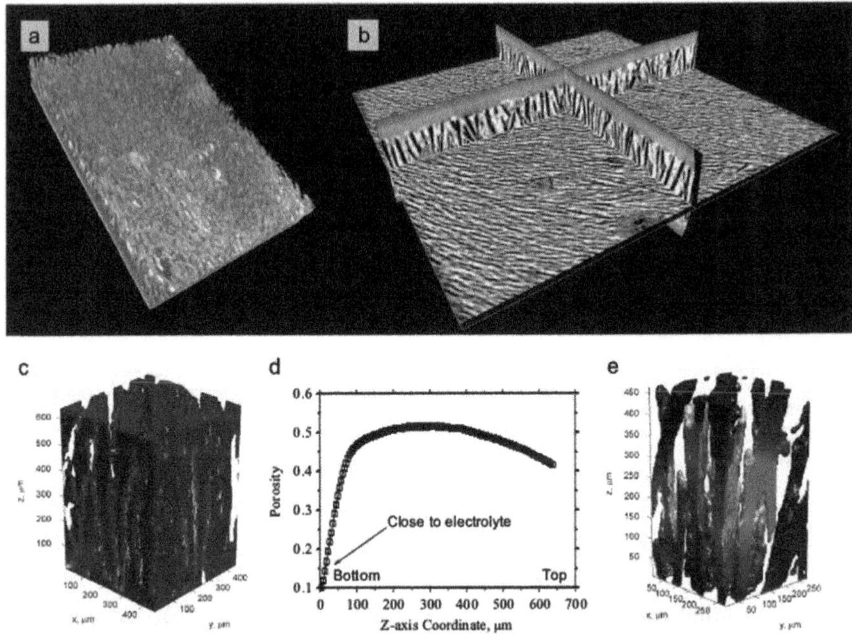

FIGURE 7.30 Some results about Ni-YSZ (SDC)/YSZ: (a) 3D X-ray microscopy image, (b) cross-sectional image, (c) reconstructed 3D microstructure (blue: solid phase, blank: pore phase), (d) porosity distribution, (e) visualization of a general field within the porous channels. Colors from red to blue proportionally represent the field potential from high to low. (Reprinted from *Nano Energ.*, 10, Chen, Y., et al. Direct-methane solid oxide fuel cells with hierarchically porous Ni-based anode deposited with nanocatalyst layer, 1–9, Copyright 2014, with permission from Elsevier.)

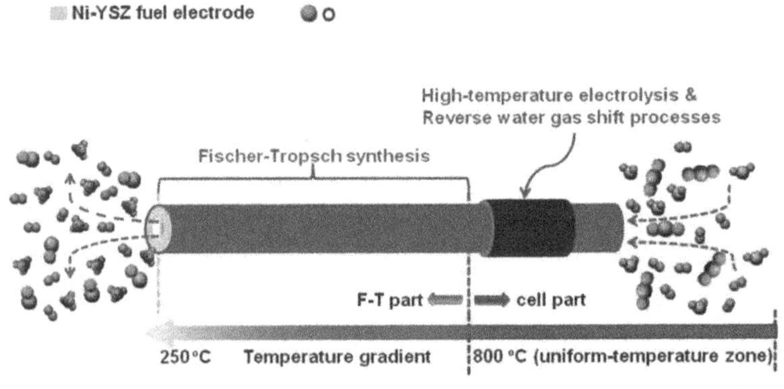

FIGURE 7.31 Schematic diagram of direct methane synthesis from HTCE to reduced temperature F-T synthesis. (Chen, L. et al., *Energ. Environ. Sci.*, 7, 4018–4022, 2014. Reproduced by permission of The Royal Society of Chemistry.)

7.5 SUMMARY

As a promising route to decrease CO_2 emission and enable scalable energy storage, HTCE with SOC has attracted many efforts in recent years. The research discoveries about material behaviors, stack performance, and process simulation were summarized in detail in this chapter. Research and development of high-temperature CO_2/H_2O co-electrolysis is advancing at rates never seen before. The fast growth is brought by the application of synthesis gas as a feedstock for chemical synthesis and the increasing demand for clean, renewable synthetic fuels. However, the development of HTCE stacks remains at the laboratory-scale stage, blocked by the degradation in the long-term operation. Therefore, more research is needed to overcome this obstacle.

REFERENCES

1. Chandler, H. W., Pollara, F. Z. Oxygen regeneration in a solid electrolyte system. *AICHE Chemical Engineering Progress Series Aerospace Life Support* **62**, 38–42 (1966).
2. Elikan, L., Archer, D. H., Zahradnik, R.L. Oxygen regeneration in solid electrolyte batteries fundamental considerations. *AICHE Chemical Engineering Progress Series Aerospace Life Support* **62**, 29–37 (1966).
3. Ash, R. L., Dowler, W. L., Varsi, G. Feasibility of rocket propellant production on Mars. *Acta Astronaut* **5**, 705–724 (1978).
4. Stancati, M. L., Niehoff, J. C., Wells, W. C., Ash, R.L. Remote automated propellant production a new potential for round trip spacecraft. *AIAA 79-0906* 262–270 (1979).
5. Richter, R. Basic investigation into the production of oxygen in a solid electrolyte. *AIAA-81-1175* 1–14 (1981).
6. Sridhar, K. R., Vaniman, B.T. Oxygen production on mars using solid oxide electrolysis. *Solid State Ionics* **93**, 321–328 (1997).
7. Sridhar, K. R., Iacomini, C. S., Finn, J. E. Combined H_2O/CO_2 solid oxide electrolysis for mars in situ resource utilization. *Journal of Propulsion and Power* **20**, 892–901 (2004).
8. Stoots, C. M. High-temperature co-electrolysis of H_2O and CO_2 for syngas production, *Fuel Cell Seminar*, Honolulu, HI (2006).
9. Hawkes, G., O'Brien, J., Stoots, C., Herring, S., Jones, R. Three-dimensional CFD model of a planar solid oxide electrolysis cell for co-electrolysis of steam and carbon dioxide, *2006 Fuel Cell Seminar*, Honolulu, HI, November 13 (2006). INL/CON-06-11720.
10. Hartvigsen, J., Elangovan, S., Elwell, J., Larsen, D., Clark, L., Meaders, T. Mechanical, structural, and thermal qualification of solid oxide electrolysis for oxygen production from Mars atmosphere carbon dioxide. *ECS Transactions* **78**, 3317–3327 (2017).
11. Elikan, L., Archer, D. H., Zahradnik, R. L. Oxygen regeneration in solid electrolyte batteries: Fundamental considerations. *AICHE Chemical Engineering Progress Series Aerospace Life Support*, **62**, 29 (1966).
12. Elikan, L., Morris, J. P. Solid oxide electrolyte system for oxygen regeneration. *NASA CR-1359* 1–166 (1969).
13. Elikan, L., Morris, J. P., Wu, C. K. Development of a solid oxide electrolyte carbon dioxide and water reduction system for oxygen recovery. *NASA CR-2014* 1–169 (1972).
14. Stancati, M. L., Niehoff, J. C., Wells, W. C., Ash, R. L. Remote automated propellant production a new potential for round trip spacecraft. *AIAA 79-0906* 262–270 (1979).
15. Hawkes, G. L., O'Brien, J., Stoots, C. M., Jones, R. 3D CFD Model of High Temperature H_2O/CO_2 Co-electrolysis, ANS Summer Meeting, Boston, 06242007.

16. Smith, J. D., Reed, M. Innovative management of carbon emission from fossil plants. *24th Annual ACERC Conference*, Prove, UT (2010).
17. Hawkes, G. L., McKellar, M. G., Wood, R., Plum, M. M. Process modeling results of bio-syntrolysis converting biomass to liquid fuel with high temperature steam electrolysis. *Clean Technology Conference & Expo 2010* (2010).
18. O'Brien, R.C., Stoots, C.M., McKellar, M.G. In-situ resource utilization and cabin atmosphere revitalization via the use of nuclear process heat and electrical power generation. *Nuclear and Emerging Technologies for Space* 2012, 1–2 (2012).
19. Jun-Young Park, E. D. W. Lower temperature electrolytic reduction of CO_2 to O_2 and CO with high-conductivity solid oxide bilayer electrolytes. *Journal of the Electrochemical Society* **152**, A1654–A1659 (2005).
20. Bidrawn, F., et al. Efficient reduction of CO_2 in a solid oxide electrolyzer. *Electrochemical and Solid-State Letters* **11**, B167 (2008).
21. Zhan, Z., et al. Syngas production by coelectrolysis of CO_2/H_2O: The basis for a renewable energy cycle. *Energy & Fuels* **23**, 3089–3096 (2009).
22. Bierschenk, D. M., Wilson, J. R., Barnett, S. A. High efficiency electrical energy storage using a methane – oxygen solid oxide cell. *Energy & Environmental Science* **4**, 944–951 (2011).
23. Bierschenk, D. M., Wilson, J. R., Miller, E., Dutton, E., Barnett, S. A. A proposed method for high efficiency electrical energy. *ECS Transactions* **35**, 2969–2978 (2011).
24. Graves, C. R. Recycling CO_2 into sustainable hydrocarbon fuels: Electrolysis of CO_2 and H_2O. PhD dissertation, Columbia University (2010).
25. Wang, T. Fuel synthesis with CO_2 captured from atmosphere: Thermodynamic analysis. *ECS Transactions* **41**, 13–24 (2012).
26. Vivek Inder Sharma, B. Y. Degradation mechanism in $La_{0.8}Sr_{0.2}CoO_3$ as contact layer on the solid oxide electrolysis cell anode. *Journal of the Electrochemical Society* **157**, B441–B448 (2010).
27. Cai, Z., Kuru, Y., Han, J.W., Chen, Y., Yildiz, B. Surface electronic structure transitions at high temperature on perovskite oxides: The case of strained $La_{0.8}Sr_{0.2}CoO_3$ thin films. *Journal of the American Chemical Society* **133**, 17696–17704 (2011).
28. Chen, Y., et al. Electronic activation of cathode superlattices at elevated temperatures – source of markedly accelerated oxygen reduction kinetics. *Advanced Energy Materials* **3**, 1221–1229 (2013).
29. Suntivich, J., et al. Design principles for oxygen-reduction activity on perovskite oxide catalysts for fuel cells and metal–air batteries. *Nature Chemistry* **3**, 546–550 (2011).
30. Mutoro E., Crumlin, E. J., Biegalski, M. D., Christen, H. M., Shao-Horn Y. Enhanced oxygen reduction activity on surface-decorated perovskite thin films for solid oxide fuel cells. *Energy & Environmental Science* **4**, 3689–3696 (2011).
31. Huang, Y. H., Dass, R. I., Xing, Z. L., Goodenough, J. B. Double perovskites as anode materials for solid-oxide fuel cells. *Science* **312**, 255–257 (2006).
32. Wei, T., Huang, Y., Jiang, L., Yang, J., Goodenough, R. Z. A. J. Thermoelectric solid-oxide fuel cell with $Ca_2Co_2O_5$ as cathode material. *RSC Advances* **3**, 2336–2340 (2013).
33. Yang, C., Yang, Z., Jin, C., Liu, M., Chen, F. High performance solid oxide electrolysis cells using $Pr_{0.8}Sr_{1.2}(Co, Fe)_{0.8}Nb_{0.2}O_{4+\delta}$-Co-Fe alloy hydrogen electrodes. *International Journal of Hydrogen Energy* **38**, 11202–11208 (2013).
34. Jin, C., Yang, C., Zhao, F., Cui, D., Chen, F. La0.75Sr0.25Cr0.5Mn0.5O3 as hydrogen electrode for solid oxide electrolysis cells. *International Journal of Hydrogen Energy* **36**, 3340–3346 (2011).
35. Jin, C., Yang, C., Chen, F. Characteristics of the hydrogen electrode in high temperature steam electrolysis process. *Journal of the Electrochemical Society* **158**, B1217–B1223 (2011).

36. Liu, T., et al. Steam electrolysis in a solid oxide electrolysis cell fabricated by the phase-inversion tape casting method. *Electrochemistry Communications* **61**, 106–109 (2015).
37. Wang, Y., Liu, T., Fang, S., Chen, F. Syngas production on a symmetrical solid oxide H_2O/CO_2 co-electrolysis cell with $Sr_2Fe_{1.5}Mo_{0.5}O_6$–$Sm_{0.2}Ce_{0.8}O_{1.9}$ electrodes. *Journal of Power Sources* **305**, 240–248 (2016).
38. Liu, Q., Yang, C., Dong, X., Chen, F. Perovskite $Sr_2Fe_{1.5}Mo_{0.5}O_{6-\delta}$ as electrode materials for symmetrical solid oxide electrolysis cells. *International Journal of Hydrogen Energy* **35**, 10039–10044 (2010).
39. Wang, Y., et al. A novel clean and effective syngas production system based on partial oxidation of methane assisted solid oxide co-electrolysis process. *Journal of Power Sources* **277**, 261–267 (2015).
40. Mogens Mogensen, K. H. The strategic electrochemical research center in Denmark. *ECS Transactions* **35**, 43–52 (2011).
41. Jensen, S. H., Sun, X., Ebbesen, S. D., Knibbe, R., Mogensen, M. Hydrogen and synthetic fuel production using pressurized solid oxide electrolysis cells. *International Journal of Hydrogen Energy* **35**, 9544–9549 (2010).
42. Ebbesen, S. D., Høgh, J., Nielsen, K. A., Nielsen, J. U., Mogensen, M. Durable SOC stacks for production of hydrogen and synthesis gas by high temperature electrolysis. *International Journal of Hydrogen Energy* **36**, 7363–7373 (2011).
43. Ebbesen, S.D., Sun, X., Mogensen, M.B. Understanding the processes governing performance and durability of solid oxide electrolysis cells. *Faraday Discussion* **182**, 393–422 (2015).
44. Jensen, S. H., et al. Large-scale electricity storage utilizing reversible solid oxide cells combined with underground storage of CO_2 and CH_4. *Energy & Environmental Science* **8**, 2471–2479 (2015).
45. Fu, Q., Dailly, J., Brisse, A., Zahid, M. High-temperature CO_2 and H_2O electrolysis with an electrolyte-supported solid oxide cell. *ECS Transactions* **35**, 2949–2956 (2011).
46. Fu, Q., Mabilat, C., Zahid, M., Brisse, A., Gautier, L. Syngas production via high-temperature steam/CO_2 co-electrolysis: An economic assessment. *Energy & Environmental Science* **3**, 1365–1608 (2010).
47. Blum, L., et al. Recent results in Jülich solid oxide fuel cell technology development. *Journal of Power Sources* **241**, 477–485 (2013).
48. Menzler, N. H., et al. Status of solid oxide fuel cell development at Forschungszentrum Jülich. *Procedia Engineering* **44**, 407–408 (2012).
49. Tietz, F., Sebold, D., Brisse, A., Schefold, J. Degradation phenomena in a solid oxide electrolysis cell after 9000 h of operation. *Journal of Power Sources* **223**, 129–135 (2013).
50. Schefold, J., Brisse, A., Tietz, F. Nine thousand hours of operation of a solid oxide cell in steam electrolysis mode. *Journal of the Electrochemical Society* **159**, A137–A144 (2012).
51. Menzler, N. H., Malzbender, J., Schoderböck, P., Kauert, R., Buchkremer, H. P. Sequential tape casting of anode-supported solid oxide fuel cells. *Fuel Cells* **14**, 96–106 (2014).
52. Nguyen, V. N., Deja, R., Peters, R., Blum, L. Methane/steam global reforming kinetics over the Ni/YSZ of planar pre-reformers for SOFC systems. *Chemical Engineering Journal* **292**, 113–122 (2016).
53. Products and technology in Sunfire GmbH, RSOC electrolyzer: High-efficiency and low-cost hydrogen production (2016). http://www.sunfire.de/en/.
54. Odukoya, A., et al. Progress of the IAHE nuclear hydrogen division on international hydrogen production programs. *International Journal of Hydrogen Energy* **41**, 7878–7891 (2016).

55. Ogier, T., et al. Enhanced performances of structured oxygen electrodes for high temperature steam electrolysis. *Fuel Cells* **13**, 536–541 (2013).
56. Mougin, J., et al. High temperature steam electrolysis stack with enhanced performance and durability. *Energy Procedia* **29**, 445–454 (2012).
57. Lay-Grindler, E., et al. Micro modelling of solid oxide electrolysis cell: From performance to durability. *International Journal of Hydrogen Energy* **38**, 6917–6929 (2013).
58. Cai, Q., Adjiman, C. S., Brandon, N. P. Modelling the 3D microstructure and performance of solid oxide fuel cell electrodes: Computational parameters. *Electrochimica Acta* **56**, 5804–5814 (2011).
59. Cai, Q., Adjiman, C. S., Brandon, N. P. Investigation of the active thickness of solid oxide fuel cell electrodes using a 3D microstructure model. *Electrochimica Acta* **56**, 10809–10819 (2011).
60. Cai, Q., Adjiman, C. S., Brandon, N. P. Optimal control strategies for hydrogen production when coupling solid oxide electrolysers with intermittent renewable energies. *Journal of Power Sources* **268**, 212–224 (2014).
61. Laguna-Bercero, M. A., Campana, R., Larrea, A., Kilner, J. A., Orera, V. M. Steam electrolysis using a microtubular solid oxide fuel cell. *Journal of the Electrochemical Society* **157**, B852–B855 (2010).
62. Laguna-Bercero, M. A., Hanifi, A. R., Monzón, H., Cunningham, J., Etsell, T. H., Sarkar, P. High performance of microtubular solid oxide fuel cells using $Nd_2NiO_{4+\delta}$ based composite cathodes. *Journal of Materials Chemistry A* **2**, 9764–9770 (2014).
63. Laguna-Bercero, M. A., Hanifi, A. R., Etsell, T. H., Sarkar, P., Orera, V.M. Microtubular solid oxide fuel cells with lanthanum strontium manganite infiltrated cathodes. *International Journal of Hydrogen Energy* **40**, 5469–5474 (2015).
64. Laguna-Bercero, M. A., Bayliss, R. D., Skinner, S. J. $LaNb_{0.84}W_{0.16}O_{4.08}$ as a novel electrolyte for high temperature fuel cell and solid oxide electrolysis applications. *Solid State Ionics* **262**, 298–302 (2014).
65. Zhang, W., Yu, B., Xu, J. Efficiency evaluation of high-temperature steam electrolytic systems coupled with different nuclear reactors. *International Journal of Hydrogen Energy* **37**, 12060–12068 (2012).
66. Zhang, W., Yu, B., Xu, J. Investigation of single SOEC with BSCF anode and SDC barrier layer. *International Journal of Hydrogen Energy* **37**, 837–842 (2012).
67. Yu, B., et al. Preparation and electrochemical behavior of dense YSZ film for SOEC. *International Journal of Hydrogen Energy* **37**, 12074–12080 (2012).
68. Wang, X., et al. Microstructural modification of the anode/electrolyte interface of SOEC for hydrogen production. *International Journal of Hydrogen Energy* **37**, 12833–12838 (2012).
69. Bo, Y., Wenqiang, Z., Jingming, X., Jing, C. Status and research of highly efficient hydrogen production through high temperature steam electrolysis at INET. *International Journal of Hydrogen Energy* **35**, 2829–2835 (2010).
70. Bo, Y., Wenqiang, Z., Jingming, X., Jing, C. Microstructural characterization and electrochemical properties of $Ba_{0.5}Sr_{0.5}Co_{0.8}Fe_{0.2}O_{3-\delta}$ and its application for anode of SOEC. *International Journal of Hydrogen Energy* **33**, 6873–6877 (2008).
71. Yu, B., Zhang, W., Chen, J., Xu, J., Wang, S. Advance in highly efficient hydrogen production by high temperature steam electrolysis. *Science in China Series B: Chemistry* **51**, 289–304 (2008).
72. Zhu, J. X., Yu, B. Electrochemical performance and microstructural characterization of solid oxide electrolysis cells. *Advanced Materials Research* **287–290**, 2506–2510 (2011).
73. Yu, B., Wen, M.F. Preparation and characterization of NiO/YSZ cathode and BSCF/SDC anode of SOEC for hydrogen production. *Advanced Materials Research* **287– 290**, 2494–2499 (2011).

74. Liang, M., et al. Preparation of LSM–YSZ composite powder for anode of solid oxide electrolysis cell and its activation mechanism. *Journal of Power Sources* **190**, 341–345 (2009).
75. Li, W., Shi, Y., Luo, Y., Cai, N. Elementary reaction modeling of solid oxide electrolysis cells: Main zones for heterogeneous chemical/electrochemical reactions. *Journal of Power Sources* **273**, 1–13 (2015).
76. Li, W., Shi, Y., Luo, Y., Wang, Y., Cai, N. Carbon deposition on patterned nickel/yttria stabilized zirconia electrodes for solid oxide fuel cell/solid oxide electrolysis cell modes. *Journal of Power Sources* **276**, 26–31 (2015).
77. Li, W., Shi, Y., Luo, Y., Cai, N. Theoretical modeling of air electrode operating in SOFC mode and SOEC mode: The effects of microstructure and thickness. *International Journal of Hydrogen Energy* **39**, 13738–13750 (2014).
78. Li, W., Shi, Y., Luo, Y., Cai, N. Elementary reaction modeling of CO_2/H_2O co-electrolysis cell considering effects of cathode thickness. *Journal of Power Sources* **243**, 118–130 (2013).
79. Li, W., Wang, H., Shi, Y., Cai, N. Performance and methane production characteristics of H_2O-CO_2 co-electrolysis in solid oxide electrolysis cells. *International Journal of Hydrogen Energy* **38**, 11104–11109 (2013).
80. Shi, Y., et al. Experimental characterization and modeling of the electrochemical reduction of CO_2 in solid oxide electrolysis cells. *Electrochimica Acta* **88**, 644–653 (2013).
81. Fan, H., Han, M. Electrochemical performance and stability of Sr-doped $LaMnO_3$-infiltrated yttria stabilized zirconia oxygen electrode for reversible solid oxide fuel cells. *International Journal of Coal Science & Technology* **1**, 56–61 (2014).
82. Chen, L., Chen, F., Xia, C. Direct synthesis of methane from CO_2-H_2O co-electrolysis in tubular solid oxide electrolysis cells. *Energy & Environmental Science* **7**, 4018–4022 (2014).
83. Chen, T., et al. High performance of intermediate temperature solid oxide electrolysis cells using $Nd_2NiO_{4+\delta}$ impregnated scandia stabilized zirconia oxygen electrode. *Journal of Power Sources* **276**, 1–6 (2015).
84. Li, J., et al. $Sr_2Fe_{1.5}Mo_{0.5}O_{6-\delta}$-$Zr_{0.84}Y_{0.16}O_{2-\delta}$ materials as oxygen electrodes for solid oxide electrolysis cells. *Fuel Cells* **14**, 1046–1049 (2014).
85. Weissbart, J., Smart, W. H. Study of electrolytic dissociation of CO_2-H_2O using a solid oxide electrolyte, NASA contractor report, CR-680. *National Aeronautics and Space Administration* (1967).
86. Tao, G. Study of carbon dioxide electrolysis at electrode/electrolyte interface: Part I. Pt/YSZ interface. *Solid State Ionics* **175**, 615–619 (2004).
87. Tao, G. Study of carbon dioxide electrolysis at electrode/electrolyte interface: Part II. Pt-YSZ cermet/YSZ interface. *Solid State Ionics* **175**, 621–624 (2004).
88. Singhal, S. C., Kendall, K. *High-Temperature Solid Oxide Fuel Cells: Fundamentals, Design and Applications* (2003).
89. Brien, J. E. O., McKellar, M. G., Stoots, C. M., Herring, J. S., Hawkes, G.L. Parametric study of large-scale production of syngas via high temperature co-electrolysis, *2007AIChE Annual Meeting* (2007).
90. McKellar, M. G., Brien, J. E. O., Stoots, C. M., Hawkes, G. L. Process model for the production of syngas via high temperature co-electrolysis, *2007ASME International Mechanical Engineering Congress & Exposition* (2007).
91. Stoots, C. M., O'Brien, J. E., Herring, J. S., Condie, K. G., Hartvigsen, J. J. INL experimental research in high temperature electrolysis for hydrogen and syngas production, *ASME 4th International Topical Meeting on High Temperature Reactor Technology*, Washington DC, September 28–October 1 (2008).

92. Stoots, C. M., Brien, J. E. O., Herring, J. S., Hartvigsen, J. J. Recent progress at the Idaho National Laboratory in high temperature electrolysis for hydrogen and syngas production. *IMECE 2008* (2008).
93. O'Brien, J. E., McKellar, M. G., Harvego, E. A., Stoots, C.M. High-temperature electrolysis for large-scale hydrogen and syngas production from nuclear energy – summary of system simulation and economic analyses. *International Journal of Hydrogen Energy* **35**, 4808–4819 (2010).
94. Sohal, M. S., Stoots, C. M., Sharma, V. I., Yildiz, B., Virkar, A. Degradation issues in solid oxide cells during high temperature electrolysis. *Journal of Fuel Cell Science and Technology* **9**, 0110171–10 (2012).
95. Zhang, X., et al. Improved durability of SOEC stacks for high temperature electrolysis. *International Journal of Hydrogen Energy* **38**, 20–28 (2013).
96. Zhang, X., O'Brien, J. E., O'Brien, R. C., Housley, G. K. Durability evaluation of reversible solid oxide cells. *Journal of Power Sources* **242**, 566–574 (2013).
97. Zhang, X., O'Brien, J. E., Tao, G., Zhou, C., Housley, G. K. Experimental design, operation, and results of a 4 kW high temperature steam electrolysis experiment. *Journal of Power Sources* **297**, 90–97 (2015).
98. Hughes, G. A., Yakal-Kremski, K., Barnett, S. A. Life testing of LSM-YSZ composite electrodes under reversing-current operation. *Physical Chemistry Chemical Physics* **15**, 17257–17262 (2013).
99. Railsback, J. G., Gao, Z., Barnett, S. A. Oxygen electrode characteristics of $Pr_2NiO_{4+\delta}$-infiltrated porous $(La_{0.9}Sr_{0.1})(Ga_{0.8}Mg_{0.2})O_{3-\delta}$. *Solid State Ionics* **274**, 134–139 (2015).
100. Jalili, H., Han, J. W., Kuru, Y., Cai, Z., Yildiz, B. New insights into the strain coupling to surface chemistry, electronic structure, and reactivity of $La_{0.7}Sr_{0.3}MnO_3$. *The Journal of Physical Chemistry Letters* **2**, 801–807 (2011).
101. Chroneos, A., Yildiz, B., Tarancón, A., Parfitt, D., Kilner, J.A. Oxygen diffusion in solid oxide fuel cell cathode and electrolyte materials: Mechanistic insights from atomistic simulations. *Energy & Environmental Science* **4**, 2774–2789 (2011).
102. Cai, Z., Kubicek, M., Fleig, J., Yildiz, B. Chemical heterogeneities on $La_{0.6}Sr_{0.4}CoO_{3-\delta}$ thin films—correlations to cathode surface activity and stability. *Chemistry of Materials* **24**, 1116–1127 (2012).
103. Chen, Y., Jung, W., Cai, Z., Kim, J. J., Tuller, H. L., Yildiz, B. Impact of Sr segregation on the electronic structure and oxygen reduction activity of $SrTi_{1-x}Fe_xO_3$ surfaces. *Energy & Environmental Science* **5**, 7979–7988 (2012).
104. Tsvetkov, N., Chen, Y., Yildiz, B. Reducibility of Co at the $La_{0.8}Sr_{0.2}CoO_3/(La_{0.5}Sr_{0.5})_2CoO_4$ hetero-interface at elevated temperatures. *Journal of Materials Chemistry A* **2**, 14690 (2014).
105. Ma, W., et al. Vertically aligned nanocomposite $La_{0.8}Sr_{0.2}CoO_3/(La_{0.5}Sr_{0.5})_2CoO_4$ cathodes – electronic structure, surface chemistry and oxygen reduction kinetics. *Journal of Materials Chemistry A* **3**, 207–219 (2015).
106. Lee, W., Han, J. W., Chen, Y., Cai, Z., Yildiz, B. Cation size mismatch and charge interactions drive dopant segregation at the surfaces of manganite perovskites. *Journal of the American Chemical Society* **135**, 7909–7925 (2013).
107. Chen, Y., et al. Segregated chemistry and structure on (001) and (100) surfaces of $(La_{1-x}Sr_x)_2CoO_4$ override the crystal anisotropy in oxygen exchange kinetics. *Chemistry of Materials* **27**, 5436–5450 (2015).
108. Feng, Z., et al. In situ studies of the temperature-dependent surface structure and chemistry of single-crystalline (001)-oriented $La_{0.8}Sr_{0.2}CoO_{3-\delta}$ perovskite thin films. *The Journal of Physical Chemistry Letters* **4**, 1512–1518 (2013).
109. Crumlin, E. J., et al. Oxygen reduction kinetics enhancement on a heterostructured oxide surface for solid oxide fuel cells. *The Journal of Physical Chemistry Letters* **1**, 3149–3155 (2010).

110. Lee, D., et al. Oxygen surface exchange kinetics and stability of (La, Sr)$_2$CoO$_{4\pm\delta}$/La$_{1-x}$Sr$_x$MO$_{3-\delta}$(M = Co and Fe) hetero-interfaces at intermediate temperatures. *Journal of Materials Chemistry A* **3**, 2144–2157 (2015).
111. Feng, Z., et al. Anomalous interface and surface strontium segregation in (La$_{1-y}$Sr$_y$)$_2$CoO$_{4\pm\delta}$/La$_{1-x}$Sr$_x$CoO$_{3-\delta}$ heterostructured thin films. *The Journal of Physical Chemistry Letters* **5**, 1027–1034 (2014).
112. Feng, Z., et al. Revealing the atomic structure and strontium distribution in nanometer-thick La$_{0.8}$Sr$_{0.2}$CoO$_{3-\delta}$ grown on (001)-oriented SrTiO$_3$. *Energy & Environmental Science* **7**, 1166 (2014).
113. Orikasa, Y., et al. Surface strontium segregation of solid oxide fuel cell cathodes proved by in situ depth-resolved X-ray absorption spectroscopy. *ECS Electrochemistry Letters* **3**, F23–F26 (2014).
114. Suntivich, J., May, K. J., Gasteiger, H. A., Goodenough, J. B., Shao-Horn, Y. A perovskite oxide optimized for oxygen evolution catalysis from molecular orbital principles. *Science* **334**, 1383–1385 (2011).
115. Goodenough, J. B. Oxide-ion conductors by design. *Nature* **404**, 821–823 (2000).
116. Goodenough, J. Electronic and ionic transport properties and other physical aspects of perovskites. *Reports on Progress in Physics* **67**, 1915–1993 (2004).
117. Wei, T., et al. Sr$_{3-3x}$Na$_{3x}$Si$_3$O$_{9-1.5x}$ (x=0.45) as a superior solid oxide-ion electrolyte for intermediate temperature-solid oxide fuel cells. *Energy and Environmental Science* **7**, 1680–1684 (2014).
118. Singh, P., Goodenough, J. B. Monoclinic Sr$_{1-x}$Na$_x$SiO$_{3-0.5x}$: New superior oxide ion electrolytes. *Journal of the American Chemical Society* **135**, 10149–10154 (2013).
119. Martinez-Coronado, R., Singh, P., Alonso-Alonso, J., Goodenough, J. Structural investigation of the oxide-ion electrolyte with SrMO$_3$ (M = Si/Ge) structure. *Journal of Materials Chemistry A* **2**, 4355–4360 (2014).
120. Mogensen, M. Electrochemical routes towards sustainable hydrocarbon fuels. *ECS transactions* **41**, 3–11 (2011).
121. Jensen, S. H., Larsen, P. H., Mogensen, M. Hydrogen and synthetic fuel production from renewable energy sources. *International Journal of Hydrogen Energy* **32**, 3253–3257 (2007).
122. Ebbesen, S. D., Graves, C., Mogensen, M. Production of synthetic fuels by co-electrolysis of steam and carbon dioxide. *International Journal of Green Energy* **6**, 646–660 (2009).
123. Ebbesen, S. D., Mogensen, M. Electrolysis of carbon dioxide in solid oxide electrolysis cells. *Journal of Power Sources* **193**, 349–358 (2009).
124. Ebbesen, S. D., Mogensen, M. Exceptional durability of solid oxide cells. *Electrochemical and Solid-State Letters* **13**, B106–B108 (2010).
125. Ebbesen, S. D., Graves, C., Hauch, A., Jensen, S. H., Mogensen, M. Poisoning of solid oxide electrolysis cells by impurities. *Journal of the Electrochemical Society* **157**, B1419–B1429 (2010).
126. Knibbe, R., Hauch, A., Hjelm, J., Ebbesen, S. D., Mogensen, M. Durability of solid oxide cells. *Green* **1**, 141–169 (2011).
127. Graves, C., Ebbesen, S. D., Mogensen, M. Co-electrolysis of CO_2 and H_2O in solid oxide cells: Performance and durability. *Solid State Ionics* **192**, 398–403 (2011).
128. Graves, C., Ebbesen, S. D., Jensen, S. H., Simonsen, S. B., Mogensen, M.B. Eliminating degradation in solid oxide electrochemical cells by reversible operation. *Nature Materials* **14**, 239–244 (2014).
129. Sun, X., et al. Performance and durability of solid oxide electrolysis cells for syngas production. *ECS Transactions* **41**, 77–85 (2012).

130. Tao, Y., Ebbesen, S. D., Mogensen, M. Degradation of solid oxide cells during co-electrolysis of H_2O and CO_2: Carbon deposition under high current densities. *ECS Transactions* **50**, 139–151 (2013).
131. Knibbe, R., Traulsen, M. L., Hauch, A., Ebbesen, S. D., Mogensen, M. Solid oxide electrolysis cells: Degradation at high current densities. *Journal of the Electrochemical Society* **157**, B1209–B1217 (2010).
132. Graves, C., Ebbesen, S. D., Mogensen, M., Lackner, K. S. Sustainable hydrocarbon fuels by recycling CO_2 and H_2O with renewable or nuclear energy. *Renewable and Sustainable Energy Reviews* **15**, 1–23 (2011).
133. Schiermeier, Q. Germany's Energy Gamble: An ambitious plan to slash greenhouse-gas emissions must clear some high technical and economic hurdles. *Nature* **496**, 156–158 (2013).
134. Blum, L., et al. Solid oxide fuel cell, stack and system development status at forschungszentrum julich. *ECS Transactions* **68**, 157–169 (2015).
135. Fang, Q., et al. Durability test and degradation behavior of a 2.5 kW SOFC stack with internal reforming of LNG. *International Journal of Hydrogen Energy* **38**, 16344–16353 (2013).
136. Nguyen, V. N., Fang, Q., Packbier, U., Blum, L. Long-term tests of a Jülich planar short stack with reversible solid oxide cells in both fuel cell and electrolysis modes. *International Journal of Hydrogen Energy* **38**, 4281–4290 (2013).
137. Peters, R., et al. Operation experience with a 20 kW SOFC system. *Fuel Cells* **14**, 489–499 (2014).
138. Peters, R., et al. Influence of operating parameters on overall system efficiencies using solid oxide electrolysis technology. *International Journal of Hydrogen Energy* **40**, 7103–7113 (2015).
139. Fang, Q., Blum, L., Menzler, N. H. Performance and degradation of solid oxide electrolysis cells in stack. *Journal of the Electrochemical Society* **162**, F907–F912 (2015).
140. Odukoya, A., et al. Progress of the IAHE nuclear hydrogen division on international hydrogen production programs. *International Journal of Hydrogen Energy* **41**, 7878–7891 (2015).
141. Reytier, M., et al. Stack performances in high temperature steam electrolysis and co-electrolysis. *International Journal of Hydrogen Energy* **40**, 11370–11377 (2015).
142. Chauveau, F., Mougin, J., Bassat, J. M., Mauvy, F., Grenier, J.C. A new anode material for solid oxide electrolyser: The neodymium nickelate $Nd_2NiO_{4+\delta}$. *Journal of Power Sources* **195**, 744–749 (2010).
143. Chauveau, F., Mougin, J., Mauvy, F., Bassat, J., Grenier, J. Development and operation of alternative oxygen electrode materials for hydrogen production by high temperature steam electrolysis. *International Journal of Hydrogen Energy* **36**, 7785–7790 (2011).
144. Aicart, J., Laurencin, J., Petitjean, M., Dessemond, L. Experimental validation of two-dimensional H_2O and CO_2 Co-electrolysis modeling. *Fuel Cells* **14**, 430–447 (2014).
145. De Saint Jean, M., Baurens, P., Bouallou, C., Couturier, K. Economic assessment of a power-to-substitute-natural-gas process including high-temperature steam electrolysis. *International Journal of Hydrogen Energy* **40**, 6487–6500 (2015).
146. Aicart, J., Petitjean, M., Laurencin, J., Tallobre, L., Dessemond, L. Accurate predictions of H_2O and CO_2 co-electrolysis outlet compositions in operation. *International Journal of Hydrogen Energy* **40**, 3134–3148 (2015).
147. Adjiman, C., et al. A review of progress in the UK supergen fuel cell programme. *ECS Transactions* **25**, 35–42 (2009).
148. Laguna-Bercero, M. A., et al. Performance of $La_{2-x}Sr_xCo_{0.5}Ni_{0.5}O_{4\pm\delta}$ as an oxygen electrode for solid oxide reversible cells. *Fuel Cells* **11**, 102–107 (2011).

149. Laguna-Bercero, M. A., Kilner, J. A., Skinner, S. J. Development of oxygen electrodes for reversible solid oxide fuel cells with scandia stabilized zirconia electrolytes. *Solid State Ionics* **192**, 501–504 (2011).
150. Odukoya, A., et al. Progress of the IAHE nuclear hydrogen division on international hydrogen production programs. *International Journal of Hydrogen Energy* **41**, 7878–7891 (2016).
151. Zhu, T., Yang, Z., Han, M. Metal-supported solid oxide fuel cell with. *Journal of the Electrochemical Society* **163**, F122–F125 (2016).
152. Yang, Z., et al. $La_{0.7}Sr_{0.3}Fe_{0.7}Ga_{0.3}O_{3-\delta}$ as electrode material for a symmetrical solid oxide fuel cell. *RSC Advances* **5**, 2702–2705 (2015).
153. Song, S., Zhang, P., Zhang, X., Han, M. Partial oxidation of methane reaction in $Ba_{0.9}Co_{0.7}Fe_{0.2}Nb_{0.1}O_{3-\delta}$ oxygen permeation membrane with three-layer structure. *International Journal of Hydrogen Energy* **40**, 10894–10901 (2015).
154. Zhang, P., et al. Mechanisms of methane decomposition and carbon species oxidation on the $Pr_{0.42}Sr_{0.6}Co_{0.2}Fe_{0.7}Nb_{0.1}O_{3-\delta}$ electrode with high catalytic activity. *Journal of Materials Chemistry A* **3**, 22816–22823 (2015).
155. Yang, Z., et al. Stability investigation for symmetric solid oxide fuel cell with. *Journal of the Electrochemical Society* **162**, F718–F721 (2015).
156. Zhu, T., et al. Evaluation of Li_2O as an efficient sintering aid for gadolinia-doped ceria electrolyte for solid oxide fuel cells. *Journal of Power Sources* **261**, 255–263 (2014).
157. Zhu, T., Yang, Z., Han, M. Evaluation of $La_{0.7}Ca_{0.3}Cr_{0.95}Zn_{0.05}O_{3-\delta}$-$Gd_{0.1}Ce_{0.9}O_{3-\delta}$ dual-phase material and its potential application in oxygen transport membrane. *Journal of Materials Science & Technology* **30**, 954–958 (2014).
158. Fan, H., Keane, M., Singh, P., Han, M. Electrochemical performance and stability of lanthanum strontium cobalt ferrite oxygen electrode with gadolinia doped ceria barrier layer for reversible solid oxide fuel cell. *Journal of Power Sources* **268**, 634–639 (2014).
159. Chen, Y., et al. Direct-methane solid oxide fuel cells with hierarchically porous Ni-based anode deposited with nanocatalyst layer. *Nano Energy* **10**, 1–9 (2014).

8 High-Temperature Electrochemical Process of CO_2 Conversion with SOCs 3

Key Materials

8.1 MATERIALS AND MICROSTRUCTURES OF ELECTRODES

Obviously, the design of electrodes is one of the key points for constructing a solid oxide cell (SOC). In order to achieve a high-power output (solid oxide fuel cell [SOFC]) or electrolytic efficiency (solid oxide electrolysis cell [SOEC], the materials, microstructures, and porosity should be considered, and the active surface area of electrode should be extended in design. M. B. Mogensen has summarized electrode materials and microstructures of SOC comprehensively, as shown in Figure 8.1.[1]

It is well known that ionic conduction (IC), electronic conduction (EC) and catalytic activity are the three essential functionalities for a SOC electrode; among them, the EC and IC can also be provided by a mixed ionic and electronic conductor (MIEC), as shown in Figure 8.1a. The material shown in Figure 8.1b–f, such as metal (M), fluorite (F), perovskite (P), double perovskite (DP), and Ruddlesden–Popper (RP), can be used for the one or more essential functionalities. It should be mentioned that the crystal structures of perovskite (P), double perovskite (DP), and Ruddlesden–Popper (RP) will be detailed later in this chapter. With regard to composite electrodes that consist of IC and EC phases, the interface between the IC, the EC, and the gas phase is the active reaction area and it is called the three-phase boundary (3PB), as shown in Figure 8.1g. If the functionalities of EC and IC are provided by MIEC, the active reaction area is the interface between the solid and gas phases and is called the two-phase boundary (2PB), as shown in Figure 8.1h. In addition to these two common microstructures of electrode, more elaborate electrode microstructures have been proposed to increase the areas of 2BP and/or 3PB, as illustrated in Figure 8.1i–n. For instance, a simple porous skeleton of stabilized zirconia (YSZ) can be used as the mechanical support with ionic conductivity; then the electronic pathways can be created and the electrocatalysts are deposited all using impregnation techniques. The structure is shown in Figure 8.1m. Consequently, the area of 3PB and 2PB is increased significantly.[1,2]

Other indispensable materials in solid oxide electrolysis cells or stacks, including electrode materials, electrolyte materials, interconnect materials, and cell sealing materials, will be discussed in this chapter.

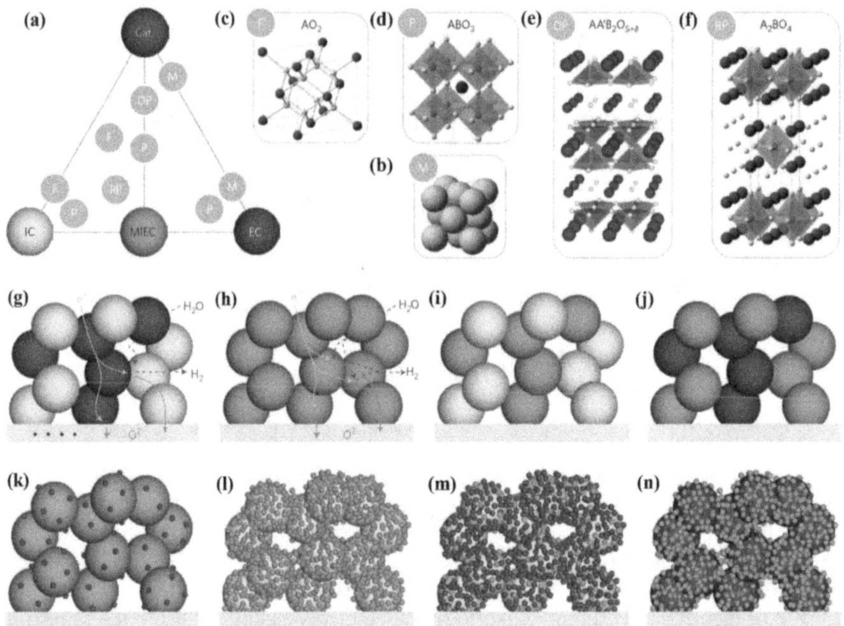

FIGURE 8.1 Materials and microstructures for SOC electrodes. (a) Three key functionalities that electrodes should exhibit to operate effectively, including ion conduction (IC), electron conduction, (EC), or combined mixed ion and electron conduction (MIEC), and catalytic activity (Cat.) Crystal structures of (b) a metal (M), (c) Fluorite (F), (d) Perovskite (P), (e) Double perovskite (DP), (f) Ruddlesden–Popper (RP) phases. Some structures of electrode includes (g) IC–EC composite structure, (h) single-phase MIEC electrode, (i) IC–MIEC composite structure, (j) MIEC–EC composite structure, (k) MIEC with dispersed catalyst particles, (l) IC coated with a percolating layer of MIEC, (m) IC coated with a percolating layer of EC and MIEC, and (n) EC coated with a percolating layer of MIEC and dispersed catalyst. (Reprinted by permission from Macmillan Publishers Ltd. *Nat. Energy*, Irvine, J.T.S. et al., 2016, Copyright 2016.)

8.2 FUEL ELECTRODE MATERIALS

Porous fuel electrodes are used to provide reaction active sites for the decomposition of H_2O and CO_2 in a SOEC. They allow reactants to be transported and products to be removed from the reaction active sites at the surface, while providing a pathway for electrons going from the interconnect to the reaction sites on the electrolyte/electrode.[3]

Ni-YSZ is widely used for solid oxide cell applications as a porous fuel electrode material due to its reasonable electrocatalytic activity, low cost, excellent chemical stability, and appropriate thermal expansion coefficient. The microstructure of the fuel electrode in SOEC is critical for cell performance, which is related to the electrode polarization resistances.[4] In a Ni-YSZ electrode, the networks of Ni and YSZ, as well as the open pores (see Figure 8.2) can contribute to the transportation of electrons, oxide ions, and gas, respectively.[5] Today, the fuel electrode (cathode) supported cell geometry has been designed to enhance cell electrochemical efficiency

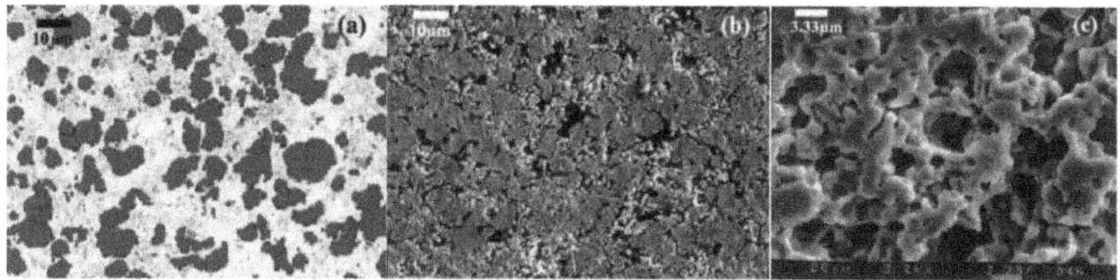

FIGURE 8.2 Microstructures of Ni-YSZ cermet: (a) nickel distribution, (b) overall cermet morphology, and (c) zirconia structural skeleton. (Reprinted by permission from Macmillan Publishers Ltd. *Nat. Mater.*, Atkinson, A. et al., 2004, Copyright 2004.)

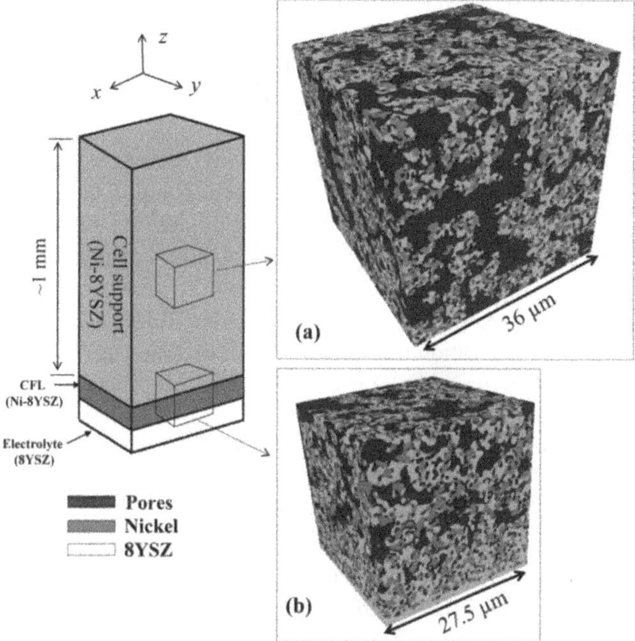

FIGURE 8.3 Three-dimensional structure of a Ni-YSZ cermet: (a) reconstruction taken in the middle of cell support; (b) reconstruction contained a part of cell support, functional layer and a part of electrolyte. (Reprinted from *J. Power Sources*, 256, Usseglio-Viretta, F. et al., Quantitative microstructure characterization of a Ni–YSZ bi-layer coupled with simulated electrode polarisation, 394–403, Copyright 2014, with permission from Elsevier.)

by limiting the ionic losses. It is composed of a porous Ni-YSZ substrate (also called support layer or current collecting layer), a functional layer (also called catalyst layer), and a thin electrolyte, as shown in Figure 8.3. Similarly, as a coarse and thick microstructure, the Ni-YSZ substrate plays an important role in gas transportation and can also induce high concentration overpotential at high current density.[6] On the contrary, exhibiting a thinner microstructure, the functional layer could be presented to lower the activation overpotential and accelerate the electrochemical process during SOEC operating. In pursuit of this, the density of triple phase boundaries (TPB) lengths, which is proportional to the active sites for electrochemical reactions, should be increased as far as possible.[7] In order to improve the ionic conduction of the YSZ, the microstructure of the functional layer should be optimized, which can also lower the YSZ tortuosity factor to enhance the oxygen ion conductivity. Therefore, the microstructure properties of both the substrate or support layer and the functional layer are directly related to the electrolysis performance of SOEC.

Aside from the tortuosity factor and density of TPB lengths, other factors, including dispersion of metal nanoparticles, particle size distribution, mean particle diameters, and even the thickness of each layer can have a significant impact on the effective transport of gas, ions, and electrons, namely, the electrochemical activity of SOEC.[6] For instance, the microstructures of the fuel electrode with different thicknesses of current collecting layer were compared.[8] The results showed that the thickness of the current collecting layer Ni-SDC (current collecting layer) cermet had a significant influence on reducing the polarization loss.

On the other hand, testing showed that the agglomeration-degradation of the Ni-YSZ-based fuel electrode materials during SOEC operation is an issue.[10] Therefore, the microstructure optimization of Ni-YSZ materials under SOEC operating conditions is of great importance.[11] Cell performance can be significantly different when cells are operated at different cell modes. As reported, short-term passivation can occur during the first few hundred hours of operation, which can be attributed to impurities.[12,13] At least part of these impurities may originate from the applied albite glass sealing used in the setup for cell testing. However, further research showed that, in addition to passivation, the main cause of long-term degradation can be induced by the sealing material as well as the decreased activity of the fuel electrode. In addition, high-current densities can also result in significant microstructural changes at the interface between the fuel electrode and the electrolyte, which is ascribed to the relocation of metal. The mechanisms behind this long-term degradation are still not fully understood. Therefore, efforts are needed to optimize electrode microstructures to avoid agglomeration.[11] More details about the degradation issues of fuel electrodes will be discussed in Section 8.6.

In addition to the common fuel electrode material Ni-YSZ, other advanced materials, such as Ni-SDC (Ni-$Sm_{0.2}Ce_{0.8}O_{2-\delta}$), SFM ($Sr_{0.5}Fe_{1.5}Mo_{0.5}O_{6-\delta}$), LSV ($La_{0.6}Sr_{0.4}VO_{3-\delta}$), LSCM ($(La_{0.75}Sr_{0.25})_{0.95}Cr_{0.5}Mn_{0.5}O_3$), STNO ($Sr_{0.94}Ti_{0.9}Nb_{0.1}O_3$), STNNO ($(Sr_{0.94})_{0.9}(Ti_{0.9}Nb_{0.1})O_3$), LST-CGO ($La_{0.35}Sr_{0.65}TiO_{3+\delta}/Ce_{0.8}Gd_{0.2}O_{2-\delta}$), LST-$CeO_2$ ($La_{0.35}Sr_{0.65}TiO_3$-$Ce_{0.5}La_{0.5}O_{2-\delta}$), are presented. Their features are shown in Table 8.1.

Although many varieties of fuel electrode materials have been proposed and developed, there is still no single compound or composite material that can meet all requirements, such as stability, activity, flexibility, and low cost, in SOEC operation simultaneously.[3] Ni-YSZ is regarded as one of the most widely used fuel electrode materials, the stability problems such as sulfur poisoning, oxidation of Ni, and so forth, are the most severe limitations. Furthermore, compared to traditional Ni-YSZ ceramics, the perovskite oxides have excellent mixed ionic and electronic conductivity and better catalytic reaction activity,[14] which are also superior for redox activity, long-term operation stability, and resistance to carbon deposition and sulfur poisoning.[15] The electrochemical reduction reaction is proved to mainly occur at the surface of perovskite-structured fuel cell materials. Unfortunately, regarding CO_2 conversion,

TABLE 8.1
Current Fuel Electrode Materials Used in Various SOEC Cells

Electrode	Electrolysis[a]	Cell	Tem. (°C)	Features	Limitations
Ni-YSZ[23]	HTCE	Ni-YSZ ∣ YSZ ∣ LSM-YSZ	850	Reasonable electro-catalytic activity Low cost	Oxidation of Ni to NiO Poisoned by H_2S
Ni-YSZ[24]	HTSE	Ni-YSZ ∣ YSZ ∣ LSM/LSF/LSCuF/LSCoF	700–850	Excellent chemically stability Appropriate thermal expansion coefficient	
Ni-SDC[25]	HTSE	Ni-Fe ∣ Ni-SDC ∣ LSGM ∣ BLC	600/1000	Decreased cathodic overpotential	
SFM[26,27]	HTSE	SFM ∣ LSGM ∣ SFM	900	Both anode and cathode Good stability Better than Ni-YSZ-LSM	
LSV[28]	HTCE	LSV ∣ YSZ ∣ Pt	900	Salient catalytic activity Long-term stability Sulfur tolerance and carbon resistance	
LSCM[29]	HTSE	LSCM(CGO/YSZ) ∣ YSZ ∣ LSM-YSZ	700–920	Atmosphere: low H_2content	Improvement of current collection and electrode microstructure is necessary
LSCM[30]	HTCE	Pd, CZY, LSCM-YSZ ∣ YSZ ∣ LSF-YSZ	800	Good conductivity in both oxidizing and reducing environments	
LSCMS-SDC[31]	HTSE	LSCM-SDC ∣ YSZ ∣ LSCM-SDC	700	Better ionic conductivity than that of LSCM	
STNO[32]	HTSE	STNO-SDC ∣ YSZ ∣ LSM-SDC	800		Insufficient electrocatalytic activity
STNNO[32]	HTSE	STNNO-SDC ∣ YSZ ∣ LSM-SDC	800	Current efficiency is 20% higher than that of STNO	
LST-CGO[33]	THE	LST-CGO ∣ YSZ ∣ LSM-CGO	700		Insufficient electrolysis performance
LST-CeO_2[24]	HTSE	LST-Ce ∣ YSZ ∣ LSM/LSF/LSCuF/LSCoF	700–850	More active than standard Ni/YSZ	

[a] High-temperature electrolysis (HTE, including HTCE and HTSE), high -emperature CO_2/H_2O co-electrolysis (HTCE), high-temperature steam electrolysis (HTSE).

the AO (A = alkaline earth or lanthanide metal, O = oxide) termination of perovskite oxides with low oxygen vacancy concentration hinders the catalysis contact between CO_2 molecular and high active B-site transition element. In addition, it's difficult to keep the micro-nano structure of the fuel electrode at a high temperature.[16–18]

To meet these challenges, some work can be focused on the following. First, as it has been proven, doping chemistry is a promising way to adjust and improve the electrochemical activity and stability of electrode materials for SOEC, as listed in Table 8.1, by doping with Mn, Cr, Fe, Nd, and so forth. As an alternative method to uniformly disperse active metal nanoparticles on the fuel electrode, in situ exsolution of transition metal in the surface/interface of fuel electrode materials not only facilitates the contact between CO_2 molecular and site-B transition meta but also increases the oxygen vacancy concentration at reaction interface, thus enhancing electrocatalytic activity and improving CO_2 conversion efficiency effectivity.[19–22] Du et al.[19] prepared a $FeNi_3$ bimetallic alloy nanoparticle catalyst-decorated ceramic fuel electrode by in situ reduction of the perovskite $Sr_2FeMo_{0.65}Ni_{0.35}O_{6-\delta}$ (SFMNi), as shown in Figure 8.4. In addition to a significant enhancement of electrochemical performance, the excellent structural stability and good coking resistance in wet methane can also be obtained. With CO_2 electrolysis using SOEC, novel in situ exsolved Fe–Ni alloy nanospheres uniformly socketed on an oxygen-deficient perovskite was proposed by Liu et al.[22] (see Figure 8.5). Compared to common fuel electrode, the stability and efficiency for CO_2 conversion has been significantly improved. Third, in view of the degradation of fuel electrode materials under high-temperature conditions and even long-term operation, the corresponding mechanisms, which may guide the designs of electrode materials, are still not fully understood. Therefore, more related research should be performed.

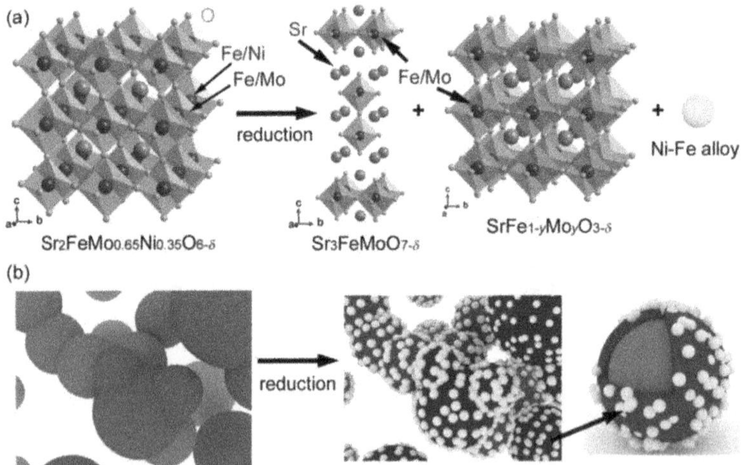

FIGURE 8.4 Illustration models for $Sr_2FeMo_{0.65}Ni_{0.35}O_{6-\delta}$ (SFMNi) fuel electrode: (a) structure transformation and (b) evolution of surface morphology. (Reprinted with permission from Myung, J. et al., *Nature*, 537, 528–531, 2016. Copyright 2016 American Chemical Society.)

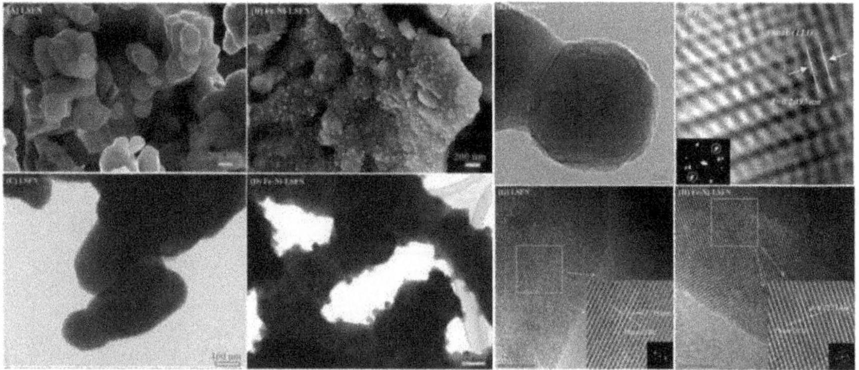

FIGURE 8.5 Microstructures of Fe-Ni-$La_{0.6}Sr_{0.4}Fe_{0.8}Ni_{0.2}O_{3-\delta}$ (LSFN) powers. (TEM: [A] before reduction, [B] after reduction), SEM images (low-resolution: [C] before reduction, [D] after reduction; high-resolution: [E] after reduction, [F] corresponding crystal lattices; the crystal lattice analyses related to fast Fourier transformation: [G] before reduction, [H] after reduction). (Reprinted with permission from Liu, S. et al., *ACS Catal.*, 6, 6219–6228, 2016. Copyright 2016 American Chemical Society.)

8.3 OXYGEN ELECTRODE MATERIALS

In order to improve the electrocatalytic performance of SOEC systems, one key approach is to improve both the activity and stability of the oxygen reduction reaction (ORR) or the oxygen evolution reaction (OER) at oxygen electrodes.[34] In relation to these topics, the materials and structures of oxygen electrodes and the oxygen ionic transport mechanisms in SOECs are discussed next.

8.3.1 Reaction Mechanism of Oxygen Electrode Process

The processes occurring on an oxygen electrode including ORR (**O_2 (gas) + $4e^-$ → $2O^{2-}$**) and OER (**$2O^{2-}$ → O_2 (gas) + $4e^-$**) are regarded as the rate determining steps of electrochemical reactions, which limit the efficiency of SOCs operation.[35] These two processes are reversed, and the detailed reaction steps of one oxygen electrode process (e.g., ORR) are shown in Figure 8.6. The detailed reaction steps of ORR are outlined as follows.[36–38]

① oxygen gas diffused towards the surface of oxygen electrode, **O_2 (*gas*) → O_2 (*electrode surface*)** and then ② absorbed on the surface, **O_2 (*electrode surface*) → O_2 (*ads.*)**; ③ the absorbed oxygen is dissociated, **O_2 (*ads.*) → 2O (*ads.*)** and ④ subsequently reduced and then incorporated into lattice with electron and oxygen vacancy ($V_o^{\cdot\cdot}$), **O (*ads.*) +$V_o^{\cdot\cdot}$ (*electrode*) + $2e^-$→O_o^x (*electrode*)**; ⑤ the lattice oxygen (O_o^x) is diffusion through bulk ⑤' or along the surface to the TPB ⑤'', which is the interface of gas, electrode and electrolyte; and then incorporate into the electrolyte at electrode/electrolyte interface ⑥ or at the TPB ⑦, **O_o^x (*electrode*) +$V_o^{\cdot\cdot}$ (*electrolyte*) →O_o^x (*electrolyte*) +$V_o^{\cdot\cdot}$ (*electrode*)** (Figure 8.6).

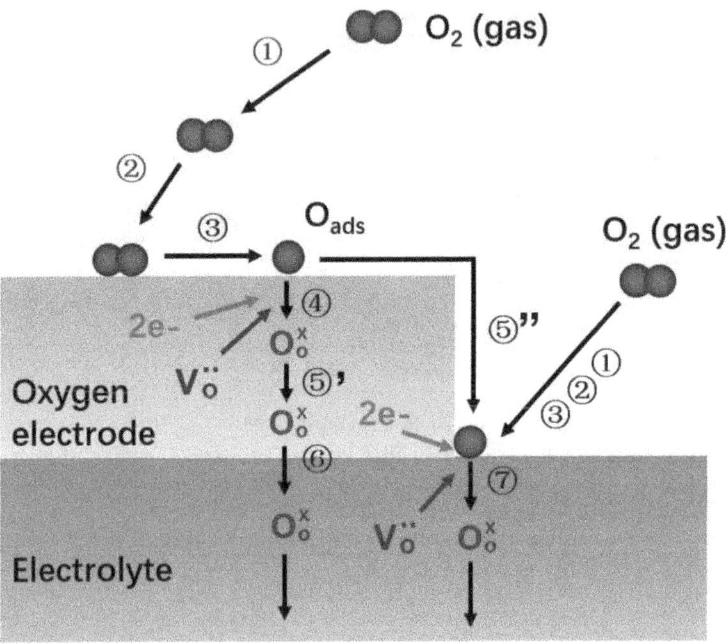

FIGURE 8.6 ORR steps on a mixed-conducting oxide electrode.

8.3.2 Mixed Ionic and Electronic Materials

It is generally recognized that ionic conduction (IC) and electronic conduction (EC) are the two essential functionalities for an oxygen electrode. EC and IC can be provided simultaneously by a mixed ionic and electronic conductor (MIEC).[16,39–41] Typically, perovskite-based MIEC oxides have been proposed as potential SOC oxygen electrode materials especially under intermediate temperatures. They include the cubic-type perovskites ($ABO_{3\pm\delta}$), double-perovskites ($AA'B_2O_{6-\delta}$), and Ruddlesden–Popper phases ($A_2BO_{4+\delta}$), as shown in Figure 8.7. Since it has been declared that ORR or OER occurring at oxygen electrodes are the rate-determining steps of the energy conversion process, the catalyst activities of oxygen electrode materials generally dominate the performance of SOC cells.[34] Thus, the materials and structures covering these three categories and the mechanisms for oxygen exchange and ion transport are discussed next in detail.

8.3.2.1 Perovskites ($ABO_{3\pm\delta}$)

8.3.2.1.1 Basic Structure

The general chemical formula for perovskites is $ABO_{3\pm\delta}$ (so-called " **113**", named from the A:B:O atomic ratios, space group: $Pm\bar{3}m$).[43,44] For many oxygen electrode candidate perovskites, the A-site is occupied by rare earth metal ions (e.g., La^{3+}, Gd^{3+}, Pr^{3+}) surrounded by 12 coordinated oxygen anions. To increase electronic and/or ionic conductivity, A-site cations are usually partially replaced by alkaline earth metal ions such

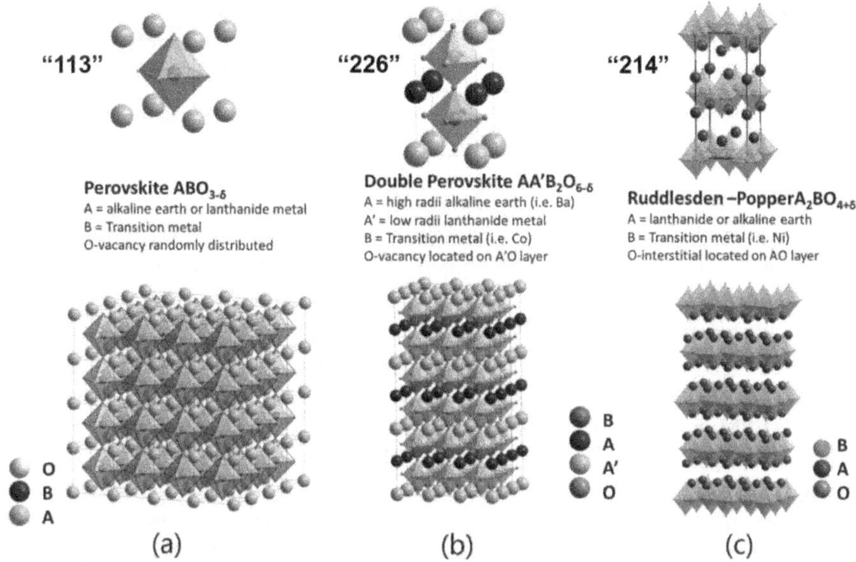

FIGURE 8.7 Crystal structures of (a) perovskite, (b) layered perovskite, and (c) Ruddlesden–Popper. (Gao, Z. et al., *Energy Environ. Sci.*, 9, 1602–1644, 2016. Reproduced by permission of The Royal Society of Chemistry.)

as Ca^{2+}, Sr^{2+}, and Ba^{2+},[45] while the B-site is often occupied by transition metal ions (e.g., Fe^{3+}, Co^{3+}, Ni^{3+}) surrounded by six coordinated oxygen anions.[46,47]

8.3.2.1.2 Oxygen Transport Mechanism

For perovskite materials, the migration paths of an oxygen anion in the cubic symmetry systems were described by Chroneos et al.[48] When the B-site is substituted by divalent ions such as Ca^{2+}, Sr^{2+}, or Ba^{2+}, an oxygen vacancy will be formed around the B-site via charge compensation, which can provide a cubic symmetrical oath for oxygen migration. As shown in Figure 8.8, oxygen anions can move from a previous equilibrium position to the adjacent oxygen vacancy, which can be called **vacancy**

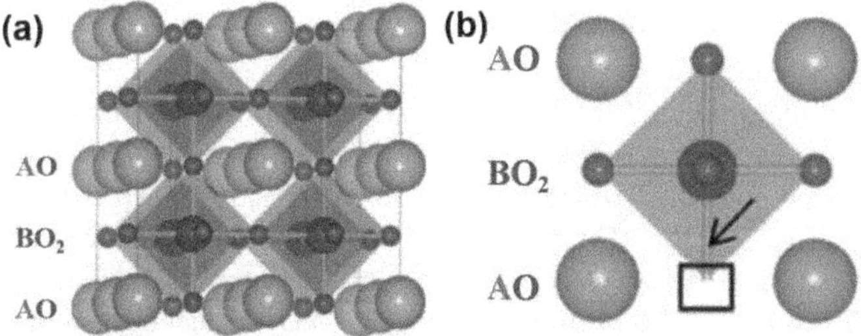

FIGURE 8.8 (a) Crystal structure (b) and oxygen migration path in the perovskite oxides. The square in (b) is vacant oxygen site. (Chroneos. A. et al., *Energy Environ. Sci.*, 4, 2774–2789, 2011. Reproduced by permission of The Royal Society of Chemistry.)

mechanism. Obviously, the transport kinetics is influenced by vacancy concentration in the lattice. Three factors can affect the energy barrier of oxygen migration: (1) the category of A-site or/and B-site ions near the migration path, (2) the configuration and distance of the surrounding vacancy, and (3) the configuration and distance of surrounding doped cations.[48]

8.3.2.1.3 *Typical Materials*

Generally, the "113" materials possess significantly superior EC (100–1000 S cm^{-1}), while their IC (0.001–0.1 S cm^{-1}) is much poorer under intermediate temperatures (e.g., 600°C).[42]

In early studies of SOCs, lanthanum strontium manganite (LSM, $La_{1-x}Sr_xMnO_3$)–based materials were the most commonly used oxygen electrode materials due to their acceptable electronic conductivities and catalytic activities. However, delamination issues related to LSM-based materials can lead to an increase in polarization resistance,[49,50] and the low ionic conductivity of LSM is also an obstacle that results in the lack of active sites for ORR/OER at the electrolyte/electrode interface. Consequently, some alternative oxygen electrode materials related to "La-Sr-Co-Fe" systems, such as $La_{0.8}Sr_{0.2}CoO_3$ (LSC),[51] $La_{0.8}Sr_{0.2}FeO_3$ (LSF),[51] and $La_{0.8}Sr_{0.2}Co_{0.2}Fe_{0.8}O_3$ (LSCF),[52] have been developed (see Table 8.2). In addition, another high-active material, $Ba_{0.5}Sr_{0.5}Co_{0.8}Fe_{0.2}O_{3-\delta}$ (BSCF), has been investigated by many researchers.[50,53–55] Suntivich et al.[54] reported a volcano-shaped relationship between OER activity and the occupancy of the 3d electron of surface B-site cations in perovskite systems. These advances in basic material science will surely help to reveal the origin of catalytic activity and provide guidance for finding more superior materials for SOC oxygen electrodes.

8.3.2.2 Double Perovskites ($AA'B_2O_{6-\delta}$)

8.3.2.2.1 *Basic Structure*

Perovskites are often modified by doping alkaline earth metal ions (A') onto the A site. If the host (A) and substitutional (A') cations are ordered in alternate "AO" planes, the structure becomes an ordered double perovskite structure $AA'B_2O_{5+\delta}$ ("**225**") (or $AA'B_2O_{6-\delta}$ ("**226**")), with an AO-BO_2-A'O-BO_2 stacking sequence (Figure 8.7b)[56]; However, if the host (A) and substitutional (A') cations mix in the same AO/A'O planes, the structure is composed of disordered doped "113" materials.

8.3.2.2.2 *Oxygen Transport Mechanism*

Similar to the vacancy mechanism discussed for perovskite structures, the appearance of oxygen vacancies located on the AO plane was confirmed by neutron diffraction studied in the double perovskites $PrBaCo_2O_{5.5}$.[57] In addition, the **anisotropic oxygen diffusion** was proposed by Parfitt et al.[58] and Seymour et al.,[59] in which the oxygen predominantly diffused in the AO (Pr-O) and BO (Co-O) layer. As shown in Figure 8.9a, the O2 (apical oxygen of BaO plane) sites are not fully occupied and thus tend to obtain oxygen ions, resulting in a decrease of occupancy in O3 (apical oxygen of GdO plane) sites. The result from the calculated oxygen density profile for $GdBaCo_2O_{5.5}$, shown in Figure 8.9b and c, indicates that the oxygen ion migration occurs primarily between the O2 and O3 sites along the a–b plane, while O1(equatorial oxygen) oxygen ions do not participate, a situation called highly anisotropic oxygen diffusion.[58]

TABLE 8.2
Several Perovskite-Based Oxygen Electrode Materials in Various SOEC Cells

Electrode	Electrolysis	Cell/Symmetric Cells	Tem. (°C)	Features	Limitations
LSM[52]	HTSE	Ni-YSZ | 10Sc1CeSZ | LSM	700–850	Good electronic conductivity	Delamination issue/unstable
LSM[101]	HTSE	Ni-YSZ | YSZ | LSM	800–900	Good catalytic activity	Relatively low ionic conductivity[100]
LSM-YSZ[51]	HTSE	Co-ceria-YSZ | YSZ | LSM-YSZ	700		
LSM-GDC[102]		LSM-GDC | YSZ | LSM-GDC	800	Better electro-catalytic activity than that of LSM based cells Can inhibit electrode delamination	
LSC-YSZ[51]	HTSE	Co-ceria-YSZ | YSZ | LSC-YSZ	700	Excellent initial performance but slowly deactivates	Cell performance can be harmed by low O_2 pressure Long-term stability issues
LSF-YSZ[51]	HTSE	Co-ceria-YSZ | YSZ | LSF-YSZ	700	High compatibility with YSZ Exhibits excellent and stable performance	
LSCF[52]	HTSE	Ni-YSZ | 10Sc1CeSZ | LSCF	700–850	Good catalytic activity Candidate as reversible oxygen electrodes	Can react with YSZ
LSCF-CGO[100]	HTCE	Ni-YSZ | YSZ | LSCF-CGO	750–850	Lower degradation rate	
LSCM[103]	HTCE	LSCF | LSGM | LSCM	750		Low electrical conductivity
LSCM-Cu[103]	HTCE	LSCF | LSGM | LSCM-Cu	750	Good electrochemical performance Increasing the proportion of fed CO_2 can result in resistance increase	
SSC-SDC[49]		SSC-SDC | YSZ | SSC-SDC	800	Co-synthesized SSC-SDC Better electrochemical activity and stability	Agglomeration of particles under a higher anodic polarization
BSCF[47]	HTSE	Ni-YSZ | YSZ | BSCF	850	Good performance (three times better than LSM)	Reacts readily with YSZ electrolyte[50]
BSCF-SDC[53]	HTSE	BSCF-SDC | YSZ | BSCF-SDC Ni-YSZ | YSZ | BSCF-SDC	850	Excellent stability	

Abbreviations: High-temperature CO_2/H_2O co-electrolysis (HTCE), high-temperature steam electrolysis (HTSE).

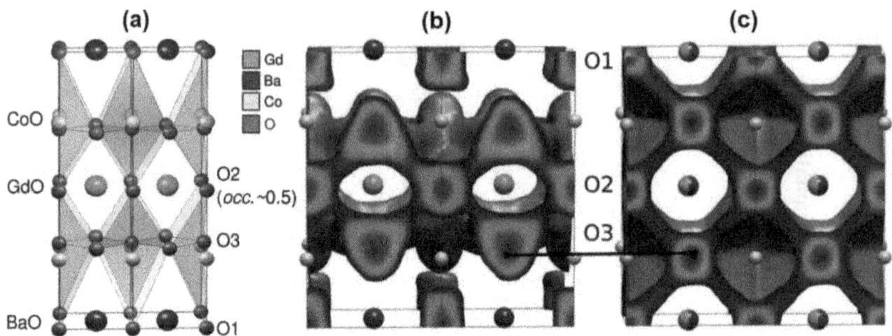

FIGURE 8.9 Typical oxygen transport mechanism in $AA'B_2O_{5+\delta}$ (e.g., $GaBa_2CoO_{5.5}$). (a) The crystal structure of $GaBa_2CoO_{5.5}$; calculated oxygen density profiles of (b) ordered and (c) disordered $GaBa_2CoO_{5.5}$ at 900 K. The O1, O2 and O3 indicate the apical oxygen of (A'O) BaO plane, apical oxygen of (AO) GdO plane, and equatorial oxygen, respectively. (Parfitt, D. et al., *J. Mater. Chem.*, 21, 2183–2186, 2011. Reproduced by permission of The Royal Society of Chemistry.)

8.3.2.2.3 Typical Materials

As typical "226" material, the $LnBaCo_2O_{5+\delta}$ family (Ln = Pr, Nd, Sm, Gd, Dy, Ho, Er, Yb) is investigated as oxygen electrode material candidates by some researchers due to their good EC, which can reach up to 1000 S cm^{-1} at high temperature, and their high oxygen surface exchange and diffusivity kinetics due to the presence of more oxygen vacancies in their crystal structures.[57,60,61] However, compared to "113" materials (e.g., $LaBaCoO_{3-\delta}$), the ordered "226" (e.g., $LaBaCo_2O_{5+\delta}$) is stable at high temperatures and low oxygen pressure, while it is metastable at low temperatures and high oxygen pressure[62] (Figure 8.9).

According to Goodenough,[63] double perovskite $Sr_2Mg_{1-x}Mn_xMoO_{6-\delta}$ showed long-term stability with a tolerance for sulfur and an outstanding single-cell performance. The maximum power density P_{max} reached 838 mW/cm^2 at 800°C in H_2 and 438 mW/cm^2 at 800°C in CH_4. Double perovskite materials such as $Sr_2Fe_{1.5}Mo_{0.5}O_{6-\delta}$,[64–67] $LaBaCo_2O_{6-\delta}$,[62,68–71] $PrBaCo_2O_{5+\delta}$,[56,72–74] and $Sr_2MgMoO_{6-\delta}$[75] were studied as oxygen electrodes in SOCs. Since many double perovskite oxides possess good EC, fast oxygen ion diffusion, and surface exchange kinetics,[61,56,76,77] they may have broad prospects for future applications, especially in the range of immediate temperatures.

8.3.2.3 Ruddlesden-Popper ($A_2BO_{4+\delta}$)

8.3.2.3.1 Basic Structure

The Ruddlesden-Popper family of $A_{n+1}B_nO_{3n+1}$ has a stacking sequence of AO-(BO_2-AO-)$_n$-, with one of its most typical members being the $A_2BO_{4+\delta}$ structure (n = 1, so-called " **214**"), as shown in Figure 8.7c.[78,79]

8.3.2.3.2 Oxygen Transport Mechanism

In addition to the vacancy mechanism, there are two additional paths for oxygen transport: **interstitial migration** (Path I) and **interstitialcy migration** (Path II), as shown in Figure 8.10a and b. In Path I, the interstitial oxygen atom A can directly

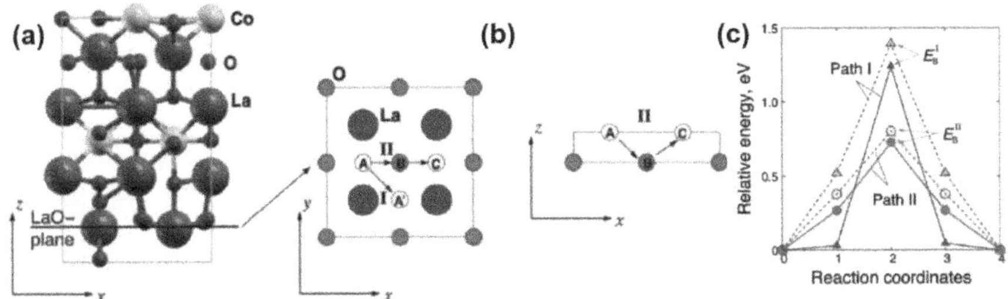

FIGURE 8.10 Typical oxygen transport mechanism in $A_2BO_{4+\delta}$ (e.g. $La_2CoO_{4+\delta}$): (a) the crystal structure of $La_2CoO_{4+\delta}$, (b) the interstitial migration mechanism (path I) and interstitial migration mechanism (path II), (c) relative energy of the system along path I and II; here the letter in the hollow circle indicates interstitial oxygen atom or interstitial site/position. (Kushima, A. et al., *Phys. Chem. Chem. Phys.*, 13, 2242–2249, 2011. Reproduced by permission of The Royal Society of Chemistry.)

move to the adjacent interstitial site A', which is Path I in Figure 8.10a. Regarding the Path II, interstitial oxygen atom A first kicks the oxygen atom at site B (an oxygen atom at the apical of octahedron) and the removed oxygen atom from site B subsequently migrates to another adjacent interstitial site C. Here, site B is on the LaO plane, while the interstitial oxygen atom and interstitial site are not on the LaO-plane.

The relative energy and the energy barriers E_B^I and E_B^{II} were also calculated and compared in Figure 8.10c.[80] It can be found that the energy barriers in Path I (1.27–1.39 eV), which is obviously higher than that of Path II (0.73–0.80 eV), thus suggesting oxygen transport via interstitialcy migration, is much more preferred in A_2BO_4 materials. These conclusions were verified by molecular dynamics (MD) simulations from $La_2CoO_{4+\delta}$, $La_2NiO_{4+\delta}$, $Pr_2NiO_{4+\delta}$, and $Nd_2NiO_{4+\delta}$.[79,81–84]

8.3.2.3.3 *Typical Materials*

In regard to the "214" materials, the oxygen IC is higher than that of perovskite materials by several orders of magnitude, while the EC (about 50–100 S cm^{-1}) is much lower that of "113." The TEC of "214" has good compatibility with some traditional electrolyte materials.

Rare-earth nickelate oxides (with K_2NiF_4-type structure) such as $La_2NiO_{4+\delta}$ and $Pr_2NiO_{4+\delta}$ can be chosen as oxygen electrode materials due to their considerable electrocatalytic activity.[85] In addition, enhanced electrochemical properties can be achieved by adding an interlayer. PNO (Pr_2NiO_4)-YDC was selected for characterizations of single SOCs for comparisons with commercially available $La_{0.6}Sr_{0.4}Fe_{0.8}Co_{0.2}O_{3-\delta}$ cells and showed similar performances. Furthermore, the substitution of Sr, Cu, or Ga in Pr_2NiO_4 can improve stability and oxygen diffusivity.[86] Another typical MIEC material $Nd_2NiO_{4+\delta}$ (NNO) shows promising electrocatalytic activity for oxygen electrode process, with a much higher oxygen diffusion coefficient of 10^{-7} cm^2 s^{-1} and a surface exchange coefficient of 10^{-6} cm s^{-1}.[84,87] Some additional Ruddlesden-Popper (RP) phase oxides in various SOEC cells are listed in Table 8.3.

TABLE 8.3
Several Ruddlesden–Popper (RP) Phase Oxides in Various SOEC Cells

Electrode	Electrolysis	Cell/Symmetric Cells	Tem. (°C)	Features
NNO[84]	HTSE	Ni-CGO ∣ TZ3Y ∣ NNO	750–850	Can be used in reversible operations Higher performance than that of LSM-containing cells
NNO-SSZ[87]	HTSE	Ni-YSZ ∣ SSZ ∣ NNO-SSZ	650–800	High catalytic activity for ORR H_2 production rate: 306.6 mL cm^{-2} h No degradation in short-term durability tests
LSCN[104]	HTSE	Ni-YSZ ∣ 10Sc1CeSZ ∣ LSCN	700–850	Cell performance in SOFC similar to that of SOEC
LSCN[104]		LSCN ∣ 10Sc1CeSZ ∣ LSCN	700–850	Stable with electrolyte 10Sc1CeSZ
PNO-YDC[85]	HTSE	Ni-YSZ ∣ YSZ ∣ PNO-YDC	650–800	Similar performances to $La_{0.6}Sr_{0.4}Fe_{0.8}Co_{0.2}O_{3-\delta}$ Long-term durability needs to be established
PNO-LSGM[78]		PNO-LSGM ∣ LSGM ∣ PNO-LSGM	650	Stability: infiltrated PNO > annealed PNO
LNO[85]		LNO ∣ YSZ ∣ LNO	800	Electrochemical properties: PNO > LNO, PNO-GDC > PNO-YDC > PNO,
LNO-GDC[85]		LNO-GDC ∣ YSZ ∣ LNO-GDC	800	LNO-YDC > LNO-GDC > LNO
LNO-YDC[85]		LNO-YDC ∣ YSZ ∣ LNO-YDC	800	Stability: LNO-GDC, LNO-YDC > LNO, PNO-GDC,
PNO[85]		PNO ∣ YSZ ∣ PNO	800	PNO-YDC > PNO
PNO-GDC[85]		PNO-GDC ∣ YSZ ∣ PNO-GDC	800	Polarization resistance:
PNO-YDC[85]		PNO-GDC ∣ YSZ ∣ PNO-YDC	800	PNO > PNO-YDC > PNO-GDC > LNO > LNO-YDC > LNO-GDC

Abbreviation: High-temperature steam electrolysis (HTSE).

In addition to the three categories of perovskite-related MIEC oxides discussed above, other traditional noble metal electrodes, such as platinum or gold, can also be used as oxygen electrode materials in SOCs.[9,88] However, each one still has some limitations in meeting the requirements of oxygen electrode materials used in SOCs, especially for further applications in long-time operation at low cost. For instance, a noble metal is costly, and the "113" materials possess excellent EC but poor IC. The "214" materials, however, have IC for oxygen migration especially, but their EC is not so good. The performance of "226" materials is superior at high temperatures, but the it is metastable at lower temperatures or in an oxidizing environment.

Therefore, the composite electrodes are developed to enhance the reactive areas and thus the electrode activity and even the stability. Some novel materials or advanced structures with good performance can also be investigated, such as the advanced electrode materials prepared with impregnation,[2,66,87,89–91] exsolution,[19,22,92–94] or heterostructure,[95–99] which may further enhance the reacting sites and activities.[1]

8.3.3 Main Issues of the Oxygen Electrode

SOCs have attracted much attention in recent years with research and development of high-temperature electrolysis advancing at unprecedented rates; however, technical hurdles must be overcome before broad commercialization is possible.[105,106] The existing SOC devices typically operate at extremely high temperatures (for instance, 800–1000°C). Although such high operating temperatures accelerate the kinetics of ORR/OER significantly, they can simultaneously lead to detrimental interactions between electrodes and electrolytes, which may damage cell compositions and material microstructures, leading to long-term durability loss and resulting in the high cost of utilization.[107] For instance, the degradation rates of a 5-cell SOC stack with 5000 hours of successful operation were approximately 15%/1000 hours (0–2000 hours) and 6%/1000 hours (2000–5000 hours) for two types of feed-gas compositions. In another case, 18% loss in H_2 production was observed after 1000 hours of operation when performed by the Idaho National Laboratory with a 25-cell SOC stack.[108]

To develop more efficient, durable, and economically feasible SOCs system, it is widely acknowledged that the key challenge is to reduce operating temperatures to the intermediate range (500°C–800°C) and even lower (300°C–500°C) while maintaining high performances.[95,109] However, the reduced operation temperatures are usually found to cause other performance degradation issues, such as sluggish ORR/OER kinetics on oxygen electrodes, low ionic conductivities in electrolytes, severe carbon deposition and sulphur poisoning on fuel electrodes, and increased polarization resistances throughout the SOCs.[110] As a result, most of the existing cell components and structures may become useless.

One of the most serious limitations for enhancing the efficiency of SOCs at reduced temperature is the sluggish ORR/OER kinetics on the oxygen electrode, and this needs to be improved accordingly. In recent studies, two state-of-the-art strategies have been explored by researchers involving the development of more active materials and the tailoring of material structures.[1] For the former, it is urgent to find more active electrode materials, such as $Ba_{0.5}Sr_{0.5}Co_{0.8}Fe_{0.2}O_{3-\delta}$ (BSCF),[55] $SmNiO_3$ (SNO),[14] $REBaCo_2O_{5+\delta}$ (RE = Pr, Nd, and Gd),[111] and other materials with perovskite type or perovskite-related type structures that have shown promise in the acceleration of ORR/OER kinetics for intermediate temperature SOCs. Regarding the tailoring of material structures, the introduction of a heterostructure possessing two phases can be an effective approach to improve the performance of electrodes in SOCs.

8.4 ELECTROLYTE

The properties of the electrolyte play an important role in electrolysis cell performance due to its contribution to the ohmic internal resistance. Although the electrolyte materials is not involved in the electrode processes directly, ideal SOC electrolyte materials should possess the following characteristics: high oxide ion conductivity (typically > 1×10^{-3} S·cm^{-1}), low electronic conductivity, good thermal stability in the reactant environment, chemical compatibility to the contacting electrode materials, an approximately matched TEC between the electrodes and contacting components, and so on.

Figure 8.11 shows the ionic conductivity of several typical solid oxide electrolytes as a function of temperature.[112] Note that the actual conductivity values for these electrolytes may also change, depending on the microstructure, exact level of doping, and fabrication and sintering processes. For example, ZrO_2 stabilized with 8 mol % Y_2O_3 (8YSZ), one of the most widely used electrolyte material for SOC, possesses relatively high ionic conductivity. In comparison, ZrO_2 stabilized with only 3 mol % Y_2O_3 (3YSZ) becomes a less fragile electrolyte with a lower ionic conductivity.

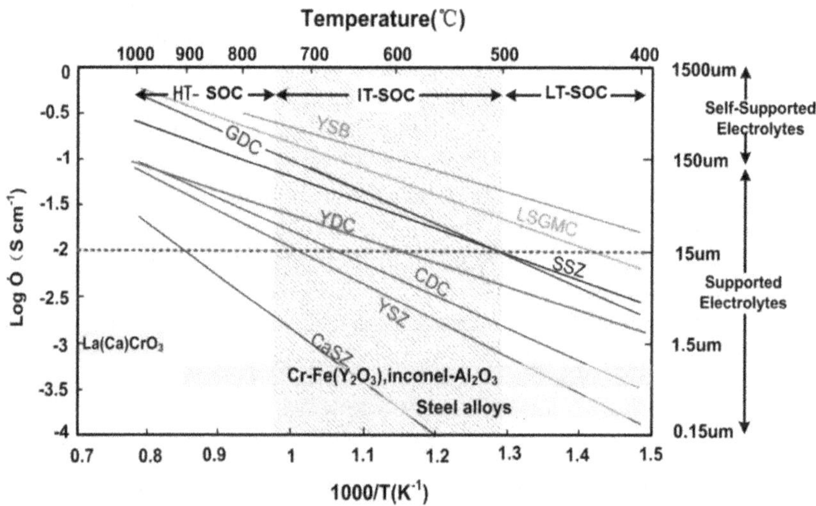

FIGURE 8.11 Variation of specific ionic conductivity of selected electrolytes with temperature. (Brett, D.J.L. et al., *Chem. Soc. Rev.*, 37, 1568, 2008. Reproduced by permission of The Royal Society of Chemistry.)

TABLE 8.4
Comparison of Electrolyte Candidate Materials

Materials	Advantages	Disadvantages
YSZ	Excellent stability in reducing and oxidizing environment Excellent mechanical stability (3YSZ) >40000 h of fuel cell operation possible High quality raw materials available	Low ionic conductivity (especially 3YSZ) Incompatible with some fuel electrode materials
ScSZ	Higher ionic conductivity Excellent stability in oxidizing and reducing environment Better long-term stability than 8YSZ	Scarce availability High price of scandium
CGO	Higher ionic conductivity Good compatibility with fuel electrode materials	Poor mechanical stability Electronic conductivity leading short circuit
LSGM	Higher ionic conductivity Good compatibility with fuel electrode materials	Ga evaporation at low PO_2 Incompatible with NiO Poor mechanical stability

Source: Yu, B. et al., *Sci. China Ser. B Chem.*, 51, 289–304, 2008.

Based on many investigations on YSZ series and other electrolyte materials such as ScYSZ and CGO,[113–116] the oxide ion transport mechanism in these ceramic electrolytes can be thermally activated.

The conductivity related to temperature yields an Arrhenius-like dependence, which is responsible for Figure 8.11 being plotted as the logarithm of conductivity versus reciprocal of temperature. Some possible electrolyte candidates for solid oxide electrolytes are listed and evaluated in Table 8.4. They include several state-of-the-art materials (e.g., scandium doped zirconia [ScSZ], lanthanum gallate with strontium doping [LSGM]) that are being developed. They show promise for application in SOCs.[113,117,118]

8.5 INTERCONNECT MATERIALS

Interconnect is a critical component in SOC, regardless whether the cell configuration is planar or tubular, the role of interconnect is twofold. It provides an electrical connection between the anode of one individual cell to the cathode of the neighboring one (in SOEC mode). It also acts as a physical barrier to protect the oxygen electrode material from the reducing environment of the cathode side, and it equally prevents the fuel electrode material from contacting with the oxidizing atmosphere of the anode side.[120,121] The criteria for the interconnect materials are the most stringent of all cell components. In particular, the chemical potential gradient stemming from considerable oxygen partial pressure differences between the cathode and anode mean severe constraints on the choice of material for the interconnect.

The use of metallic interconnection between planar cells would result in a lower ohmic loss, improved resistance to thermal and mechanical shock, and reduced manufacturing costs. However, the choice of the interconnection material is closely related to the choice of electrolyte because the ionic resistivity of the electrolyte is temperature dependent, dropping by a magnitude of two between 800°C and 900°C. Metallic interconnections would have to operate at lower temperatures than present-day ceramic interconnects.[112]

High growth rate and easy volatilization are the two main drawbacks of chromia. Therefore, the currently developed chromium-bearing metallic interconnects can be applied only for operation under 700°C. Future research should be directed on[122,123]: (1) the development of new interconnect materials with low thermal expansion and high electrical conductivity that can operate in both reducing and oxidizing atmospheres; (2) the development of new interconnect alloys from a fundamental understanding of the oxidation kinetics and oxide conductivity; (3) the development of compliant metallic interconnect designs in combination with novel and low cost stack design concepts; (4) the development of effective and low-cost protective coating materials for metallic interconnects; (5) the investigation of novel coating approaches for metallic interconnects based on chemical, physical vapor deposition and thermal spray techniques; and (6) reducing operating temperatures of SOC to 800°C without compromising the power density.

8.6 CELL SEALING MATERIALS

Sealing is a very important concern for solid oxide stacks used in SOC mode to generate hydrogen and subsequent syngas. Because the product hydrogen and/or CO is the primary valuable product for co-electrolysis of CO_2 and H_2O and due especially to hydrogen's small size, the gas molecules can quickly leak out of high-temperature collection ducts, which will negatively affect the gas purity and efficiency. Figure 8.12 shows seals typically found in a SOC stack with metallic internal gas manifolds and a metallic bipolar plate (in SOFC mode).[124] Common seals include: (1) cell to metal

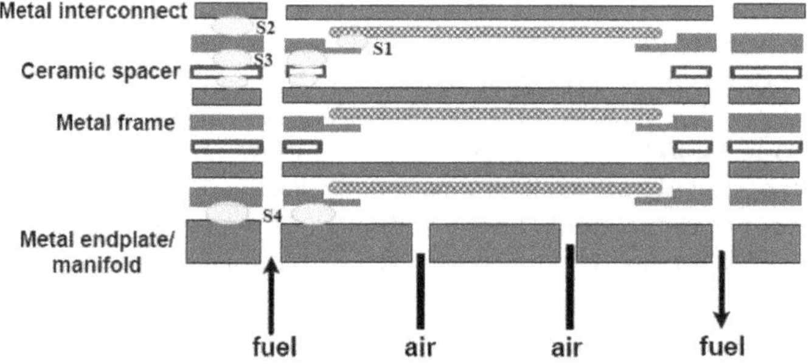

FIGURE 8.12 Schematic of seals typically found in a planar design SOC stack (in SOFC mode).

frame, (2) metal frame to metal interconnect, (3) frame/interconnect pair to electrically insulating spacer, (4) stack to base manifold plate, and (5) cell electrode edge or electrolyte to interconnect edge.

In many instances, an edge seal exists between the zirconia electrolyte and a high-temperature metal frame (or housing). The sealant must endure both the oxidizing environment of the anode area and the reducing condition of the cathode area. The cathode of SOEC has higher inlet steam content than the anode chamber of a fuel cell. Compared with the tubular design, which allows a separation between the oxidizing and reducing zones, the edge sealing in a planar configuration is more difficult. Much effort is needed to address this issue.

Possible candidate sealing materials are silicate, borate, phosphate-based glass systems, and micarex. Three general approaches are currently being pursued: rigid bonded sealing, compressive sealing, and compliant bonded sealing. Reliability is also a major issue in cell seal development because, during long periods of operation and especially during thermal cycling, the thermal profile can vary considerably, creating a variety of stresses.[125] Rather than using traditional glass-based materials, which provide a rigid seal, Pacific Northwest National Laboratory researchers are working with mica materials to construct a more flexible, compressive seal. No one sealing technique will likely satisfy all stack designs and system applications. By comparison, the development efforts on compressive sealing have been more limited in scope, but good progress has been achieved with hybrid mica seals. The concept needs to be tested on full-size components and test stacks to identify potential design and performance issues with scale-up.

Researchers are using a gasket approach, where mica materials are put between cell components and an external load is applied, making it similar to an O ring or a head gasket in a car. The mica-based seal is a more forgiving seal in terms of thermal expansion mismatch because it is made of parallel layers that de-bond from each other under high temperatures.[126,127] As a result, if a certain cell component expands significantly more than an adjacent component, the mica material can mechanically de-couple the two components, preventing the build-up of destructive stresses.

8.7 SUMMARY

The key materials, including electrode materials, electrolyte materials, interconnect materials, and cell sealing materials, were introduced in this chapter. All are the indispensable components used in solid oxide cells and even stacks. Oxygen electrodes are highlighted here. In addition to the reaction mechanism of oxygen electrode process, three main categories of perovskite-based MIEC oxides containing perovskites, double-perovskites and Ruddlesden–Popper phases oxides were highlighted. More studies for these key materials need to be performed to further enhance the efficiency and durability of SOCs. To achieve this goal, some characterization and measuring techniques is necessary, and more details are provided in the next chapter.

REFERENCES

1. Irvine, J. T. S., et al. Evolution of the electrochemical interface in high-temperature fuel cells and electrolysers. *Nature Energy* **1**, 15014 (2016).
2. Liu, Z., et al. Fabrication and modification of solid oxide fuel cell anodes via wet impregnation/infiltration technique. *Journal of Power Sources* **237**, 243–259 (2013).
3. Kan, W. H., Thangadurai, V. Challenges and prospects of anodes for solid oxide fuel cells (SOFCs). *Ionics* **21**, 301–318 (2015).
4. Jiang, S. P. Challenges in the development of reversible solid oxide cell technologies: A mini review. *Asia-Pacific Journal of Chemical Engineering* **11**, 386–391 (2016).
5. Boukamp, B. Anodes sliced with ions. *Nature Materials* **5**, 517–518 (2006).
6. Usseglio-Viretta, F., et al. Quantitative microstructure characterization of a Ni–YSZ bi-layer coupled with simulated electrode polarisation. *Journal of Power Sources* **256**, 394–403 (2014).
7. Lay-Grindler, E., et al. Degradation study by 3D reconstruction of a nickel-yttria stabilized zirconia cathode after high temperature steam electrolysis operation. *Journal of Power Sources* **269**, 927–936 (2014).
8. Uchida, H., et al. Effect of microstructure on performances of hydrogen and oxygen electrodes for reversible SOEC/SOFC. *ECS Transactions* **68**, 3307–3313 (2015).
9. Atkinson, A., et al. Advanced anodes for high-temperature fuel cells. *Nat Mater* **3**, 17–27 (2004).
10. Keane, M., Fan, H., Han, M., Singh, P. Role of initial microstructure on nickel-YSZ cathode degradation in solid oxide electrolysis cells. *International Journal of Hydrogen Energy* **39**, 18718–18726 (2014).
11. Hauch, A., Ebbesen, S. D., Jensen, S. H., Mogensen, M. Solid oxide electrolysis cells: microstructure and degradation of the Ni/Yttria-stabilized zirconia electrode. *Journal of the Electrochemical Society* **155**, B1184–B1193 (2008).
12. Hauch, A., Jensen, S. H., Ramousse, S., Mogensen, M. Performance and durability of solid oxide electrolysis cells. *Journal of the Electrochemical Society* **153**, A1741–A1747 (2006).
13. Hauch, A., Jensen, S. H., Bilde-Sørensen, J. B., Mogensen, M. Silica segregation in the Ni/YSZ electrode. *Journal of the Electrochemical Society* **154**, A619–A626 (2007).
14. Zhou, Y., et al. Strongly correlated perovskite fuel cells. *Nature* **534**, 231–234 (2016).
15. Shao, Z., et al. A thermally self-sustained micro solid-oxide fuel-cell stack with high power density. *Nature* **435**, 795–798 (2005).
16. Druce, J., et al. Surface termination and subsurface restructuring of perovskite-based solid oxide electrode materials. *Energy and Environmental Science* **7**, 3593–3599 (2014).
17. Crumlin, E. J., et al. Oxygen reduction kinetics enhancement on a heterostructured oxide surface for solid oxide fuel cells. *The Journal of Physical Chemistry Letters* **1**, 3149–3155 (2010).
18. Mizusaki, J., Mima, Y., Yamauchi, S., Fueki, K., Tagawa, H. Nonstoichiometry of the perovskite type oxides $La_{1-x}Sr_xCoO_{3-\delta}$. *Journal of Solid State Chemistry* **80**, 102–111 (1989).
19. Du, Z., et al. High-performance anode material $Sr_2FeMo_{0.65}Ni_{0.35}O_{6-\delta}$ with in situ exsolved nanoparticle catalyst. *ACS Nano* **10**, 8660–8669 (2016).
20. Myung, J., Neagu, D., Miller, D. N., Irvine, J. T. S. Switching on electrocatalytic activity in solid oxide cells. *Nature* **537**, 528–531 (2016).
21. Zhou, J., et al. In situ growth of nanoparticles in layered perovskite $La_{0.8}Sr_{1.2}Fe_{0.9}Co_{0.1}O_{4-\delta}$ as an active and stable electrode for symmetrical solid oxide fuel cells. *Chemistry of Materials* **28**, 2981–2993 (2016).

22. Liu, S., Liu, Q., Luo, J. Highly stable and efficient catalyst with in situ exsolved Fe–Ni alloy nanospheres socketed on an oxygen deficient perovskite for direct CO_2 electrolysis. *ACS Catalysis* **6**, 6219–6228 (2016).
23. Ebbesen, S. D., Graves, C., Mogensen, M. Production of synthetic fuels by Co-electrolysis of steam and carbon dioxide. *International Journal of Green Energy* **6**, 646–660 (2009).
24. Marina, O. A.,et al. Electrode performance in reversible solid oxide fuel cells. *Journal of the Electrochemical Society* **154**, B452–B459 (2007).
25. Ishihara, T., Kannou, T. Intermediate temperature steam electrolysis using $LaGaO_3$-based electrolyte. *Solid State Ionics* **192**, 642–644 (2011).
26. Liu, Q., Yang, C., Dong, X., Chen, F. Perovskite $Sr_2Fe_{1.5}Mo_{0.5}O_{6-\delta}$ as electrode materials for symmetrical solid oxide electrolysis cells. *International Journal of Hydrogen Energy* **35**, 10039–10044 (2010).
27. Liu, Q., Dong, X., Xiao, G., Zhao, F., Chen, F. A novel electrode material for symmetrical SOFCs. *Advanced Materials* **22**, 5478–5482 (2010).
28. Ge, X., et al. Robust solid oxide cells for alternate power generation and carbon conversion. *RSC Advances* **1**, 715–724 (2011).
29. Yang, X., Irvine, J. T. S. $(La_{0.75}Sr_{0.25})_{0.95}Mn_{0.5}Cr_{0.5}O_3$ as the cathode of solid oxide electrolysis cells for high temperature hydrogen production from steam. *Journal of Materials Chemistry* **18**, 2349 (2008).
30. Bidrawn, F., et al. Efficient reduction of CO_2 in a solid oxide electrolyzer. *Electrochemical and Solid-State Letters* **11**, B167 (2008).
31. Chen, S., et al. A composite cathode based on scandium-doped chromate for direct high-temperature steam electrolysis in a symmetric solid oxide electrolyzer. *Journal of Power Sources* **274**, 718–729 (2015).
32. Yang, L., et al. Redox-reversible niobium-doped strontium titanate decorated with in situ grown nickel nanocatalyst for high-temperature direct steam electrolysis. *Dalton Transactions* **43**, 14147 (2014).
33. Li, S., Li, Y., Gan, Y., Xie, K., Meng, G. Electrolysis of H_2O and CO_2 in an oxygen-ion conducting solid oxide electrolyzer with a $La_{0.2}Sr_{0.8}TiO_{3+\delta}$ composite cathode. *Journal of Power Sources* **218**, 244–249 (2012).
34. Liu, S. M., et al. Atomic-scale insights into the oxygen ionic transport mechanisms of oxygen electrode in solid oxide cells: A review. *Progress in Chemistry* **26**, 1570–1585 (2014).
35. Sase, M., et al. Enhancement of oxygen exchange at the hetero-interface of (La, Sr) CoO_3/(La, $Sr)_2CoO_4$ in composite ceramics. *Solid State Ionics* **178**, 1843–1852 (2008).
36. Zhao, C., et al. Measurement of oxygen reduction-evolution kinetics enhanced (La, Sr)CoO_3/(La, $Sr)_2CoO_4$ hetero-structure oxygen electrode in operating temperature for SOCs. *International Journal of Hydrogen Energy* (2018). doi:10.1016/j.ijhydene.2018.04.128.
37. Ringuedé, A., Fouletier, J. Oxygen reaction on strontium-doped lanthanum cobaltite dense electrodes at intermediate temperatures. *Solid State Ionics* **139**, 167–177 (2001).
38. Kawada, T., et al. Microscopic observation of oxygen reaction pathway on high temperature electrode materials. *Solid State Ionics* **177**, 3081–3086 (2006).
39. Adler, S. B. Mechanism and kinetics of oxygen reduction on porous $La_{1-x}Sr_xCoO_{3-\delta}$ electrodes. *Solid State Ionics* **111**, 125–134 (1998).
40. Adler, S. B. Limitations of charge-transfer models for mixed-conducting oxygen electrodes. *Solid State Ionics* **135**, 603–612 (2000).
41. Ebbesen, S. D., Jensen, S. H., Hauch, A., Mogensen, M. B. High temperature electrolysis in alkaline cells, solid proton conducting cells, and solid oxide cells. *Chemical Reviews* **114**, 10697–10734 (2014).

42. Gao, Z., Mogni, L. V., Miller, E. C., Railsback, J. G., Barnett, S. A. A perspective on low-temperature solid oxide fuel cells. *Energy & Environmental Science* **9**, 1602–1644 (2016).
43. Lacorre, P., Goutenoire, F. O., Bohnke, O., Retoux, R., Laligant, Y. Designing fast oxide-ion conductors based on $La_2Mo_2O_9$. *Nature* **40**, 856–858 (2000).
44. Steele, B. C., Heinzel, A. Materials for fuel-cell technologies. *Nature* **414**, 345–352 (2001).
45. Knibbe, R., Hauch, A., Hjelm, J., Ebbesen, S. D., Mogensen, M. Durability of solid oxide cells. *Green* **1**, 141–169 (2011).
46. Adler, S.B. Factors governing oxygen reduction in solid oxide fuel cell cathodes. *Chemical Reviews* **104**, 4791–4844 (2004).
47. Bo, Y., et al. Microstructural characterization and electrochemical properties of $Ba_{0.5}Sr_{0.5}Co_{0.8}Fe_{0.2}O_{3-\delta}$ and its application for anode of SOEC. *International Journal of Hydrogen Energy* **33**, 6873–6877 (2008).
48. Chroneos, A., Yildiz, B., Tarancón, A., Parfitt, D., Kilner, J. A. Oxygen diffusion in solid oxide fuel cell cathode and electrolyte materials: mechanistic insights from atomistic simulations. *Energy & Environmental Science* **4**, 2774–2789 (2011).
49. Jiang, W., et al. Performance and stability of co-synthesized $Sm_{0.5}Sr_{0.5}CoO_3$–$Ce_{0.8}Sm_{0.2}O_{1.9}$ composite oxygen electrode for solid oxide electrolysis cells. *International Journal of Hydrogen Energy* **40**, 561–567 (2015).
50. Kim-Lohsoontorn, P., Brett, D. J. L., Laosiripojana, N., Kim, Y., Bae, J. Performance of solid oxide electrolysis cells based on composite $La_{0.8}Sr_{0.2}MnO_{3-\delta}$ – yttria stabilized zirconia and $Ba_{0.5}Sr_{0.5}Co_{0.8}Fe_{0.2}O_{3-\delta}$ oxygen electrodes. *International Journal of Hydrogen Energy* **35**, 3958–3966 (2010).
51. Wang, W., Huang, Y., Jung, S., Vohs, J. M., Gorte, R. J. A comparison of LSM, LSF, and LSCo for solid oxide electrolyzer anodes. *Journal of the Electrochemical Society* **153**, A2066–A2070 (2006).
52. Laguna-Bercero, M.A., Kilner, J.A. & Skinner, S.J. Development of oxygen electrodes for reversible solid oxide fuel cells with scandia stabilized zirconia electrolytes. *Solid State Ionics* **192**, 501–504 (2011).
53. Zhang, W., Yu, B., Xu, J. Investigation of single SOEC with BSCF anode and SDC barrier layer. *International Journal of Hydrogen Energy* **37**, 837–842 (2012).
54. Suntivich, J., May, K. J., Gasteiger, H. A., Goodenough, J. B., Shao-Horn, Y. A perovskite oxide optimized for oxygen evolution catalysis from molecular orbital principles. *Science* **334**, 1383–1385 (2011).
55. Shao, Z., Haile, S. M. A high-performance cathode for the next generation of solid-oxide fuel cells. *Nature* **431**, 170–173 (2004).
56. Kim, G., et al. Rapid oxygen ion diffusion and surface exchange kinetics in $PrBaCo_2O_{5+x}$ with a perovskite related structure and ordered A cations. *Journal of Materials Chemistry* **17**, 2500 (2007).
57. Streule, S., et al. High-temperature order-disorder transition and polaronic conductivity in $PrBaCo_2O_{5.48}$.*Physical Review B* **73**, 094203–094205 (2006).
58. Parfitt, D., Chroneos, A., Tarancón, A., Kilner, J. A. Oxygen ion diffusion in cation ordered/disordered $GdBaCo_2O_{5+d}$. *Journal of Materials Chemistry* **21**, 2183–2186 (2011).
59. Seymour, I. D., et al. Anisotropic oxygen diffusion in $PrBaCo_2O_{5.5}$ double perovskites. *Solid State Ionics* **216**, 41–43 (2012).
60. Tarancón, A., et al. Effect of phase transition on high-temperature electrical properties of $GdBaCo_2O_{5+x}$ layered perovskite. *Solid State Ionics* **179**, 611–618 (2008).
61. Tarancon, A., et al. Stability, chemical compatibility and electrochemical performance of $GdBaCo_2O_{5+x}$ layered perovskite as a cathode for intermediate temperature solid oxide fuel cells. *Solid State Ionics* **179**, 2372–2378 (2008).

62. Bernuy-Lopez, C., Høydalsvik, K., Einarsrud, M., Grande, T. Effect of A-site cation ordering on chemical stability, oxygen stoichiometry and electrical conductivity in layered $LaBaCo_2O_{5+\delta}$ double perovskite. *Materials* **9**, 154 (2016).
63. Yun-Hui, H., Dass, R. I., Xing, Z. L., Goodenough, J. B. Double perovskites as anode materials for solid-oxide fuel cells. *Science* **312**, 255–257 (2006).
64. Bugaris, D. E., et al. Investigation of the high-temperature redox chemistry of Sr2Fe1.5Mo0.5$O_{6-\delta}$via in situ neutron diffraction. *Journal of Materials Chemistry A* **2**, 4045 (2014).
65. Suthirakun, S., et al. Theoretical investigation of H_2O oxidation on the $Sr_2Fe_{1.5}Mo_{0.5}O_6$(001) perovskite surface under anodic solid oxide fuel cell conditions. *Journal of the American Chemical Society* **136**, 8374–8386 (2014).
66. Rath, M. K., Lee, K. Superior electrochemical performance of non-precious Co-Ni-Mo alloy catalyst-impregnated $Sr_2FeMoO_{6-\delta}$ as an electrode material for symmetric solid oxide fuel cells. *Electrochimica Acta* **212**, 678–685 (2016).
67. Li, H., Zhao, Y., Wang, Y., Li, Y. $Sr_2Fe_{2-x}Mo_xO_{6-\delta}$ perovskite as an anode in a solid oxide fuel cell: Effect of the substitution ratio. *Catalysis Today* **259**, 417–422 (2016).
68. Liu, J., Collins, G., Liu, M., Chen, C. Superfast oxygen exchange kinetics on highly epitaxial $LaBaCo_2O_{5+\delta}$ thin films for intermediate temperature solid oxide fuel cells. *APL Materials* **1**, 031101 (2013).
69. Choi, S., et al. Electrochemical properties of an ordered perovskite $LaBaCo_2O_{5+\delta}$–$Ce_{0.9}Gd_{0.1}O_{2-\delta}$ composite cathode with strontium doping for intermediate-temperature solid oxide fuel cells. *Electrochemistry Communications* **34**, 5–8 (2013).
70. Pang, S., et al. Characterization of cation-ordered perovskite oxide $LaBaCo_2O_{5+\delta}$ as cathode of intermediate-temperature solid oxide fuel cells. *International Journal of Hydrogen Energy* **37**, 6836–6843 (2012).
71. Li, R., Ge, L., Chen, H., Guo, L. Preparation and performance of triple-layer graded $LaBaCo_2O_{5+\delta}$–$Ce_{0.8}Sm_{0.2}O_{1.9}$ composite cathode for intermediate-temperature solid oxide fuel cells. *Electrochimica Acta* **85**, 273–277 (2012).
72. Liu, J., et al. Ultrafast oxygen exchange kinetics on highly epitaxial $PrBaCo_2O_{5+\delta}$ thin films. *Applied Physics Letters* **100**, 193903 (2012).
73. Meng, F., Xia, T., Wang, J., Shi, Z., Zhao, H. Praseodymium-deficiency $Pr_{0.94}BaCo_2O_{6-\delta}$ double perovskite: A promising high performance cathode material for intermediate-temperature solid oxide fuel cells. *Journal of Power Sources* **293**, 741–750 (2015).
74. Burriel, M., et al. Anisotropic oxygen ion diffusion in layered $PrBaCo_2O_{5+\delta}$. *Chemistry of Materials* **24**, 613–621 (2012).
75. Bernuy-Lopez, C., Allix, M., Bridges, C.A., Claridge, J. B., Rosseinsky, M. J. $Sr_2MgMoO_{6-\delta}$: Structure, phase stability, and cation site order control of reduction. *Chemistry of Materials* **19**, 1035–1043 (2007).
76. Maignan, A. et al. Structural and magnetic studies of ordered oxygen-deficient perovskites $LnBaCo_2O_5$, closely related to the "112" structure. *Journal of Solid State Chemistry* **142**, 247–260 (1999).
77. Kim, G., et al. Oxygen exchange kinetics of epitaxial $PrBaCo_2O_{5+\delta}$ thin films. *Applied Physics Letters* **88**, 024103 (2006).
78. Railsback, J. G., Gao, Z., Barnett, S. A. Oxygen electrode characteristics of $Pr_2NiO_{4+\delta}$-infiltrated porous $(La_{0.9}Sr_{0.1})(Ga_{0.8}Mg_{0.2})O_{3-\delta}$. *Solid State Ionics* **274**, 134–139 (2015).
79. Laguna-Bercero, M. A., et al. High performance of microtubular solid oxide fuel cells using Nd_2NiO_{4+d}-based composite cathodes. *Journal of Materials Chemistry A* **2**, 9764–9770 (2014).
80. Kushima, A., et al. Interstitialcy diffusion of oxygen in tetragonal $La_2CoO_{4+delta}$. *Physical Chemistry Chemical Physics* **13**, 2242–2249 (2011).

81. Chroneos, A., Parfitt, D., Kilner, J. A., Grimes, R. W. Anisotropic oxygen diffusion in tetragonal La2NiO$_{4+d}$: molecular dynamics calculations. *Journal of Materials Chemistry* **20**, 266–270 (2010).
82. Perrichon, A., et al. Lattice dynamics modified by excess oxygen in $Nd_2NiO_{4+\delta}$: triggering low-temperature oxygen diffusion. *The Journal of Physical Chemistry C* **119**, 1557–1564 (2015).
83. Montenegro-Hernández, A., Vega-Castillo, J., Mogni, L., Caneiro, A. Thermal stability of $Ln_2NiO_{4+\delta}$ (Ln: La, Pr, Nd) and their chemical compatibility with YSZ and CGO solid electrolytes. *International Journal of Hydrogen Energy* **36**, 15704–15714 (2011).
84. Chauveau, F., Mougin, J., Bassat, J. M., Mauvy, F., Grenier, J. C. A new anode material for solid oxide electrolyser: The neodymium nickelate $Nd_2NiO_{4+\delta}$. *Journal of Power Sources* **195**, 744–749 (2010).
85. Ogier, T., et al. Enhanced performances of structured oxygen electrodes for high temperature steam electrolysis. *Fuel Cells* **13**, 536–541 (2013).
86. Ferchaud, C., et al. High performance praseodymium nickelate oxide cathode for low temperature solid oxide fuel cell. *Journal of Power Sources* **196**, 1872–1879 (2011).
87. Chen, T., et al. High performance of intermediate temperature solid oxide electrolysis cells using $Nd_2NiO_{4+\delta}$ impregnated scandia stabilized zirconia oxygen electrode. *Journal of Power Sources* **276**, 1–6 (2015).
88. Bius, H. M. On the history of solid electrolyte fuel cells. *Journal of Solid State Electrochemistry* **1**, 2–16 (1997).
89. Zhang, X., et al. Perovskite LSCM impregnated with vanadium pentoxide for high temperature carbon dioxide electrolysis. *Electrochimica Acta* **212**, 32–40 (2016).
90. Chen, T., et al. High performance solid oxide electrolysis cell with impregnated electrodes. *Electrochemistry Communications* **54**, 23–27 (2015).
91. Wei, B., et al. Performance evaluation of an anode-supported solid oxide fuel cell with $Ce_{0.8}Sm_{0.2}O_{1.9}$ impregnated $GdBaCo_2O_{5+\delta}$ cathode. *International Journal of Hydrogen Energy* **37**, 13491–13498 (2012).
92. Zhu, Y., et al. Promotion of oxygen reduction by exsolved silver nanoparticles on a perovskite scaffold for low-temperature solid oxide fuel cells. *Nano Letters* **16**, 512–518 (2016).
93. Neagu, D., et al. Nano-socketed nickel particles with enhanced coking resistance grown in situ by redox exsolution. *Nature Communications* **6**, 8120 (2015).
94. Cassidy, M., Gamble, S., Irvine, J. T. S. Application of exsolved structures as a route to more robust anodes for improved biogas utilisation in SOFCs. *ECS Transactions* **68**, 2029–2036 (2015).
95. Ma, W., et al. Vertically aligned nanocomposite $La_{0.8}Sr_{0.2}CoO_3/(La_{0.5}Sr_{0.5})_2CoO_4$cathodes – electronic structure, surface chemistry and oxygen reduction kinetics. *Journal of Materials Chemistry A* **3**, 207–219 (2015).
96. Lee, D., et al. Oxygen surface exchange kinetics and stability of $(La, Sr)_2CoO_{4\pm\delta}$/ La1−xSrx $MO_{3-\delta}$(M = Co and Fe) hetero-interfaces at intermediate temperatures. *Journal of Materials Chemistry A* **3**, 2144–2157 (2015).
97. Feng, Z., et al. Anomalous interface and surface strontium segregation in $(La_{1-y}Sr_y)_2$ $CoO_{4\pm\delta}$ /La_{1-x} Sr_x $CoO_{3-\delta}$ heterostructured thin films. *The Journal of Physical Chemistry Letters* **5**, 1027–1034 (2014).
98. Tsvetkov, N., Chen, Y., Yildiz, B. Reducibility of Co at the $La_{0.8}Sr_{0.2}CoO_3$/ $(La_{0.5}Sr_{0.5})_2CoO_4$ hetero-interface at elevated temperatures. *Journal of Materials Chemistry A* **2**, 14690 (2014).
99. Han, J. W., Yildiz, B. Mechanism for enhanced oxygen reduction kinetics at the (La, Sr) CoO_3/(La, Sr)$_2CoO_{4+\delta}$ hetero-interface. *Energy & Environmental Science* **5**, 8598–8607 (2012).

100. Hjalmarsson, P., Sun, X., Liu, Y., Chen, M. Influence of the oxygen electrode and inter-diffusion barrier on the degradation of solid oxide electrolysis cells. *Journal of Power Sources* **223**, 349–357 (2013).
101. Yang, C., Coffin, A., Chen, F. High temperature solid oxide electrolysis cell employing porous structured $(La_{0.75}Sr_{0.25})_{0.95}MnO_3$ with enhanced oxygen electrode performance. *International Journal of Hydrogen Energy* **35**, 3221–3226 (2010).
102. Kongfa, C., Ai, N., San Ping, J. Development of (Gd, Ce)O_2-impregnated (La, Sr)MnO_3 anodes of high temperature solid oxide electrolysis cells. *Journal of the Electrochemical Society* **157**, P89–P94 (2010).
103. Xing, R., Wang, Y., Zhu, Y., Liu, S., Jin, C. Co-electrolysis of steam and CO_2 in a solid oxide electrolysis cell with $La_{0.75}Sr_{0.25}Cr_{0.5}Mn_{0.5}O_{3-\delta}$-Cu ceramic composite electrode. *Journal of Power Sources* **274**, 260–264 (2015).
104. Laguna-Bercero, M. A., et al. Performance of $La_{2-x}Sr_xCo_{0.5}Ni_{0.5}O_{4\pm\delta}$ as an oxygen electrode for solid oxide reversible cells. *Fuel Cells* **11**, 102–107 (2011).
105. Zheng, Y., et al. A review of high temperature co-electrolysis of H_2O and CO_2 to produce sustainable fuels using solid oxide electrolysis cells (SOECs): Advanced materials and technology. *Chemical Society Reviews* **46**, 1427–1463 (2017).
106. Zheng, Y., et al. Energy related CO_2 conversion and utilization advanced materials/nanomaterials, reaction mechanisms and technologies. *Nano Energy* **40**, 512–539 (2017).
107. Moçoteguy, P., Brisse, A. A review and comprehensive analysis of degradation mechanisms of solid oxide electrolysis cells. *International Journal of Hydrogen Energy* **38**, 15887–15902 (2013).
108. Sohal, M. S., Stoots, C. M., Sharma, V. I., Yildiz, B., Virkar, A. Degradation issues in solid oxide cells during high temperature electrolysis. *Journal of Fuel Cell Science and Technology* **9**, 0110171-10 (2012).
109. Choi, S., et al. Exceptional power density and stability at intermediate temperatures in protonic ceramic fuel cells. *Nature Energy* **3**, 202–210 (2018).
110. Lee, K. T., Manthiram, A. Comparison of $Ln_{0.6}Sr_{0.4}CoO_{3-\delta}$ (Ln = La, Pr, Nd, Sm, and Gd) as cathode materials for intermediate temperature solid oxide fuel cells. *Journal of the Electrochemical Society* **4**, A794–A798 (2006).
111. Liu, S., Zhang, W., Li, Y., Yu, B. $REBaCo_2O_{5+d}$ (RE ¼ Pr, Nd, and Gd) as promising oxygen electrodes for intermediate-temperature solid oxide electrolysis cells. *RSC Advances* **7**, 16332–16340 (2017).
112. Brett, D. J. L., Atkinson, A., Brandon, N. P., Skinner, S. J. Intermediate temperature solid oxide fuel cells. *Chemical Society Reviews* **37**, 1568 (2008).
113. Osada, N., Uchida, H., Watanabe, M. Polarization behavior of SDC cathode with highly dispersed Ni catalysts for solid oxide electrolysis cells. *Journal of the Electrochemical Society* **153**, A816 (2006).
114. Stoots, C. M., Herring, J. S., Hartvigsen, J. Hydrogen production performance of a 10-Cell planar solid-oxide electrolysis stack. *Journal of Fuel Cell Science and Technology* **3**, 213–219 (2006).
115. Eguchi, K., et al. Power generation and steam electrolysis characteristics of an electrochemical cell with a zirconia- or ceria-based electrolyte. *Solid State Ionics* **86– 88**, 1245–1249 (1996).
116. Zhu, B., et al. Electrolysis studies based on ceria-based composites. *Electrochemistry Communications* **8**, 495–498 (2006).
117. Ni, M., Leung, M. K., Leung, D. Y. Technological development of hydrogen production by solid oxide electrolyzer cell (SOEC). *International Journal of Hydrogen Energy* **33**, 2337–2354 (2008).
118. Lessing, P. A. Materials for hydrogen generation via water electrolysis. *Journal of Materials Science* **42**, 3477–3487 (2007).

119. Yu, B., Zhang, W., Chen, J., Xu, J., Wang, S. Advance in highly efficient hydrogen production by high temperature steam electrolysis. *Science in China Series B: Chemistry* **51**, 289–304 (2008).
120. Molenda, J. High-temperature solid-oxide fuel cells. New trends in materials research. *Materials Science* **24**, 5–11 (2006).
121. Haile, S. M. Fuel cell materials and components. *Acta Materialia* **51**, 5981–6000 (2003).
122. Zhu, W. Z., Deevi, S. C. Development of interconnect materials for solid oxide fuel cells. *Materials Science and Engineering* **A348**, 227–243 (2003).
123. Singh, P., Minh, N. Q. Solid oxide fuel cells: Technology status. *International Journal of Applied Ceramic Technology* **1**, 5–15 (2004).
124. Stevenson, J. W. *SOFC Seals: Materials Status, SECA Core Technology Program-SOFC Seal Meeting* (Sandia National Laboratory, Albuquerque, NM, 2003).
125. Steele, B. C. H. Materials for IT-SOFC stacks 35 years R&D: The inevitability of gradualness? *Solid State Ionics* **134**, 3–20 (2000).
126. Lessing, P. A. A review of sealing technologies applicable to solid oxide electrolysis cells. *Journal of Materials Science* **42**, 3465–3476 (2007).
127. Weil, K. S. The state-of-the-art in sealing technology for solid oxide fuel cells. *JOM* **58**, 37–44 (2006).

9 High-Temperature Electrochemical Process of CO_2 Conversion with SOCs 4

Measurement, Characterization, and Simulation

9.1 ELECTROCHEMICAL MEASUREMENT

First, with regards to electrochemical research, the current-voltage (I–V) measurement may be the most fundamental and important method, especially in the research of solid oxide cell (SOC). The measurement can be conducted using an electrochemical interface in solid oxide electrolysis cell (SOEC) or solid oxide fuel cell (SOFC) conditions.[1] The voltages, including open circuit voltages (OCVs), can be observed and compared to study properties of some electrochemical processes, and resistance such as area-specific resistance (ASR) can be calculated for further investigation. For example, the basic electrochemical performance at various temperatures for H_2O/CO_2 co-electrolysis is shown in Figure 9.1, and the influence of humidity for H_2O electrolysis is also studied with I-V measurement.

There is no denying that electrochemical impedance spectroscopy (EIS) is one of the earliest and most insightful methods used in SOC characterization. With it, the impedance of a SOC over a wide range of frequencies (typically 10^{-2} to 10^{7} Hz) can be measured with alternating current or voltage.[3–7] Meanwhile, different electrochemical processes can be separated according to electrical relaxation times, and the electrochemical properties can then be studied. For example, the electrochemical processes and corresponding electrochemical properties can be investigated systematically using EIS, as shown in Figure 9.2a.[8]

In comparison with bulk materials, the heterostructured interface materials have attracted increasing attention in SOC because of their inherent attractive

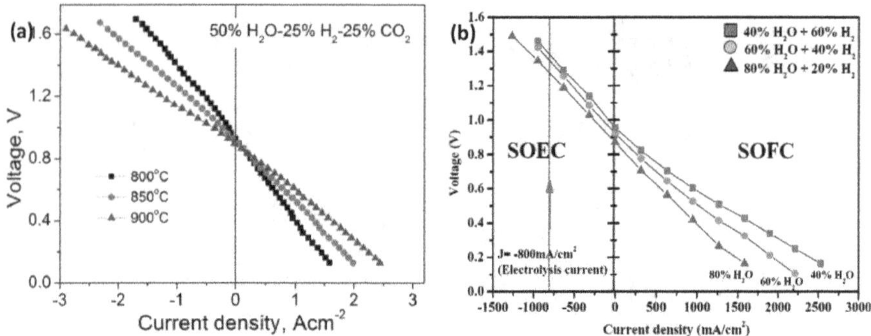

FIGURE 9.1 (a) Current-voltage curves of the Ni-SDC-YSZ/YSZ/LSM-SDC-YSZ SOEC recorded in SOEC and SOFC modes at various temperatures. (Yang, C. et al., *J. Mater. Chem. A*, 3, 15913–15919, 2015. Reproduced by permission of The Royal Society of Chemistry); (b) current-voltage curves of the Ni-YSZ cathodes exposed to various humidity at 800°C in SOEC and SOFC modes. (Reproduced from Kim, S.J. et al., *Int. J. Hydrogen Energ.*, 40, 9032–9038, 2015 with permission from Pergamon-Elsevier Science Ltd., Copyright 2018.)

physical, chemical, and electrochemical properties, especially the enhanced oxygen electrocatalytic activity and electronic/ionic conductivity.[9–18] The heterostructured interfaces are investigated using EIS to compare electrochemical performance of different materials, as shown in Figure 9.2b.[19] EIS can be performed in situ and in operando, and this technique is more insightful and accurate for electrochemical

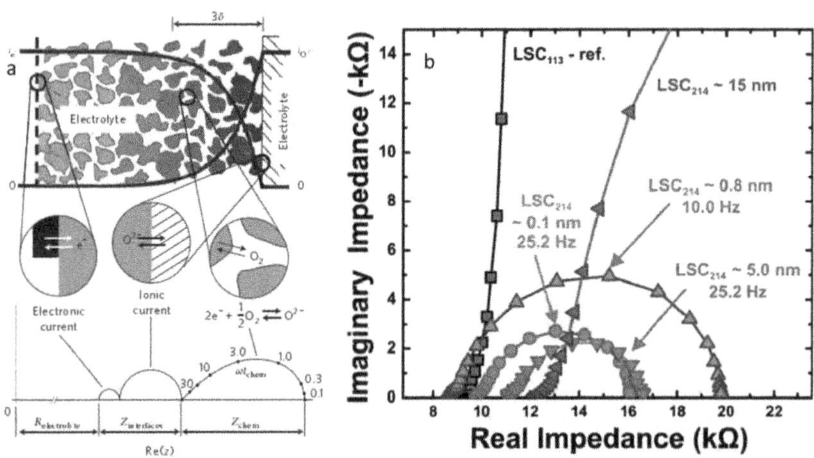

FIGURE 9.2 (a) Schematic of the ALS model for a porous mixed-conducting oxygen electrode. (Reproduced from *Solid State Ion.*, 135, Adler, S.B., Limitations of charge-transfer models for mixed-conducting oxygen electrodes, 603–612, Copyright 2000, with permission from Elsevier.) (b) EIS results of microelectrodes for the research of heterostructured interfaces by comparing with LSC_{113} and LSC_{214}. (Reprinted with permission from Crumlin, E.J. et al., *J. Phys. Chem. Lett.*, 1, 3149–3155, 2010. Copyright 2010 American Chemical Society.)

research of SOC.[20,21] Note that, in order to better understand materials' properties, the information from EIS is always complemented or better interpreted in conjunction with information on the structure (especially the microstructure) of materials.

9.2 MICROSTRUCTURE CHARACTERIZATION

9.2.1 SEM, TEM, AND STEM

The scanning electron microscope (SEM) is one of the most common methods to observe the microstructure of materials for SOC,[2,7,22] such as the microstructures of SOCs with the configuration Ni-YSZ|YSZ|LSM-YSZ, as shown in Figure 9.3. The surface features of materials can be presented clearly and directly using SEM at different scales.

There are some differences between SEM and transmission electron microscopy (TEM). First, TEM is based on transmitted electrons, while SEM is based on scattered electrons. TEM provides details about the internal composition of SOC material with a three-dimensional image, whereas SEM specializes two-dimensional depictions of the surface of material. For an analogy, TEM and SEM are similar to a microscope and a camera, respectively. In addition, TEM can show many characteristics such as morphology, crystallization, and even magnetic domains, whereas SEM can show only the morphology of SOC material.

Figure 9.3 shows the surface/interface structures of perovskite. The figure includes a SEM image of an A-site-deficient, O-stoichiometric $La_{0.52}Sr_{0.28}Ni_{0.06}Ti_{0.94}O_3$ surface and two TEM images of exsolved particle-substrate interface. It can be seen clearly that TEM has much higher resolution than SEM; the former is able to analyze at nano level. More information can be achieved from TEM by contrast, but SEM is still indispensable for the study of SOC materials in general.

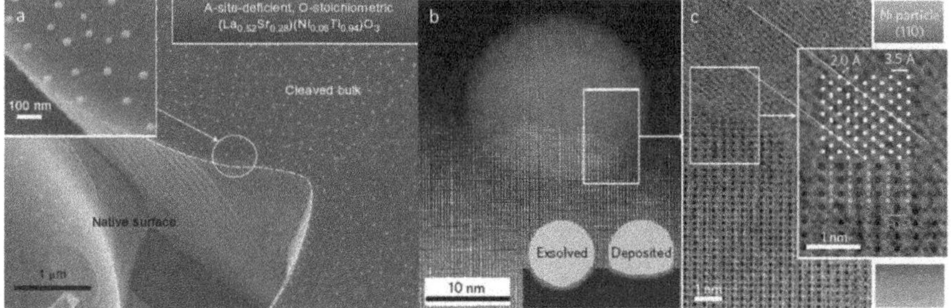

FIGURE 9.3 (a) SEM image shows surface structure of perovskite in the formation of exsolutions; (b) TEM images show interface of exsolved particle-substrate: a Ni particle exsolved on (110) native surface facet and (c) interface of metal-perovskite with outstanding atomic planes and orientations. (Reprinted by permission from Macmillan Publishers Ltd. *Nat. Chem.*, Neagu, D. et al., 2013, Copyright 2013; *Nat. Commun.*, Neagu, D. et al., 2015, Copyright 2015.)

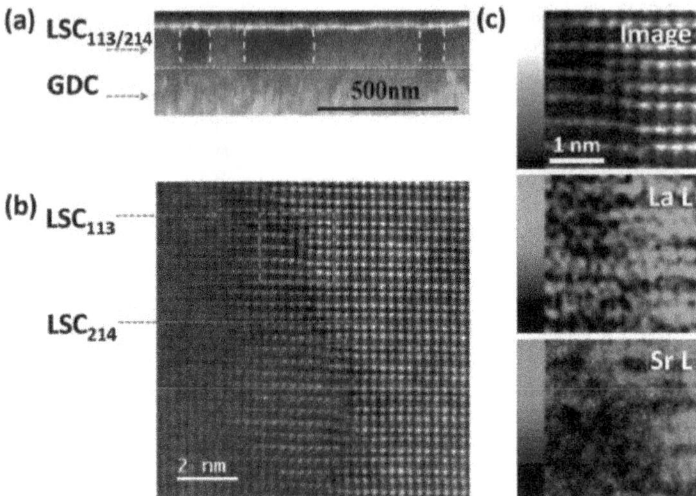

FIGURE 9.4 (a) SEM image of the $LSC_{113/214}$ vertically aligned nanocomposite film and (b) STEM image of the interface near LSC_{113} and LSC_{214}, (c) STEM images of the regions which is marked in (b) with red square. (From Ma, W. et al., *J. Mater. Chem. A*, 3, 207–219, 2015. Reproduced by permission of The Royal Society of Chemistry.)

Another technique for exploiting the high spatial resolution possible with thin specimens is use of the scanning transmission electron microscope (STEM). A finely focused beam is scanned, as in the scanning microscope, and the image is generated by electrons transmitted through the sample.[24] This technique is occasionally used for the study of SOC.[25,26] An example is the research of hetero-interfaces between the perovskite $La_{1-x}Sr_xCoO_3$ (LSC_{113}) and the Ruddlesden-Popper $(La_{1-x}Sr_x)_2CoO_4$ (LSC_{214}) phases, as shown in Figure 9.4. Note that energy-dispersive X-ray spectroscopy (EDX) is sometimes coupled with TEM or STEM to get more information about SOC such as elemental analysis of electrode/electrolyte interface (SEM-EDX or STEM-EDX).[26–28]

9.2.2 FIB-SEM and XCT

A microstructure overview can be achieved by SEM and three-dimensional (3D) reconstruction based on focused ion beam (FIB) tomography.[20,29] The advent of dual-beam FIB-SEM can significantly facilitate the process of 3D microstructure reconstruction by presenting high-quality volumetric data. FIB is first performed to ion mill a trench in the SOC, as shown in Figure 9.5.[29] This technique provides a way to make quantitative correlations between processing and microstructure.[25]

Although FIB-SEM is not suitable for the in situ study of SOC on account of its destructive nature, an alternative, X-ray computed tomography (XCT), is

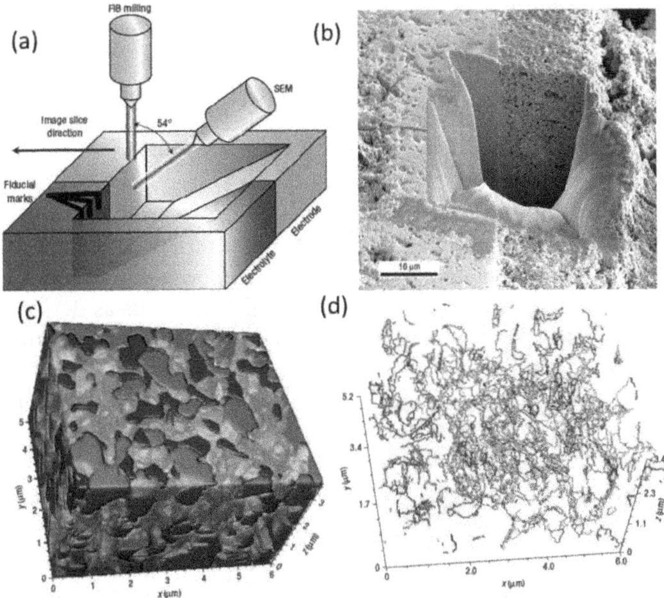

FIGURE 9.5 Images of TPB region in Ni-YSZ composite electrodes of solid oxide cells taken using FIB-SEM: (a) schematic diagram of FIB-SEM, (b) SEM image of the etched region in the electrode, (c) 3D electrode reconstruction (green for Ni, gray for YSZ-translucent/, and blue for the pore phases), (d) distribution of TPB boundaries in the electrode. (Reprinted by permission from Macmillan Publishers Ltd. *Nat. Mater*, Wilson, J.R. et al., 2006, Copyright 2006.)

presented for in situ and possibly even in operando studies due to the features of nondestructive, sub-50-nm resolution and nonvacuum operation.[30] It can be seen clearly from Figure 9.6 that the XCT technique is applied to show the 3D microstructure of SOC, which includes YSZ electrolyte and Ni-YSZ electrode. The collected data form XCT system can be post-processed to enable visualization

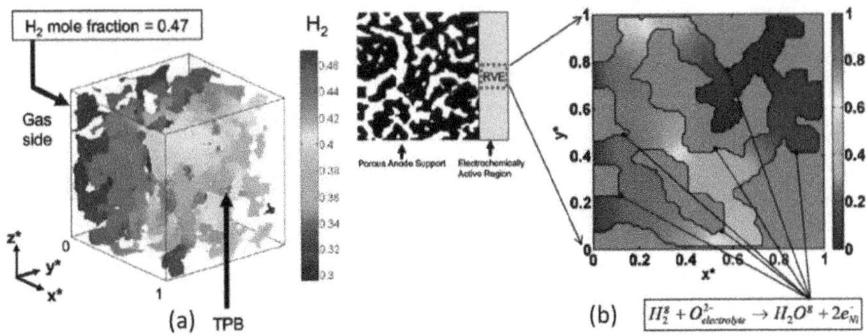

FIGURE 9.6 Application of XCT in solid oxide cells: (a) pore-scale species distribution of H_2 in electrode, (b) variation of H_2 concentration. (Reproduced from John, R. et al., *J. Electrochem. Soc.*, 155, B504–B508, 2008 with permission from ECS, Copyright 2008.)

of the actual porous microstructure, and the structural parameters such as the porosity and tortuosity of the porous SOC electrodes can also be calculated using this technique.[30]

9.2.3 STM/STS and AFM

The scanning tunneling microscope (STM) was invented by Binnig and Rohrer, who shared the 1986 Nobel Prize in physics.[31,32] The spectroscopy taken using STM is correspondingly called scanning tunneling spectroscopy (STS). It can be used to see individual atoms distinctly and arrange the atoms as desired.[33,34] STM images of perovskite LSC_{113} vertically aligned nanocomposite structure surfaces are presented in Figure 9.7, this measurement is performed by a modified variable temperature scanning tunneling microscope (VT-STM). The screw dislocations can be seen clearly in Figure 9.7a and b.

The invention of STM was quickly followed by the development of a family of related techniques. Of the later techniques, the atomic force microscope (ATM) is the most important, even for the studies of SOC.[25,35,36] The two modes of AFM are non-contact (NC-AFM) and intermittent contact (IC-AMF).[33] Aside from STM, the AFM can also be used to see individual atoms distinctly. The grain structure of surface can be observed obviously and the roughness can be measured.[25] For example, the Ni-YSZ electrode has been investigated to understand the influence of the segregation of impurities on electrode/electrolyte interface performance.[36] In addition, the relationship between surface chemistry and surface oxygen exchange kinetics was studied and the (001) and (100) surfaces' topography of $(La_{1-x}Sr_x)_2CoO_4$ (LSC) was assessed using AFM, as shown in Figure 9.8.[37]

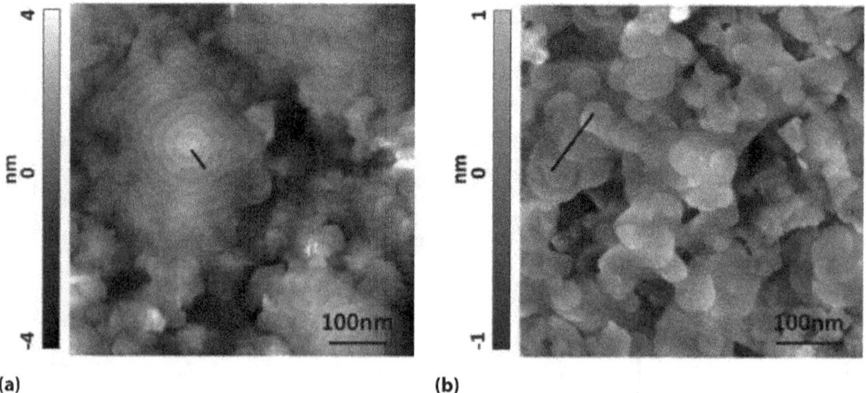

FIGURE 9.7 STM images of perovskite LSC_{113} vertically aligned nanocomposite structure surface on (a) STO and (b) GDC/YSZ. (Ma, W. et al., *J. Mater. Chem. A*, 3, 207–219, 2015. Reproduced by permission of The Royal Society of Chemistry.)

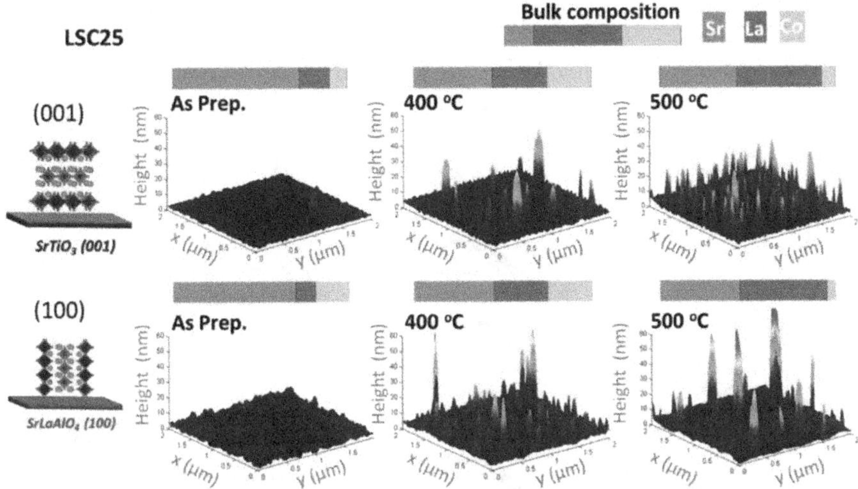

FIGURE 9.8 Surface topography (3D image) of AFM for LSC25. (Reprinted with permission from Chen, Y. et al., *Chem. Mater.*, 27, 5436–5450, 2015. Copyright 2015 American Chemical Society.)

9.3 SURFACE ANALYSIS

Various techniques are used to probe the surface at various depths, including X-ray photoelectron spectroscopy or electron spectroscopy for chemical analysis (XPS/ESCA), secondary ion mass spectroscopy (SIMS), auger electron spectrometry (AEM), and low-energy ion scattering spectroscopy (LEIS). The inspected depths from interface are 1–10 nm, several to hundreds of nm, 1–3 nm, and the outer atomic monolayer, respectively. The applications of these methods on SOC studies are detailed below.

In these surface analysis methods, XPS is the most common technique used in the characterization of SOC materials.[27,38,39] It is also known as ESCA. When an X-ray or electron beam of energy impinges on sample atoms, inner shell electrons are ejected, and the ejected electrons can be measured. Therefore, the surface chemistry of a material (such as the valence state of elements or phase state of material) in its as-received state or after some treatment can be analyzed using XPS. For instance, to validate the elemental valence change, XPS was performed to test the reduced and oxidized $Sr_{0.94}Ti_{0.9}Nb_{0.1}O_3$ samples, and the oxidation state was confirmed by comparing the binding energy of different states.[27] Some other cases of XPS spectra for SOC materials are shown in Figure 9.9.

In many respects, the process of AES is similar to that of XPS, except for one fundamental step. In the AES process, a certain proportion of inner-shell ionization gives rise to the ejection of a bound electron rather than the emission of a characteristic X-ray photon. Compared with XPS, the spatial resolution of AES is much higher,

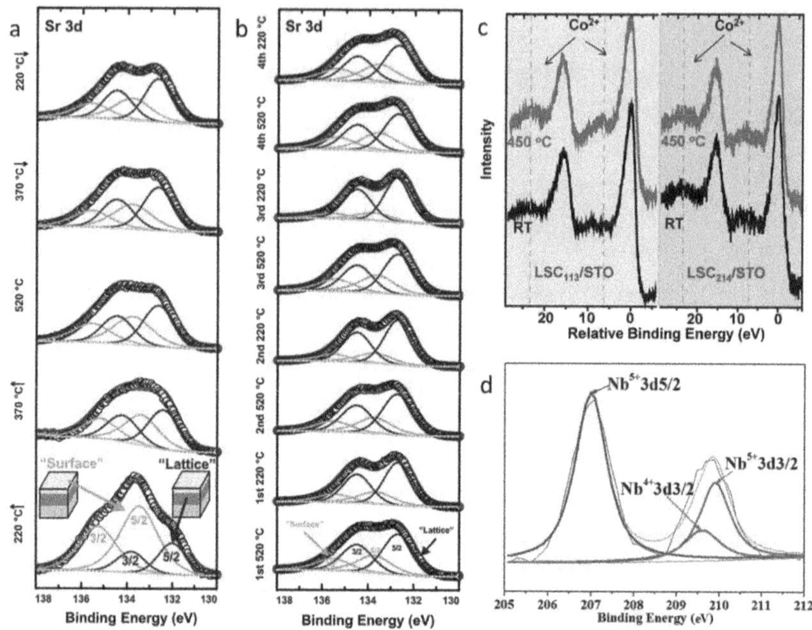

FIGURE 9.9 Some XPS spectra for SOC materials: (a) and (b) Sr 3d XPS of LSC_{113} under different temperature condition (From Chen, Y., et al., *Adv. Energy Mater.,* 3, 1221–1229, 2013; Feng, Z., et al., *J. Phys. Chem. Lett.,* 4, 1512–1518, 2013); (c) Co 2 p XPS on the LSC_{113}/STO and LSC_{214}/STO. (From Chen, Y., et al., *Adv. Energy Mater.,* 3, 1221–1229, 2013); (d) XPS results of Nb in the oxidized STNNO ($(Sr_{0.94})_{0.9}(Ti_{0.9}Nb_{0.1})_{0.9}Ni_{0.1}O_3$) sample. (Feng, Z., et al., In situ studies of the temperature-dependent surface structure and chemistry of single-crystalline (001)-oriented $La_{0.8}Sr_{0.2}CoO_{3-\delta}$ perovskite thin films. *J. Phys. Chem. Lett.*, 2013, 4, 1512–1518; Copyright 2013 Wiley-VCH. Reproduced with permission; Chen, Y., et al., Electronic activation of cathode superlattices at elevated temperatures—Source of markedly accelerated oxygen reduction kinetics. *Adv. Energy Mater.*, 2013, 3, 1221–1229. Copyright 2014, Wiley-VCH. Reproduced with permission; Yang, L., et al., Redox-reversible niobium-doped strontium titanate decorated with in situ grown nickel nanocatalyst for high-temperature direct steam electrolysis. *Dalton Trans.*, 2014, 43, 14147. Copyright 2013 Wiley-VCH. Reproduced with permission.)

and Auger spectra may be more sensitive to the chemical state than that of XPS at times. Figure 9.10 compares Auger spectrum mapping and the SEM image of Co cation at the LSC surface.[25] The intensity of the Co peak can indicate its concentration, as shown in Auger spectrum.

In general, SIMS is a relatively new surface analysis technique for SOC research.[41–44] The composition of the sample surface can be analyzed by sputtering with a focused primary ion beam, then collecting and analyzing ejected secondary ions. The elemental, isotopic, or molecular composition of the surface can be determined by measuring the mass/charge ratios of these secondary ions with a mass spectrometer such as sector, quadrupole, or time-of-flight. With regard to SOC studies, in order to investigate the oxygen exchange at the hetero-interface of

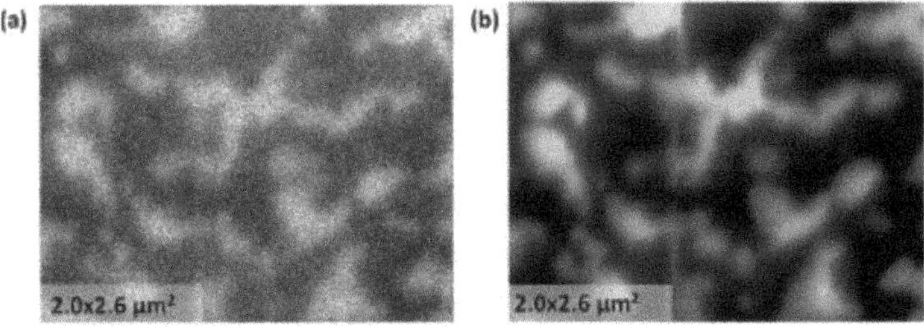

FIGURE 9.10 Comparison of Auger spectrum mapping (a) and SEM image (b) of Co cation at the LSC surface. (Ma, W. et al., *J. Mater. Chem. A*, 3, 207–219, 2015. Reproduced by permission of The Royal Society of Chemistry.)

(La, Sr)CoO_3/(La, $Sr)_2CoO_4$ in composite ceramics,[41,42] 3D SIMS was performed to visualize the isotope distribution; fast oxygen-incorporation paths was determined with SIMS images. Another case is the study of boron deposition and poisoning of $La_{0.8}Sr_{0.2}MnO_3$ oxygen electrodes of SOC, as shown in Figure 9.11. The SIMS images show the elements' distribution and confirm the separation of the electrode/electrolyte interface region.[43]

LEIS is also known as ion scattering spectroscopy (ISS). It is similar to SIMS but its inspected depth is just the outer atomic monolayer. LEIS is sensitive to both structure and composition of surfaces. The material information such as relative positions of atoms in a surface lattice and the elemental identity of those atoms can be achieved by data collecting and deducing. This technique is an emerging method for SOC studies, especially in the research of segregation.[44,45] For example, surface chemistry

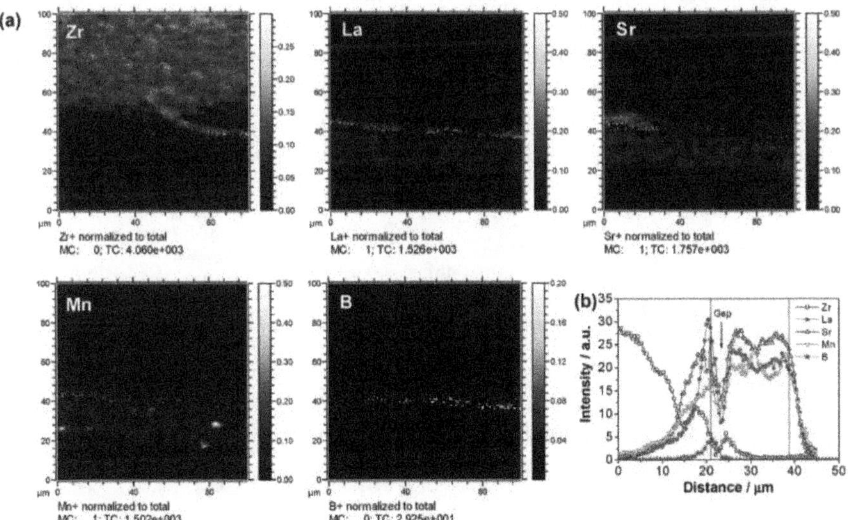

FIGURE 9.11 (a) SIMS elemental images and (b) line scan profiles of the cross-section LSM electrode. (Reproduced from Chen, K. et al., *Int. J. Hydrogen Energy*, 41, 1419–1431, 2016 with permission from Pergamon-Elsevier Science LTD, Copyright 2016.)

and oxygen surface exchange in $LnBaCo_2O_{5+\delta}$ air electrodes was investigated. LEIS was used for the analysis of the surface and near surface chemistry because it provides information from the first mono-atomic layer of the materials.[44]

9.4 SIMULATION AND CALCULATION METHOD

Computational methods such as density functional theory (DFT) have been employed to complement the above-mentioned techniques to probe and predict the structure and properties of local interfaces at an atomic scale. Such methods may provide information about bulk and surface structure, transport and mobility of oxygen and electrons, and ultimately electrode reaction mechanisms. One example is included in Figure 9.12, which illustrates the mechanism and energy barrier involved during oxygen migration in a perovskite structure.[5,20]

Using DFT-based calculations, Jeong Woo Han and Bilge Yildiz[46] proved that the hetero-interface region of perovskite-based oxides was a dynamic SOC subsystem with significant high oxygen surface exchange and migration kinetics. The activation at the hetero-interface was thought to be caused by anisotropy lattice strain. According to Milind J. Gadre,[47] the chemical and structural asymmetry leads to cation segregation and lattice reconstruction at interfacial region, and therefore induced many unique properties in this region.

Molecular dynamics (MD) is an advanced computational simulation method to learn the physical movements of atoms and molecules. The atoms and molecules are simulated to interact within a certain period of time and thus give a systematic view of the dynamical evolution process. MD simulation proposes that trajectories of atoms and molecules are controlled by interatomic potentials or molecular mechanics force. Newton's equations are applied to mathematically solve motion of interacting particles. The method was originally developed to detect theoretical physics in the late 1950s. MD simulation is widely applied today in calculation of chemical physics, materials science, and the modeling of biomolecules.

In the field of SOC, MD simulation is used to identify the mechanism of oxygen ion diffusion. For example, to detect the oxygen diffusion in Pr_2NiO_{4+d}, Alexander Chroneos[48] employed MD simulations with effective pair potentials to replicate the

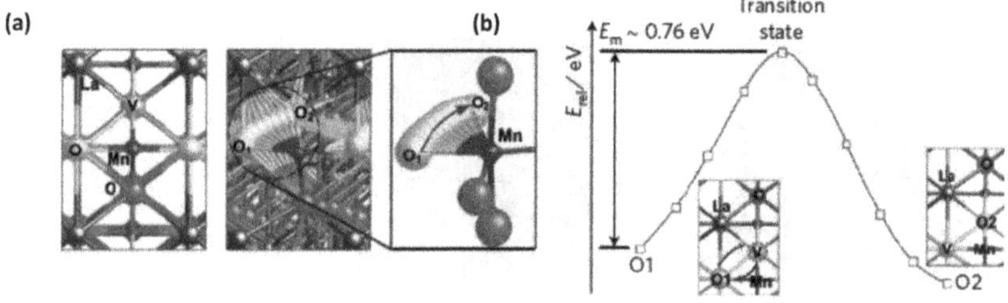

FIGURE 9.12 Understanding active interfaces using density functional theory: (a) trajectory (top and side view); (b) corresponding energy of oxygen migration in a perovskite structure (LSM) by computational methods. (Reproduced from Irvine, J.T.S. et al., *Nat. Energy*, 1, 15014, 2016 with permission from Pergamon-Elsevier Science LTD, Copyright 2010.)

interaction between ions. The results of the simulation indicate that oxygen diffusion in $Pr_2NiO_{4+\delta}$ is highly anisotropic, with migration chiefly via an interstitialcy mechanism in the a–b plane. In addition, the calculated oxygen diffusivity has a weak dependence on the concentration of oxygen interstitials, similar to previous experimental observations. The anisotropic oxygen diffusion in tetragonal $La_2NiO_{4+\delta}$ was also studied by molecular dynamics calculations. In addition, MD simulation was used to calculate the oxygen ion diffusion in cation ordered/disordered $GdBaCo_2O_{5+\delta}$ and reveals that disorder on the cation sublattice led to a drop in the total oxygen diffusivity of the material.[49]

9.5 PRODUCT ANALYSIS

In regard to the products of CO_2 electrochemical conversion, high-temperature electrolysis using SOEC normally has better selectivity when compared to low-temperature electrolysis. Therefore, product analysis for CO_2/H_2O co-electrolysis is relatively easier. The main products in co-electrolysis are CO (g) and H_2 (g) according to Equations (9.1) and (9.2).

$$H_2O + 2e^- \rightarrow CO + O^{2-} \tag{9.1}$$

$$CO_2 + 2e^- \rightarrow H_2 + O^{2-} \tag{9.2}$$

However, CO_2 (g) and H_2O (g/l) can also be detected sometimes in the gas mixture downstream.

Gas chromatograph (GC) is the most fundamental and important method for product analysis for CO_2/H_2O co-electrolysis. Measurements are performed starting from OCV up to a certain applied voltage such as 1.4 V.[50] One important factor is to control the stabilization of the voltage and temperature before GC measurements. Mass spectrum (MS) is also used to measure and calculate experimental H_2 or/and CO productions.[51] The content of residual CO_2 in the gas mixture downstream of the stack can be measured by a dual gas analyzer.[50] The mass flow rates of gases should be controlled using mass flow controllers.[2,52,53]

Alenazey et al.[50] investigated the influence of fuel composition and operating parameters such as current density on the relative amounts of products, including H_2 and CO. The composition of downstream gases after water condensation was also analyzed using GC. Ebbesen et al.[54] studied the reaction mechanisms of the electrochemical reduction of CO_2 and H_2O, and investigated the reaction mechanisms during co-electrolysis. The gases at both the inlet and the outlet were analyzed by GC with a Hayesep N column and a thermal conductivity detector (TCD). The results showed that the water gas shift (WGS) reaction can reach an equilibrium in the process of co-electrolysis, thus suggesting that one of the products CO was not only produced via electrode reaction (Equation [9.2]) but also via WGS reaction.

Syngas (H_2O and CO_2) production though co-electrolysis with a Cu-based SOEC was studied by Gaudillere et al.[51] The expected syngas production was calculated based on Faraday's equation, while the experimental syngas production was measured by both GC and MS. The H_2 production in the fuel electrode exhaust gas with time and current and the CO production can be calculated and shown graphically for comparison. Through these comparisons, the reaction mechanisms of CO_2 electrolysis were investigated, and the feasibility of syngas production via co-electrolysis was further confirmed. The authors proposed that, in order to reach higher yields and efficiencies, both the system fabrication and the electrocatalysis of SOECs needed to be further improved.

9.6 SUMMARY

Measurement, characterization, and simulation for materials or electrochemical process in solid oxide cell operation were summarized in this chapter. Electrochemical measurements such as I–V or EIS are usually preformed for the test of half-cell or full-cell in SOCs, while the characterizations about microstructure and surface analysis were carried out for further insight about specific materials. The simulation and calculation methods are often used to reaction mechanisms in atomic scale.

Aside from those widely used methods, the development of other advanced characterization techniques, such as in situ scanning tunneling microscopy (STM), hard X-ray photoelectron spectroscopy (HAXPES), near edge X-ray absorption fine structure (NEXAFS), and so on, makes it possible to detect local composition and structure of materials with very high resolution, thus achieving additional understanding about materials used in SOCs.

REFERENCES

1. Kim, S. J., Kim, K. J., Choi, G. M. A novel solid oxide electrolysis cell (SOEC) to separate anodic from cathodic polarization under high electrolysis current. *International Journal of Hydrogen Energy* **40**, 9032–9038 (2015).
2. Yang, C., et al. Co-electrolysis of H_2O and CO_2 in a solid oxide electrolysis cell with hierarchically structured porous electrodes. *Journal of Materials Chemistry A* **3**, 15913–15919 (2015).
3. Irvine, J. T. S., et al., Electroceramics: Characterization by impedance spectroscopy. *Advanced Materials* **2**, 132–138 (1990).
4. Zhan, Z. An octane-fueled solid oxide fuel cell. *Science* **308**, 844–847 (2005).
5. Yang, L., et al. Enhanced sulfur and coking tolerance of a mixed ion conductor for SOFCs: $BaZr_{0.1}Ce_{0.7}Y_{0.2-x}Yb_xO_{3-\delta}$. *Science* **326**, 126–129 (2009).
6. Selman, J. R. Poison-tolerant fuel cells. *Science* **326**, 52–53 (2009).
7. Suzuki, T., et al. Impact of anode microstructure on solid oxide fuel cells. *Science* **325**, 852–855 (2009).
8. Adler, S. B. Limitations of charge-transfer models for mixed-conducting oxygen electrodes. *Solid State Ionics* **135**, 603–612 (2000).
9. Feng, Z., et al. Anomalous interface and surface strontium segregation in $(La_{1-y}Sr_y)_2CoO_{4\pm\delta}$ /$La_{1-x}Sr_xCoO_{3-\delta}$ heterostructured thin films. *The Journal of Physical Chemistry Letters* **5**, 1027–1034 (2014).
10. Sata, N., Eberman, K., Maier, K.E.J. Mesoscopic fast ion conduction in nanometre-scale planar heterostructures. *Nature* **408**, 946–949 (2000).
11. Adler, S. B. Factors governing oxygen reduction in solid oxide fuel cell cathodes. *Chemical Reviews* **104**, 4791–4844 (2004).
12. Lee, Y., Kleis, J., Rossmeisl, J., Morgan, D. Ab initio energetics of $LaBO_3(001)$ (B=Mn, Fe, Co, and Ni) for solid oxide fuel cell cathodes. *Physical Review* **80**, 224101–224120 (2009).

13. Suntivich, J., et al. Design principles for oxygen-reduction activity on perovskite oxide catalysts for fuel cells and metal–air batteries. *Nature Chemistry* **3**, 546–550 (2011).
14. Jin Suntivich, K. J. M. H. A perovskite oxide optimized for oxygen evolution catalysis from molecular orbital principles. *Science* **334**, 1383–1385 (2011).
15. Kim, C. H., Qi, G., Dahlberg, K., Li, W. Strontium-doped perovskites rival platinum catalysts for treating NO_x in simulated diesel exhaust. *Science* **327**, 1624–1627 (2010).
16. Gorte, R. J., Vohs, J. M. Catalysis in solid oxide fuel cells. *Annual Review of Chemical and Biomolecular Engineering* **2**, 9–30 (2011).
17. Chen, Y., et al. Electronic activation of cathode superlattices at elevated temperatures—Source of markedly accelerated oxygen reduction kinetics. *Advanced Energy Materials* **3**, 1221–1229 (2013).
18. Goodenough, J. Electronic and ionic transport properties and other physical aspects of perovskites. *Reports on Progress in Physics* **67**, 1915–1993 (2004).
19. Crumlin, E. J., et al. Oxygen reduction kinetics enhancement on a heterostructured oxide surface for solid oxide fuel cells. *The Journal of Physical Chemistry Letters* **1**, 3149–3155 (2010).
20. Irvine, J. T. S., et al. Evolution of the electrochemical interface in high-temperature fuel cells and electrolysers. *Nature Energy* **1**, 15014 (2016).
21. Ebbesen, S. D., Jensen, S. H., Hauch, A., Mogensen, M. B. High temperature electrolysis in alkaline cells, solid proton conducting cells, and solid oxide cells. *Chemical Reviews* **114**, 10697–10734 (2014).
22. Neagu, D., Tsekouras, G., Miller, D. N., Ménard, H., Irvine, J. T. S. In situ growth of nanoparticles through control of non-stoichiometry. *Nature Chemistry* **5**, 916–923 (2013).
23. Neagu, D., et al. Nano-socketed nickel particles with enhanced coking resistance grown in situ by redox exsolution. *Nature Communications* **6**, 8120 (2015).
24. Reed, S. J. B. *Electron Microprobe Analysis* (second edition), Cambridge University Press, Cambridge, UK (1993).
25. Ma, W., et al. Vertically aligned nanocomposite $La_{0.8}Sr_{0.2}CoO_3$ /$(La_{0.5}Sr_{0.5})_2CoO_4$ cathodes—Electronic structure, surface chemistry and oxygen reduction kinetics. *Journal of Materials Chemistry A* **3**, 207–219 (2015).
26. Graves, C., Ebbesen, S. D., Jensen, S. H., Simonsen, S. B., Mogensen, M. B. Eliminating degradation in solid oxide electrochemical cells by reversible operation. *Nature Materials* **14**, 239–244 (2014).
27. Yang, L., et al. Redox-reversible niobium-doped strontium titanate decorated with in situ grown nickel nanocatalyst for high-temperature direct steam electrolysis. *Dalton Transactions* **43**, 14147 (2014).
28. Xu, S., et al. Composite cathode based on Fe-loaded LSCM for steam electrolysis in an oxide-ion-conducting solid oxide electrolyser. *Journal of Power Sources* **239**, 332–340 (2013).
29. Wilson, J. R., et al. Three-dimensional reconstruction of a solid-oxide fuel-cell anode. *Nature Materials* **5**, 541–544 (2006).
30. John, R., Izzo, J. A. S. J. Nondestructive reconstruction and analysis of SOFC anodes using X-ray computed tomography at sub-50 nm resolution. *Journal of the Electrochemical Society* **155**, B504–B508 (2008).
31. Binnig, G., Rohrer, H., Gerber, C., Weibel, E. Surface studies by scanning tunneling microscopy. *Physical Review Letters* **49**, 57–61 (1982).
32. Chen, C. J. *Introduction to Scanning Tunneling Microscopy* (second edition), Oxford University Press, New York, (2008).

33. Binnig, G., Rohrer, H. In touch with atoms. *Reviews of Modern Physics* **71**, S324–S330 (1999).
34. Binnig, G., Rohrer, H. Scanning tunneling microscopy from birth to adolescence. *Reviews of Modern Physics* **59**, 615–625 (1987).
35. Jensen, K. V., Primdahl, S., Chorkendorff, I., Mogensen, M. Microstructural and chemical changes at the Ni/YSZ interface. *Solid State Ionics* **144**, 197–209 (2001).
36. Hansen, K., Norrman, K., Mogensen, M. H_2-H_2O-Ni-YSZ electrode performance-effect of segregation to the interface. *Journal of the Electrochemical Society* **151**, A1436–A1444 (2004).
37. Chen, Y., et al. Segregated chemistry and structure on (001) and (100) surfaces of $(La_{1-x}Sr_x)_2CoO_4$ override the crystal anisotropy in oxygen exchange kinetics. *Chemistry of Materials* **27**, 5436–5450 (2015).
38. Chen, S., et al. A composite cathode based on scandium-doped chromate for direct high-temperature steam electrolysis in a symmetric solid oxide electrolyzer. *Journal of Power Sources* **274**, 718–729 (2015).
39. Kan, W. H., Thangadurai, V. Challenges and prospects of anodes for solid oxide fuel cells (SOFCs). *Ionics* **21**, 301–318 (2015).
40. Feng, Z., et al. In situ studies of the temperature-dependent surface structure and chemistry of single-crystalline (001)-oriented $La_{0.8}Sr_{0.2}CoO_{3-\delta}$ perovskite thin films. *The Journal of Physical Chemistry Letters* **4**, 1512–1518 (2013).
41. Sase, M., et al., Enhancement of oxygen exchange at the hetero interface of (La, Sr) CoO_3/(La, $Sr)_2CoO_4$ in composite ceramics. *Solid State Ionics* **178**, 1843–1852 (2008).
42. Sase M., et al., Promotion of oxygen surface reaction at the hetero-interface of (La, Sr) CoO_3/(La, $Sr)_2CoO_4$. *ECS Transactions* **7**, 1055–1060 (2007).
43. Chen, K., Hyodo, J., Ai, N., Ishihara, T., Jiang, S. P. Boron deposition and poisoning of $La_{0.8}Sr_{0.2}MnO_3$ oxygen electrodes of solid oxide electrolysis cells under accelerated operation conditions. *International Journal of Hydrogen Energy* **41**, 1419–1431 (2016).
44. Téllez, H., Druce, J., Kilner, J. A., Ishihara, T. Relating surface chemistry and oxygen surface exchange in $LnBaCo_2O_{5+d}$ air electrodes. *Faraday Discuss* **182**, 145–157 (2015).
45. Druce, J., et al. Surface termination and subsurface restructuring of perovskite-based solid oxide electrode materials. *Energy & Environmental Science* **7**, 3593–3599 (2014).
46. Han, J. W., Yildiz, J. W. Mechanism for enhanced oxygen reduction kinetics at the (La, Sr)CoO_{3-d}/(La, $Sr)_2CoO_{4+d}$ hetero-interface. *Energy & Environmental Science* **5**, 8598–8607 (2012).
47. Gadre, M. J., Lee, Y. L., Morgan, D. Cation interdiffusion model for enhanced oxygen kinetics at oxide heterostructure interfaces. *Physical Chemistry Chemical Physics* **14**, 2606–2616 (2012).
48. Parfitt, D., Chroneos, A., Kilner, J. A., Grimes, R. W. Molecular dynamics study of oxygen diffusion in $Pr_2NiO_{4+\delta}$. *Physical Chemistry Chemical Physics* **12**, 6834 (2010).
49. Parfitt, D., Chroneos, A., Tarancón, A., Kilner, J. A. Oxygen ion diffusion in cation ordered/disordered $GdBaCo_2O_{5+d}$. *Journal of Materials Chemistry* **21**, 2183–2186 (2011).
50. Alenazey, F., et al. Production of synthesis gas (H_2 and CO) by high-temperature Co-electrolysis of H_2O and CO_2. *International Journal of Hydrogen Energy* **40**, 10274–10280 (2015).
51. Gaudillere, C., Navarrete, L., Serra, J. M. Syngas production at intermediate temperature through H_2O and CO_2 electrolysis with a Cu-based solid oxide electrolyzer cell. *International Journal of Hydrogen Energy* **39**, 3047–3054 (2014).

52. Wang, Y., Liu, T., Fang, S., Chen, F. Syngas production on a symmetrical solid oxide H_2O/CO_2 co-electrolysis cell with $Sr_2Fe_{1.5}Mo_{0.5}O_6$–$Sm_{0.2}Ce_{0.8}O_{1.9}$ electrodes. *Journal of Power Sources* **305**, 240–248 (2016).
53. Xing, R., Wang, Y., Zhu, Y., Liu, S., Jin, C. Co-electrolysis of steam and CO_2 in a solid oxide electrolysis cell with $La_{0.75}Sr_{0.25}Cr_{0.5}Mn_{0.5}O_{3-\delta}$ –Cu ceramic composite electrode. *Journal of Power Sources* **274**, 260–264 (2015).
54. Ebbesen, S. D., Knibbe, R., Mogensen, M. Co-electrolysis of steam and carbon dioxide in solid oxide cells. *Journal of the Electrochemical Society* **159**, F482–F489 (2012).

10 High-Temperature Electrochemical Process of CO_2 Conversion with SOCs 5

Advanced Fabrication Methods (Infiltration and Freeze Casting)

10.1 INFILTRATION FOR NANO-STRUCTURED $Ln_{1-x}Sr_xMO_{3-\delta}$ (Ln=La, Sm; B=Mn, Co, Fe) SOC ELECTRODE

10.1.1 Introduction of Infiltration Used in SOCs

Despite numerous advantages and the broad application prospects, the large-scale application of solid oxide cells (SOCs) devices is strongly limited by high operating temperatures (typically close to 1000°C). Such high operating temperatures accelerate the reaction kinetics significantly; on the other hand, they lead to detrimental interactions between cell components, which may damage cell compositions and material microstructures, leading to long-term durability loss and high costs of utilization.[1,2] Reduced operation temperatures can lead to many performance degradation issues, however, such as sluggish oxygen reduction/evolution reaction (ORR/OER) kinetics on oxygen electrodes. Most of the existing cell components and structures may become no longer applicable for intermediate temperate solid oxide cells (IT-SOCs).[3–8] In addition, degradation of electrodes,[9] especially the delamination of oxygen electrodes,[10,11] is widely observed in both solid oxide fuel cells (SOFCs) and solid oxide electrolytic cells (SOECs), especially in large cells and SOC stacks. Because the thermal expansion coefficient (TEC) of traditional electrode materials is inconsistent with that of electrolytes, gaps can be generated between them when a drastic oxygen reduction reaction occurs.[12] An effective method for mitigating these problems is utilizing the electrode/electrolyte material as an electrode scaffold and then infiltrating a highly active component into it, a process known as infiltration or impregnation. For example, the TEC of LSC is as high as $20.5 \times 10^{-6}\,K^{-1}$ (30°C–1000°C),[13] while the value is reduced to $12.6 \times 10^{-6}\,K^{-1}$ (25°C–800°C) after infiltrating into YSZ scaffold, which is much closer to that of YSZ ($11.8 \times 10^{-6}\,K^{-1}$).

On the other hand, as a reliable, mature and effective strategy, the infiltration method is initially used in the industrial catalysis field. Nevertheless, in the field of SOCs, the feasibility of infiltration is hindered due to the lack of highly active and durable electrode scaffolds.[14–16] With the development of electrodes with an advanced micro-nano structure, oxygen electrodes prepared by chemical infiltration process can be fabricated successfully and efficiently.

In comparison to traditional sponge-structured oxygen electrodes, the electrodes obtained by the infiltration method possesses significant advantages[16–19]: (1) it effectively extends the area of three-phase boundaries (TPBs), which is conducive to improving electrocatalytic activity for ORR; (2) the active catalyst is deposited in prefabricated porous scaffolds, thus avoiding its direct contact with electrolytes and suppressing many side reactions; (3) the heat treatment temperature of infiltration can be reduce to as low as 350°C–850°C, which can further suppress the possible detrimental reactions and cation diffusions between electrode and electrolyte materials. In addition, severe coarsening of electrode particles can be restrained, and the preparation cost can be reduced by infiltration. As a result, electrodes fabricated by infiltration possess lower thermal shock resistance, and excellent anti-oxidative and anti-reductive activities.

This chapter reviews recent progress in infiltration technology. It starts with the basic principle and typical process. The performance enhancement of infiltrated oxygen electrodes (compared to some traditional perovskite-based electrodes) in SOCs are then discussed. Finally, the unique advantages of infiltration technology are discussed, and the future development of this technique is highlighted.

10.1.2 Process of Infiltration

In general, the process of infiltration includes the deposition, desiccation, and pyrolysis of solution precursors inside a porous scaffold.[20] Figure 10.1 shows several typical infiltration stages. The first is the *fabrication of a porous electrode scaffold.* As shown in Figure 10.1a, an ionic and/or electronic conducting scaffold is sintered at high temperatures to ensure an intimate connection with the electrolyte and excellent structural stability of the electrode. The second is *infiltration of the catalytically active phase.* As shown in Figure 10.1b, metal nitrate solution with surfactants is infiltrated into the porous scaffold due to capillary force. The infiltrated electrode is subsequently calcined at relative low temperatures to form the products, which involve two possible morphologies: discrete distribution (Figure 10.1c) or thin and continuous film (Figure 10.1d).[21]

There are two approaches for infiltration: multiple-step infiltration[22] and one-step infiltration.[23] The multiple-step infiltration process continuously accomplishes the enhancement of electrode performance, while one-step infiltration can provide an easier approach for surface modification and avoid degradation during the calcination process. However, how to optimize various parameters for the desired nanostructure is still a challenge for one single step.[21] Also, it is of great importance to design and optimize the morphology of the electrode scaffold before infiltration.

In addition, many studies have shown that the viscosity of the infiltrate/catalyst plays an important role in controlling surface morphology during the infiltration process.[24–29]

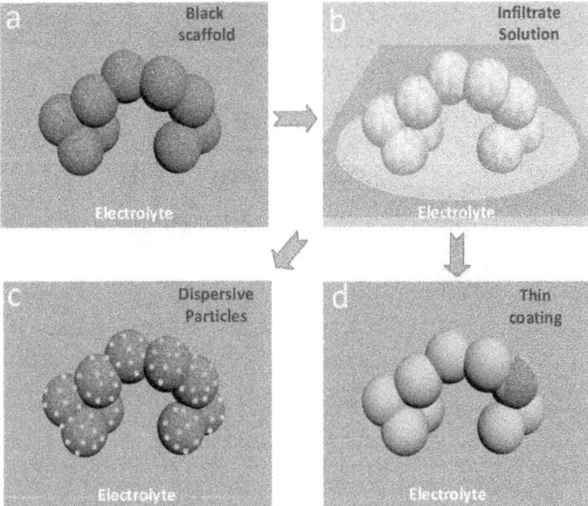

FIGURE 10.1 Simplified flow diagram of infiltration: (a) a pre-sintering electrode scaffold; (b) infiltrate solution is added into the scaffold; (c) dispersive particles; and (d) thin coating, two different morphologies are formed on the scaffold surface. (Reprinted from *Nano Lett.*, 7, Sholklapper, T.Z. et al., Nanostructured solid oxide fuel cell electrodes, 2136–2141, Copyright 2007, with permission from Elsevier.)

The infiltration precursor is composed of metal nitrate, solvent, and various surfactants. The influence factors of viscosity refer to the amount of metal cations, the qualities of the solvent, and the types of surfactants. High viscosity hinders the formation of a uniform distributed coating inside a well-sintered scaffold, while low viscosity, generates a united one.[29] It has also been noted that the choice of surfactant has great effects on the porous electrode fabrication, such as citric acid,[30,31] urea,[32] polymeric dispersant[23,33] and Triton-X100,[34–36] and so on. And various solvents are utilized to improve the stability and wettability of the infiltrate/catalyst solution.[37,38] Furthermore, electrodeposition and vacuum have also been used to facilitate the infiltration process.[35,39]

10.1.3 Infiltration with Various $Ln_{1-x}Sr_xMO_{3-\delta}$ SOC Electrodes

10.1.3.1 LSM-YSZ Electrode

$La_{1-x}Sr_xMnO_{3-\delta}$ (LSM) has good compatibility with YSZ at typical SOC operating conditions because the TEC of LSM (12.0×10^{-6} K^{-1}, 30°C–1000°C) approximately matches with YSZ (10.0×10^{-6} K^{-1}, 30°C–1000°C). As one of the most commonly used electrochemical catalysts, LSM exhibits high electronic conductivity, reported to be 200 S cm^{-1} at 800°C.[40] However, given the lack of oxygen vacancies, LSM possesses poor oxide ion conductivity, only about 4×10^{-8} S cm^{-1}.[41]

Therefore, when applying an LSM-based electrode, it is necessary to combine the ionic conductivity of YSZ with the electronic conductivity of LSM to improve the performance of the single cell. However, conventional LSM-YSZ composite electrodes still possess very high polarization resistance, which makes it difficult to meet the qualifications of single-cell performance. On the contrary, the polarization impedance of infiltrated LSM-YSZ electrodes is significantly lower. Figure 10.2 is a typical SEM

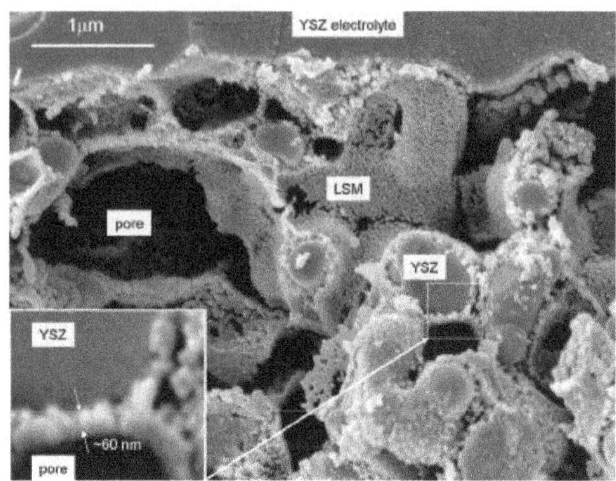

FIGURE 10.2 Micrograph of the LSM- infiltrated YSZ electrode. (Reproduced from Sholklapper, T.Z. et al., *Electrochem. Solid-State Lett.*, 9, A376–A378, 2006. With permission from ECS, Copyright 2006.)

micrograph of infiltrated LSM-YSZ electrodes. It shows that LSM nanoparticles are uniformly deposited on the surface of the YSZ scaffold. In previous studies, Hojberg et al. infiltrated $Ce_{0.8}Sm_{0.2}O_2$ (SDC)-LSM into a porous LSM-YSZ scaffold, and found that the polarization resistance before and after infiltration were reduced from 2.17 to 0.39 Ω cm^2 at 600°C, and from 0.19 to 0.039 Ω cm^2 at 750°C, respectively.[42] Armstrong et al. reported a power density up to 1.2 W cm^2 at 800°C in H_2 for a cell with infiltrated LSM-YSZ electrodes by depositing LSM nanoparticles into a porous YSZ scaffold.[43] Sholklapper et al. fabricated a similar infiltrated LSM-YSZ electrode and tested a power density of 0.3 W cm^{-2} at 650°C.[23] Sholklapper et al. reported that the infiltrated LSM-YSZ composites were stable for 500h at 650°C.[44]

Figure 10.3 illustrates the structure of different LSM-YSZ electrodes, and it further demonstrates the essential relationship between the microstructure and the performance for infiltrated electrodes.[45] As shown in Figure 10.3a, although the conventional LSM-YSZ cathode (electrode I) provides the available paths for both electronic and oxygen ionic transport, its maximum power density is still unsatisfied (Figure 10.3c). Unlike electrode I, the similar anode-supported LSM+YSZ (electrode II) cell with infiltrated LSM-YSZ cathode can achieve a power density that is four times higher (0.83W cm^{-2} at 750°C).

Electrode III with continuous Pd particles deposited on the LSM-YSZ bone surface by infiltrating palladium nitrate solution possesses the highest power density (1.42 W cm^{-2}). This can be explained as follows: although the infiltrated LSM-YSZ (electrode II) and the Pd-infiltrated LSM-YSZ (electrode III) exhibit a similar microstructure, the electrochemical reaction paths are different. In the case of LSM-infiltrated YSZ, the YSZ scaffold only provides the ionic conductivity, while the infiltrated LSM coating provides triple phase boundaries (TPBs) for ORR. For Pd-infiltrated LSM-YSZ, first the LSM-YSZ scaffold provides a mixed oxygen ionic and electronic conductivity;

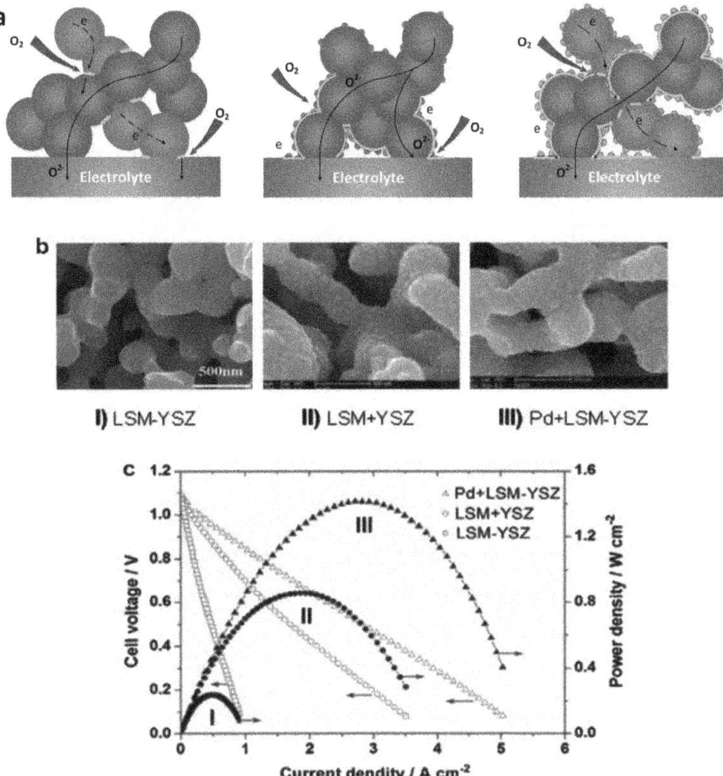

FIGURE 10.3 Structure and performance diagram of Pd-infiltrated YSZ and Pd-infiltrated LSM-YSZ electrodes: (a) structure schematic of LSM-YSZ (electrodes I), LSM+YSZ (electrodes II) and Pt+LSM-YSZ (electrodes III), respectively, demonstrating transfer paths of oxygen ions and electrons; (b) micrographs of electrodes I, II, and III; (c) performance of the cells with electrode I, II, and II at 750°C in H_2. Electrodes: (I) conventional LSM-YSZ composite cathode, (II) infiltrated LSM-YSZ composite cathode, and (III) Pt infiltrated LSM-YSZ composite cathode.(Reprinted from *Electrochem. Commun.*, 11, Liang, F. et al., High performance solid oxide fuel cells with electrocatalytically enhanced (La, Sr)MnO_3 cathodes, 1048–1051, Copyright 2009, with permission from Elsevier.)

second, the Pd nanoparticles significantly extend the areas of TPBs and efficiently promote the reaction rate of oxygen species (Figure 10.3a).

To find the effects of infiltration on surface and interface chemistry, a model about samaria doped ceria (SDC) infiltrated LSM-YSZ electrodes was established (see Figure 10.4)[34] after adding a metal nitrate solution and infiltrating adequately, the solvent is evaporated by vacuum or heating. A thin layer consisting of concentrated metal cations, nitrates, and surfactant is formed on the scaffold surface. Meantime the interactions among the different components become stronger. The possible interactions with the YSZ-SDC scaffold need to be considered. It is proposed that the presence of nitrate ions creates an acid environment. LSM seems to be significantly affected with respect to surface chemistry and can be locally dissolved, which may result in a recombination of the surface.

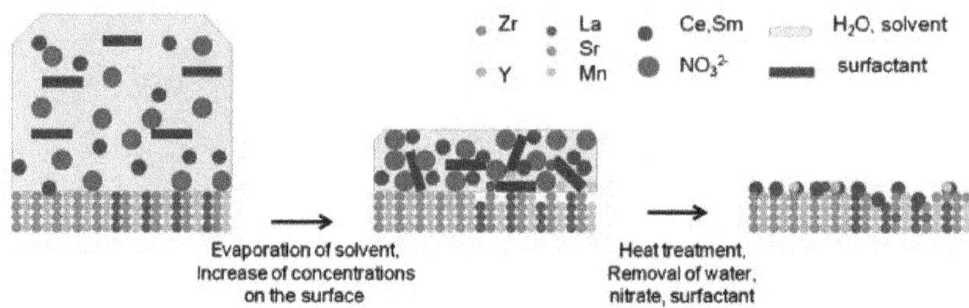

FIGURE 10.4 Schematic diagram of the interactions during infiltration process. (Reprinted from *Sol. State Ion.*, 195, Knöfel, C. et al., Modifications of interface chemistry of LSM–YSZ composite by ceria nanoparticles, 36–42, Copyright 2011, with permission from Elsevier.)

10.1.3.2 LSC-YSZ Electrode

It is known that a transition metal such as Co and Fe can have various valences. For example, Co is mainly in states of Co(II), Co(III), and Co(IV). In a pure $LaCoO_3$ system without doping, the equal numbers of n-type and p-type carriers are formed according to following disproportionation reaction:

$$2Co^{3+} \rightarrow Co^{2+} + Co^{4+} \tag{10.1}$$

This can lead to enhanced electron mobility and conductivity at high temperatures.[46] Simultaneously, replacing La with divalent Sr can generate oxygen vacancies, which promotes ionic conductivity of Sr-doped $LaCoO_3$ (LSC).[47] Therefore, LSC has superior mixed ionic-electronic conductivity. The electronic conductivity can reach up to 1584 S cm^{-1}, and ionic conductivity is as high as 0.22 S cm^{-1} at 800°C in air.[48] In addition, the oxygen surface-exchange coefficient of LSC is also reported to be excellent (10^{-5}–10^{-7} cm s^{-1}), while the value of LSM is only 10^{-8}–10^{-7} cm s^{-1}. However, LSC is still not widely used in SOCs because it may react with YSZ, which may form an insulating $La_2Zr_2O_7$ and $SrZrO_3$ phase at the YSZ-LSC interface at elevated temperatures and degrade the performance.[49,50]

On the other hand, much research indicates that LSCo-YSZ cathodes fabricated by infiltration exhibit even higher activities and appear fewer degradation issues. According to Huang et al., the polarization impedance of LSCo-infiltrated YSZ cathode was as low as 0.03 Ω cm^2 at 700°C.[51] In another study by Samson et al., the ASR of infiltrated LSCo-$Ce_{0.9}Gd_{0.1}O_{1.95}$ at 600°C and 400°C was measured to be 0.044 and 2.3 Ω cm^2 in air, respectively.[52] Armstrong et al.[53] pointed out that a cell with infiltrated LSC-YSZ cathode (approximately 30 vol% LSC) exhibited a peak power density of 2.1 W cm^2 at 800°C. The infiltration method enables the fabrication of LSC-YSZ composites at low temperatures; thus, the detrimental interfacial reactions and TEC mismatch between LSCo and YSZ can be diminished. As discussed in previous sections in this chapter, LSC exhibits high TECs ($20.5 \times 10^{-6} K^{-1}$, 30°C–1000°C) due to the low-spin to high-spin transition associated with the Co^{3+} ions, therefore inducing an obvious mismatch when coupling with YSZ electrolyte.[13] Fortunately, the infiltrated composite (LSC-YSZ) has a

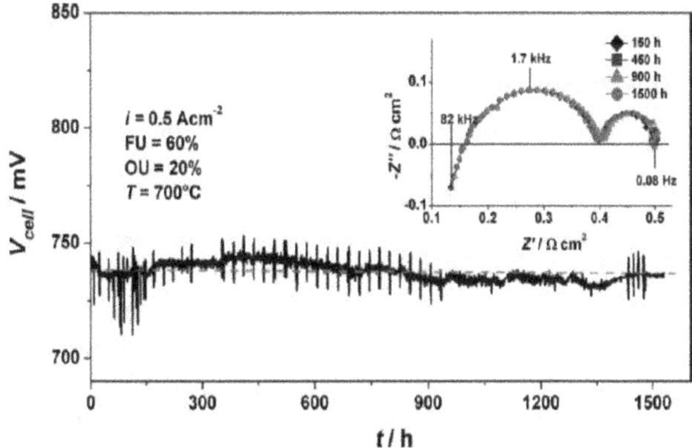

FIGURE 10.5 The test of long-term stability for the cell with infiltrated LSC-CGO cathode. (Reprinted from *Fuel Cells*, 13, Samson, A.J. et al., Durability and performance of high performance infiltration cathodes, 511–519, Copyright 2013, with permission from Elsevier.)

much lower TEC (12.6×10^{-6} K^{-1}, 25°C–800°C), which indicates the enhanced thermal stability of the whole single cell.

However, it was also discovered by Y. Huang et al. that LSC-YSZ cathodes may exhibit some performance degradation at 700°C.[51,55] Such degradation can be mainly attributed to the increase in ohmic losses. Since Sase et al. declared that solid-state reactions between $LaCoO_3$ and YSZ may occur even at 700°C,[56] the poor compatibility makes LSC not suitable in YSZ-based systems, even at relatively low operating temperatures. Thus, Samson et al. proposed to replace the YSZ scaffold by doped ceria (CGO) to effectively avoid those solid-state reactions and further improve the durability. They prepared a cell with infiltrated LSC-CGO cathode and found no obvious degradation even after 1500 h at 700°C, as shown in Figure 10.5.[54]

10.1.3.3 LSF-YSZ Electrode

The infiltrated Sr-doped $LaFeO_3$ (LSF)-YSZ electrode has advanced compatibility and stability compared to that of LSC-YSZ cathodes because there is no evidence to show the significant interface reaction between LSF and YSZ below 1200°C,[57,58] and only the formation of insulating phases can be observed above 1400°C.[47] In addition, compared with LSM-YSZ electrodes, LSF-YSZ composites exhibit superiority in electrochemical performance.[59–61] In principle, partial replacement of La^{3+} in $LaFeO_3$ with Sr^{2+} results in the perovskite structure changing from orthorhombic in $LaFeO_3$ to nearly cubic in $La_{0.8}Sr_{0.2}FeO_3$, at the same time inducing the partial oxidation of Fe^{3+} to Fe^{4+} for charge compensation effects and generating oxygen vacancies.[62,63] Therefore, LSF gave excellent reducibility to oxygen reduction reaction.

However, LSF performs slightly poorly in electrochemistry: the electronic conductivity of LSF (50 S cm^{-1}, 800°C) is much lower than that of LSC. And the ionic conductivity of LSF is reported to be 8×10^{-4} $S{\cdot}cm^{-1}$ at 800°C in air,[64] which is lower than LSC but is many orders of magnitude higher than LSM. Gorte et al.[58,65] have done a lot of research on the infiltrated LSF-YSZ cathode. The impedance of an LSF-YSZ composite prepared by infiltrating 40 wt % LSF is approximately 0.1 Ω cm^2 at 700°C in air. In symmetric-cell measurements, electrodes calcined at 850°C show a linear increase of the area-specific resistance (ASR) from initial 0.13 to 0.55 Ω cm^2 after 2500 h at 700°C. In addition, the impedance spectra of the deactivated LSF-YSZ electrode is found to be the impedance spectra shows strong current dependence of resistivity in LSF-YSZ electrode. In order to determine the source of deactivation, Gorte has examined the structure of the electrodes by SEM, as shown in Figure 10.6.

Figure 10.6a is a micrograph of a blank, porous YSZ backbone. Figure 10.6b shows that infiltrated LSF nanoparticles exist as a layer on the surfaces of a YSZ scaffold after sintering at 850°C. The LSF particles are smaller than 0.1 μm in size, and the covering layer with high porosity is conducive to diffuse gas-phase molecules to the TPB sites on the YSZ surface. Figure 10.6c is a SEM image of the same electrode shown in Figure 10.6b, but it has been tested for 1000 h at 700°C and by an additional 700 h at 800°C, indicating that the average size of LSF particles grow to approximately 0.2 μm on account of sintering caused by aging and form a dense film over the YSZ. Finally, the micrograph in Figure 10.6d shows the structure of an LSF-YSZ composite heated to 1000°C and incurring a similar coarsening of LSF particles.

Figure 10.7 is a schematic of a nano-structured LSF-infiltrated YSZ electrode based on the SEM micrograph results. It illustrates the reason of deactivation. As shown in Figure 10.7a, after sintering at 850°C, the LSF particle deposited above the YSZ scaffold and ensured the transfer of oxygen species. However, when these nanoparticles were sintered at 1000°C, a dense coating of LSF formed and the transmission of oxygen ions was restricted. Taken together, the growth in LSF particles

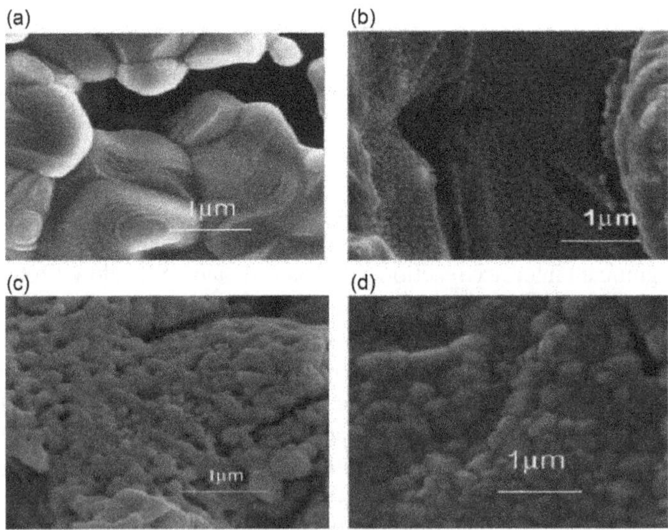

FIGURE 10.6 Microphones of (a) blank YSZ substrate, (b) thin LSF coating above the YSZ backbone after sintering at 850°C, (c) LSF coating after testing for 1000 h at 700°C, and (d) LSF film after sintering at 1100°C. (Reproduced from Wang, W. et al., *J. Electrochem. Soc.*, 154, B439–B445, 2007. With permission from ECS, Copyright 2007.)

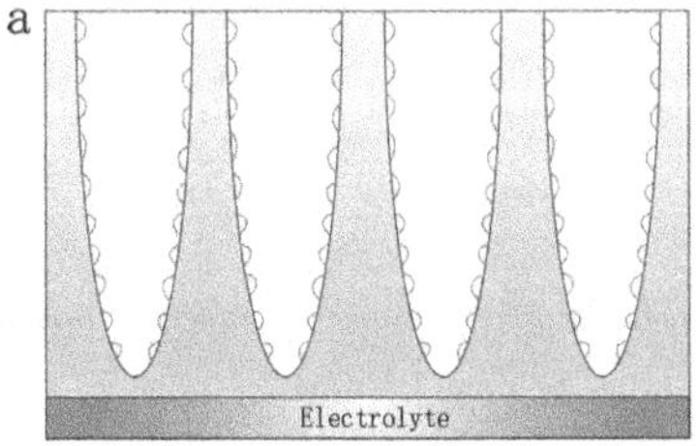

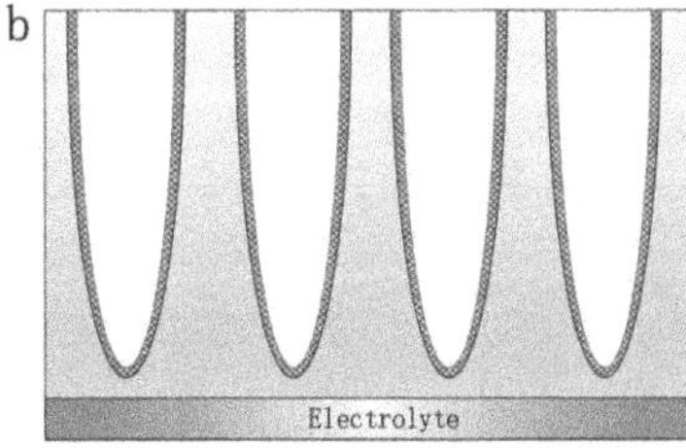

FIGURE 10.7 Schematic showing the deactivation of LSF-infiltrated YSZ electrode: (a) dispersive particles after sintering at 850°C and (b) a thin coating heated to 1000°C.

revealed that the surface area of LSF decreased, the corresponding rate of oxygen reduction reaction decreased, and the electrochemical properties for LSF-YSZ composites also decreased. These observations lead to the conclusion that the formation of a dense polycrystalline layer on the YSZ surface contributes significant deactivation of LSF-YSZ electrodes after high-temperature sintering or long-term operation, which is quite different from the inactivation caused by interfacial reactions for LSM-YSZ.

10.1.3.4 LSCF-YSZ Electrode

As mentioned above, LSC has excellent mixed conductivity, but at the same time, its exhibits a high TEC due to the low-spin to high-spin transition associated with the Co^{3+} ions. In order to improve the match with electrolyte materials and maintain catalytic activity in the meantime, the common trade-off is partially substituting Co with Fe in $La_{1-x}Sr_xCo_{1-y}Fe_yO_{3-\delta}$. For an LSCF system, the bond energy of Fe-O is stronger than that of Co-O; therefore, there exists a tighter combination among the atoms that alleviates the lattice expansion caused by heating. In addition, the doping of Fe means a decrease in the Co content. Therefore, the effect on TEC arising from the spin of Co^{3+} ions can also decline accordingly[66]. Due to the presence of a large amount of oxygen vacancies, LSCF has a superior ionic conductivity (approximately 0.18 S cm^{-1}, 900°C), and simultaneously owns a desired electronic conductivity.[40,46,67–69] Therefore, LSCF performs well in property and is commonly used as a SOC electrode.[70,71]

A later study prepared nano-structured LSCF-YSZ electrodes by infiltration of LSCF into a pre-sintered porous YSZ scaffold.[73] Chen et al. pointed out that the impedance of an infiltrated $La_{0.8}Sr_{0.2}Co_{0.5}Fe_{0.5}O_3$–YSZ composite is as low as 0.047 Ω cm^2 at 750°C.[74] As for fabricating LSCF-YSZ electrodes by wet impregnation, high-temperature calcination and adverse solid-state reaction with YSZ electrolyte were prevented because of a low temperature in cell fabrication.[22] Liu et al.[72,75] tested the stability of infiltrated LSCF–YSZ electrodes for 120 h at 750°C in air. The polarization resistance of the cathode increased from 0.17 to 0.30 Ω cm^2, indicating a rapid deactivation. The degradation of an LSCF-infiltrated YSZ electrode mainly comes from the high polarization resistance as a result of the ORR. The micrograph in Figure 10.8 shows the LSCF-YSZ cathode before and after the 120 h at 750°C. Figure 10.8a shows an untested cathode. Continuous LSCF particles are dispersive above the YSZ scaffold. Subsequently, the morphology of LSCF particles turned out

FIGURE 10.8 SEM images of the infiltrated LSCF-YSZ electrode: (a) before the stability test, (b) after testing at 750°C for 120 hours. (Reprinted from *Int. J. Hydrogen Energ.*, 37, Liu, Y.H. et al., Performance degradation of impregnated $La_{0.6}Sr_{0.4}Co_{0.2}Fe_{0.8}O_3$+$Y_2O_3$ stabilized ZrO_2 composite cathodes of intermediate temperature solid oxide fuel cells, 4388–4393, Copyright 2012, with permission from Elsevier.)

to be more distributed and flattened after a long-term stability test at 750°C. Also, the coarsening leads to a reduced surface area due to sintering, as shown in Figure 10.8b. The shape of the LSCF is similar to the LSF nanoparticles in the LSF-infiltrated YSZ electrode.[58]

Similar deactivation occurred in the stability test for the impregnated LSCF–GDC cathodes for 500 h at 750°C. The polarization and ohmic resistance increased from 0.38 to 0.83 Ω cm^2 and 1.79 to 2.14 Ω cm^2, respectively. The reasons for the degradation in electrochemical activity were twofold. On the one hand, the formation of the insulating phase, $SrCoO_x$, was largely responsible for the attenuation of the LSCF electrode. On the other hand, the agglomeration and coarsening of the LSCF-GDC electrode also caused inactivation of the O_2 reduction reaction. Fortunately, experiments found that the introduction of MgO or LNF proved to be effective in inhibiting the growth of LSCF particles, thus improving the stability and maintaining catalytic activity for the cell.[76]

10.1.3.5 SSC-Infiltrated Electrodes

As mentioned, although the mixed conductivity of LSC is desired, the solid-state reaction and incompatibility between the LSC and YSZ electrolytes mean that the Sr-doped $LaCoO_3$ system is unsuitable for SOFC cathode materials. Tu et al. substituted Sm for La and found that there was no interfacial reaction between $SmCoO_3$ and YSZ at high temperatures. Therefore, the issue of degradation was avoided.[77] Compared with the LSC electrode, the rate of oxygen absorption and dissociation for SSC is much higher than that of LSC, and the overpotential is half that of LSC.[78,79] The $Sm_{0.7}Sr_{0.3}CoO_3$ electrode exhibits a relatively high electronic conductivity, reported to be 500 S/cm at 1000°C.[80]

A study reported that the electronic conductivity of infiltrated SSC-SDC is 15 S cm^{-1} at 700°C, cathode polarization resistance is 0.05 Ω cm^2, and peak power density reaches up to 0.936 Wcm^{-2}. However, for infiltrating SSC into the YSZ, the solid-state

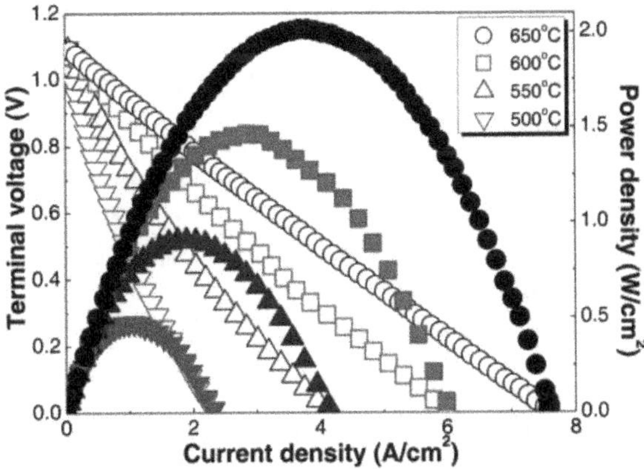

FIGURE 10.9 Characteristics of a cell with infiltrated 12.9vol% SSC into LSGM. Plots of voltage and power density versus current density at 500°C–650°C. (Reprinted by permission from Macmillan Publishers Ltd. *Sci. Rep.*, Han, D. et al., 2012, copyright 2012.)

reaction in the stability test was at 700°C. Therefore, in order to ensure long-term stability, SDC is preferred over YSZ as a backbone when infiltrating SDC.[81] Da Han et al. pointed out that infiltrating SSC catalyst into a high-porosity LSGM backbone resulted in superior catalytic activity and low interfacial resistances. A fuel cell with that infiltrated cathode produced power densities up to 2.02 W cm^{-2} at 650°C, shown as Figure 10.9.[82]

10.1.3.6 Comparison of Infiltrated Electrodes' Performance

As discussed, most infiltrated oxygen electrodes exhibit superior electrochemical performance. Table 10.1 summarizes the primary characteristic of the $Ln_{1-x}Sr_xMO_{3-\delta}$-infiltrated (Ln=La, Sm; B=Mn, Co, Fe) electrode system in SOCs. It can be seen that, at temperatures ranging from 600°C to 700°C, in all infiltrated electrodes listed in the table, SSC-infiltrated LSGM cathode exhibits the most desirable electrochemical performance because the power density is as high as 2.02 Wcm^{-2} and the polarization impedance is only 0.021 $\Omega\cdot cm^2$ at 650°C. However, if SDC is substituted for LSGM as the scaffold, the power density (0.936 Wcm^{-2} at 700°C) is significantly decreased. The Pt-infiltrated LSM/YSZ electrode also exhibits a superior power density of 1.42 Wcm^{-2} at 750°C.

In the high-temperature region (700°C–800°C), the infiltrated LSC-YSZ electrodes are extremely excellent, with a power density that reaches up to 2.1 Wcm^{-2} at 800°C. Despite all this, LSC is readily reactive with the YSZ scaffold. Replacing the scaffold of YSZ with GDC could effectively avoid the solid-sate reaction at high temperature. But for the infiltrated LSC-GDC electrode, the stability at temperatures higher than 700°C still needs to be studied. The power density of infiltrated LSM-YSZ cathodes (1.2 Wcm^{-2} at 800°C) is higher than that of LSF-YSZ (0.85 Wcm^{-2} at 800°C), while the latter has the advantage of low polarization resistance (0.13 $\Omega\cdot cm^2$ at 700°C). The deactivation mechanisms of infiltrated LSF-YSZ

TABLE 10.1
Primary Characteristic of the $Ln_{1-x}Sr_xMO_{3-\delta}$-Infiltrated (Ln = La, Sm; B = Mn, Co, Fe) Electrodes System

Infiltrate	Scaffold	Resistance (Ω·cm²)	Performance of Full Cell	References
LSM	YSZ	Conventional: R_{ASR}[a] = 0.3 Ω·cm² at 700°C	Conventional: 0.26 Wcm⁻² at 850°C	83
		Infiltrated: R_p[b] = 1.6 Ω·cm² at 600°C	Infiltrated: 1.2 Wcm⁻² at 800°C	45
LSC	GDC	Infiltrated: R_p = 0.062 Ω·cm² at 600°C		54
LSC	SDC	Infiltrated: R_p = 0.36 Ω·cm² at 600°C	Infiltrated: 0.815 Wcm⁻² at 600°C	84
LSCF	GDC	Infiltrated: R_p = 0.24 Ω·cm² at 600°C		85
Pt	LSM/YSZ	Infiltrated: R_p = 0.9 Ω·cm² at 600°C	Infiltrated: 1.42 Wcm⁻² at 750°C	45
SSC	LSGM	Infiltrated: R_p = 0.021 Ω·cm² at 650°C	Infiltrated: 2.02 Wcm⁻² at 650°C	82
LSF	YSZ	Infiltrated: R_P = 0.13 Ω·cm² at 700°C	Infiltrated: 0.85 Wcm⁻² at 800°C	58,86
SSC	SDC	Conventional: R_i[c] = 2.5 Ω·cm² at 600°C	Conventional: 0.12 Wcm⁻² at 500°C	87
		Infiltrated: R_P = 0.05 Ω·cm² at 700°C	Infiltrated: 0.936 Wcm⁻² at 700°C	88
SDC-LSM	LSM/YSZ	Infiltrated: R_p = 0.039 Ω·cm² at 750°C		42
LSCF	YSZ	Infiltrated: R_p = 0.047 Ω·cm² at 750°C	Infiltrated: 0.473 Wcm⁻² at 750°C	74,89
LSC	YSZ	Infiltrated: R_{ARS} = 0.25 Ω·cm² at 800°C	Infiltrated: 2.1 Wcm⁻² at 800°C	53

[a] R_{ASR} = area specific resistance
[b] R_p = polarization resistance
[c] R_i = interfacial resistance

electrode are different from those of the LSM-YSZ. It is more remarkable that the power density of the infiltrated LSCF-YSZ electrode is significantly unsatisfactory, as low as 0.473 Wcm^{-2} at 750°C.

The nano-structured electrodes prepared by infiltration have a unique advantage. However, the poor long-term stability caused by the solid-state reaction or coarsening still need to be resolved. Nevertheless, the data provided in Table 10.1 demonstrate that the infiltration process is a promising technology in the preparation of electrodes, such as infiltrated LSC-YSZ and SSC-LSGM electrodes, especially with regard to cell stacks, for providing high electrochemical activity and reducing the mismatched TEC.

10.1.4 Conclusions and Future Prospects

Up to this point in the chapter, diverse nano-structured infiltrated oxygen electrodes have been reviewed. Extensive attention has been devoted to the enhancement of the performance and catalytic activity of SOC composite electrodes by using the method of infiltration, especially in solving the mismatch problems. Compared to conventional electrode fabrication techniques, the impregnation approach allows for a lower sintering temperature, greater control of microstructure, more diminished TECs, and a more extensive choice of electrode materials.

In spite of stabilities being tested for hundreds and even thousands of hours in button cells, further study on the long-term stability for larger cells or stacks is still necessary to meet industrial demands. Some research groups anticipate that cells with infiltrated electrodes will be more suitable for industrial application if the infiltration process is continuously improved and the reaction mechanism of the infiltrated electrode is studied more extensively. It is believed that infiltration method has bright prospect for the design and fabrication of high-performance SOC electrodes, with its high catalytic activity and long-term stability at lower temperatures.

10.2 An Electrolyte-Electrode Interface Structure with Directional Micro-Channel Fabricated by Freeze Casting

10.2.1 Freeze Casting Technology Used in SOCs

Generally, solid oxide cells are fabricated with electrolyte supports or electrode supports. Compared to the electrolyte support, the electrode support has relatively reduced ohmic resistance due to the thinner electrolyte.[90,91] However, they both have a sponge-like structure whose irregular pores would greatly limit the oxygen release kinetics. Freeze casting is an advanced near-net-shape casting technique that has been widely used in biology and medicine for fabricating scaffolds with directional graded micro-channels.[92–94] Therefore, freeze casting can be used for preparing electrode/electrolyte supports in SOCs for more directional pores to release gas. Freeze casting can be divided into two processes: casting and vacuum freezing, both with the goal of forming and controlling graded directional microstructures (5–100 μm). In addition, these supports have many so-called arms acting as self-supporting bridges among the pores. These arms have both high strength and great toughness that contribute to the enhancement of high mechanical strength.[95,96] As a result, the performance of the cells with directional supports can be improved significantly.

In 2007, NASA[97] and Sofie et al.[92] first synthesized Ni-YSZ hydrogen electrode supports with graded micro-channel structures via freeze-casting technology. The traditional fabrication methods of electrolyte/electrode supports include dry-pressing[98–101] and tape-casting.[102,103] However, these supports usually have low porosity, a high tortuosity factor, low pore number, and TPBs. Consequently the OER/ORR kinetics are reduced due to the sluggish gas generation and release. In the freeze casting method,[93,104] the prepared supports have high porosity,[95] a low tortuosity factor, and extended TPBs. If the electrolyte material (typically yttria stabilized zirconia, YSZ[105,106]) was chosen as a support, the delamination between electrode and electrolyte caused by the unmatched coefficient of thermal expansion (CTE) at high temperatures can be effectively avoided.[9,12,107–110] For example, the CTE of Sr-doped $LaCoO_3$ (LSC) is 20.5×10^{-6} K^{-1}(30°C–1000°C),[111] which is far more than that of YSZ (10.5×10^{-6} K^{-1}, 25°C–1000°C).[112]

Although the freeze casting method has many advantages, the application has been limited to the medicine and bionic fields so far.[113–120] For example, hydroxy-apatite as the most commonly used material in freeze casting; it has high mechanical strength and can be used for tooth and bone,[121–123] for example, 145 MPa at 47% porosity and 65 MPa at 56% porosity. The applications and developments of this technology in other areas such as the chemical industry are rarely promoted.[94,124]

In this section, to facilitate the development of freeze casting in electrode fabrication, we focus on the novel YSZ electrode scaffolds with graded directional micro-channels. The processes and the related principle of freeze casting are summarized in detail. The critical factors that have an effect on the pore morphology, such as the freezing temperature, freeze rate, dispersion media, additives, and solid loading, are discussed in detail. In addition, the future development and prospects for freeze casting technology also are proposed.

10.2.2 The Process and Critical Factors of Freeze Casting Technology

Freeze casting can be divided into freezing and vacuum drying. The effect of critical factors on the scaffold morphologies, such as the freezing temperature, freeze rate, dispersion media, additives, and solid loading, are discussed in detail in this section.

10.2.2.1 The Process of Freeze Casting

The process of freeze casting can be divided into four steps (see Figure 10.10):

① Preparing the YSZ slurry. The YSZ slurry is prepared by sufficiently mixing YSZ powders (Tosoh Company, Japan, average particle size: 0.3 μm), solvents (e.g., water or camphene),[125–127] and additives (e.g., glycerol and polyvinyl alcohol)[96,128] (Figure 10.10a).

② Casting the YSZ slurry. Take the water-based solvent as an example, casting the well-dispersed YSZ slurry into the nylon mold,[129–131] which was fixed on a cooper plate near the cool source at −196°C to 0°C (Figure 10.10b). The solvent (water) in the YSZ slurry is solidified and forms an ice nucleus below zero degrees, and then the ice crystals grow along the direction of temperature gradient (the bottom slurry near the cold source when the slurry surface is exposed to air) and push away the YSZ particles, resulting in the formation of straight graded pores.[132]

③ Drying in vacuum. Shift the frozen scaffold to the vacuum drier (Figure 10.10c) and dry in the vacuum for 24 hours. A honeycomb-like scaffold with graded micro-channels is obtained after sublimation, where the pore is a replica of the ice crystal.

④ Sintering the scaffold. The scaffold is sintered at high temperatures to improve mechanical strength (Figure 10.10d).

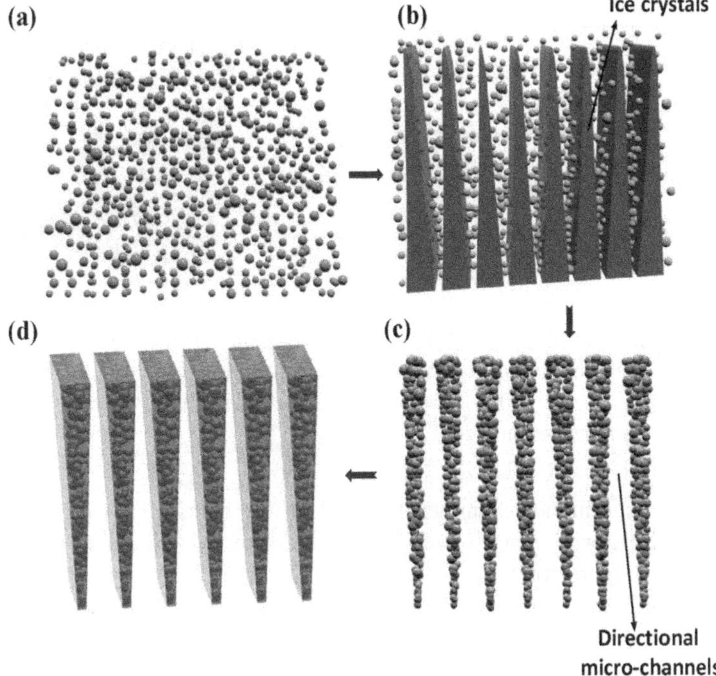

FIGURE 10.10 The process of freeze casting: (a) Preparing a YSZ slurry; (b) freezing; (c) sublimation; and (d) calcination.

10.2.2.2 Effect of Critical Factors on Morphologies

10.2.2.2.1 Effect of Dispersion Medium

In general, water,[133] camphene,[125–127] and tert-butyl alcohol[134,135] are widely used as dispersion media in freeze casting. Among them, water is the most commonly used solvent because it is clean, available, and low cost. The morphology of ice crystals during the freeze casting is determined by the competitive relation between ice nucleation and ice growth.[136–140] The morphologies of ice crystals are different under different freezing condition, including needle, columnar, layered, and branched pores. The common ice molecule structure is shown in Figure 10.11a.[95] If the freezing temperature gradually decreases, the whole slurry system can be assumed to be in thermal equilibrium. In this case, ice crystals first nucleate on the surface of slurry and then grow downward along the direction of the temperature gradient.

As a result, homogeneous columnar pores are formed. However, if the slurry is frozen at a supercool condition, the ice crystals initially nucleate on the bottom of the slurry and grow quickly, which will entrap the YSZ particles and make the ice arrangement disordered. Then, the growth rate declines along the direction of the temperature gradient, and the repulsion force from ice crystals to the particles will gradually strengthen. In this case, it is easier for crystals to grow along the direction of the temperature gradient (a-axis) than the c-axis. Studies have shown that the ice growth rate along the a-axis is 10^2 to 10^3 times faster than that of the c-axis, which

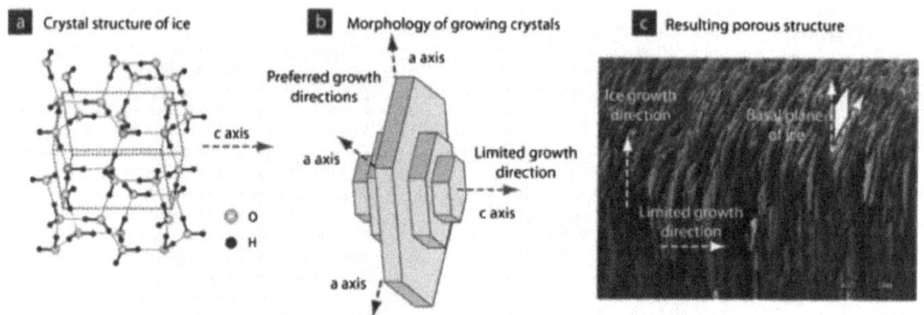

FIGURE 10.11 Ice crystals growth during freeze casting. (From Deville, S.: Freeze-casting of porous ceramics: A review of current achievements and issues. *Adv. Eng. Mater.*, 10, 155–169, 2008. Copyright Wiley-VCH Verlag GmbH & Co KGaA. Reproduced with permission.)

is attributed to the higher chemical potential on the c-axis.[141] As a result, the ice morphology transforms from a disordered three-dimensional net to well-oriented lamellar pores (Figure 10.11b).

In theoretical terms, the repulsion to YSZ particles from the ice is the premise of pore formation. From a thermodynamic and dynamic view, it is necessary to satisfied two criteria in the meantime. In thermodynamics, the interfacial energy (δ_{PS}) between the solidification front (the front of the ice crystals) and the YSZ particles should be larger than the sum of the interfacial energies (δ_{PL}) of particle-liquid and (δ_{SL}) of solid-liquid, as shown in Equation (10.2)[142]:

$$\delta_{PS} > \delta_{PL} + \delta_{SL} \tag{10.2}$$

In dynamics, the repulsive force (F_δ) and the attractive drag force (F_η) on the YSZ particles are defined in Equations (10.3) and (10.4).[142] When the repulsive force is larger than the drag force, the YSZ particles will be repulsed from the ice crystals:

$$F_\delta = 2\pi\Delta\delta_0\left(\frac{a_0}{d}\right)^n \tag{10.3}$$

$$F_\eta = \frac{6\pi\eta v R^2}{d} \tag{10.4}$$

where a_0 is the average intermolecular distance, R is the radius of YSZ particle, d is the distance between particle and ice crystals, a_0 is the average intermolecular distance, v is the crystal growth rate, n is a constant from 1 to 4, and η is the viscosity of suspension.

Here, we introduce a critical rate (υ_c), which represents the YSZ particles. It is exactly not enough to be entrapped by ice crystals, and it can be estimated as follows[143,144]:

$$\upsilon_c = \frac{\rho_l}{9\eta\rho_s}\left[-\frac{A}{2\pi Dd_0} - gDd_0\left(\rho_P - \rho_l\right)\right] \tag{10.5}$$

where ρ_P, ρ_l, and ρ_s are the density of particle, liquid (water), and solid (ice crystal), respectively; D is the YSZ particle diameter; d_0 is the minimal distance between the ice crystals and the YSZ particle; A is the Hamaker constant; g is the gravitational constan; and η is the viscosity of the slurry. When the ice growth rate (v) is faster than υ_c, the particles are entrapped by the ice; otherwise, the YSZ particles will be repelled and form the well-oriented pores.

The solidification behavior of YSZ particles is similar to other materials, such as alumina,[96,141] titanium dioxide,[145] and silica.[146,147] Deville et al.[148,149] researched the freezing process of alumina in water, as shown in Figure 10.12. The first the ice growth rate (v) is too fast to reject the particles, resulting in the formation of a dense net structure at the bottom. With ice advancing along the temperature gradient, v decreases.

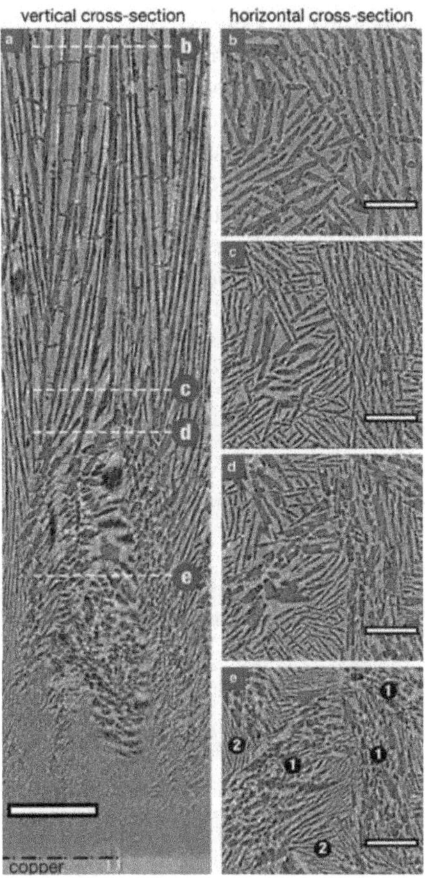

FIGURE 10.12 SEM of various zones in the alumina scaffold prepared by freeze casting with 28 vol% solid loading. Scale bars: (a) 250 mm, (b)–(e) 150 mm (vertical and horizontal cross-sections).

When v is less than υ_c, the alumina particles are gradually repelled by the ice crystals and form columnar to lamellar morphologies in the steady state.

In addition, the freeze casting method has the potential to synthesize specific morphology by controlling the freezing conditions.[129,150–157]

Amphene is another commonly used dispersion medium.[158–161] Compared with water, camphene has smaller volume changes (–3.1%) and thus prevents cracks.[95,132] Camphene's melting point is only about 44°C–48°C[162]; therefore, it can be removed via sublimation at room temperature. In general, the camphene-based slurry is heated at 55°C–60°C and then solidified in a mold below 44°C. Kah et al. prepared Ni-YSZ scaffolds through freeze casting with a camphene suspension.[126] Figure 10.13 shows the redistribution of the Ni-YSZ particles. Unlike the lamellar ice crystals in water-based systems, camphene crystals usually exist as a honeycomb after freeze casting. In addition, some bridges structures are formed among the pores.[163]

10.2.2.2.2 Effect of Additives

Additives can modify suspension viscosity and interaction among particles. The pore structures can be controlled by introducing various additives. Glycerol, polyacrylic acid, and polyvinyl alcohol (PVA) are common additives in aqueous systems.[96,128] Among them, glycerol has many advantages, such as a low freezing point, no toxicity, and low cost; thus, it is the most frequently used additive.[164–166] Because glycerol has polar groups, water molecules are the easiest to concentrate around them to form hydrogen bonds. When adding glycerol, the viscosity will be enhanced effectively and the diffusion of the solvent is limited, which generates smaller ice crystals.[167,168] Some bridge structures among the ceramic walls are generated by adding glycerol.

PVA is another common additive used for modifying the pore structure.[153,169,170] The viscosity of the solution is increased because the particles tend to be adsorbed by PVA molecules. Therefore, smaller lamellar or columnar pores are generated in this case, and some bridge structures are also generated by the interconnected YSZ ceramic walls.[153,170] SEM images of YSZ scaffolds with and without PVA are shown in Figure 10.14.[170] Both the YSZ scaffolds with 2 and 6 wt% PVA have smaller pores, and the pore size decreases as the PVA content increases.

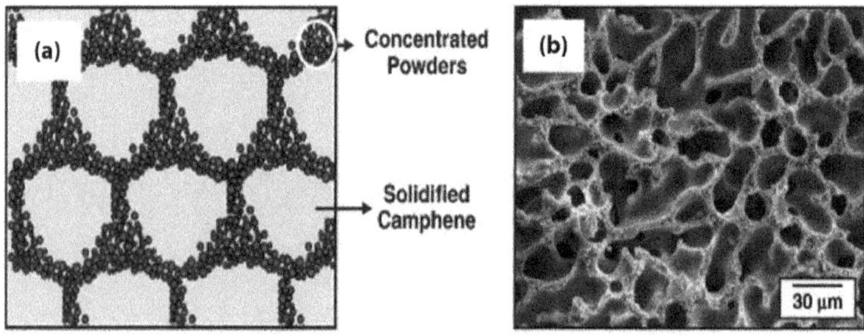

FIGURE 10.13 (a) Schematic illustration of the solidified camphene in Ni-YSZ slurry and (b) SEM of the Ni-YSZ composite sintered at 1200°C for 3 h in air. (Reprinted from *Mater. Lett.*, 61, Koh, Y.H. et al., Freeze casting of porous Ni–YSZ cermets, 61, 1283–1287, Copyright 2007, with permission from Elsevier.)

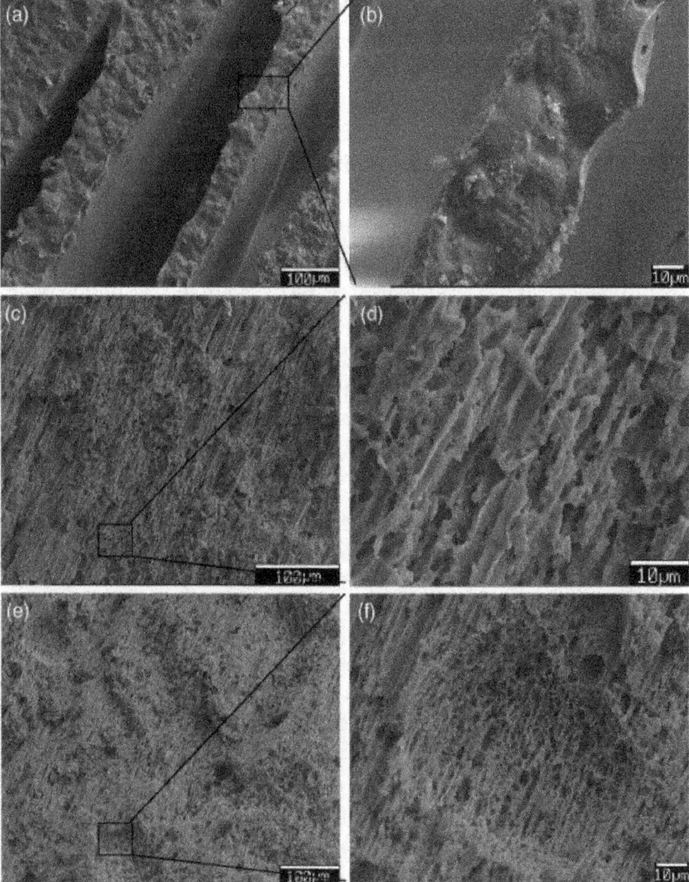

FIGURE 10.14 SEM images of YSZ scaffolds sintered at 1300°C: (a) and (b) without PVA; (c) and (d) with 2 wt% PVA; (e) and (f) with 6 wt% PVA, vertical section. (From Zuo, K.H. et al., Properties of microstructure-controllable porous yttria-stabilized ziroconia ceramics fabricated by freeze casting, *Int. J. Appl. Ceram. Tec.*, 2008, 5, 198–203. Copyright Wiley-VCH Verlag GmbH & Co KGaA. Reproduced with permission.)

10.2.2.2.3 Effect of the Freeze Conditions

During freeze casting, the freeze conditions, such as freeze temperature and freeze rate, have important effects on the pore morphologies.[132,171] The typical freezing temperatures of an aqueous suspension ranges from −18°C to −196°C,[117] which decides the freeze rate, and the freeze rate affects the dominance between the rate of nucleation and growth of ice crystals.[172] If the freeze temperature is too low, the nucleation rate is far quicker than the growth rate; therefore, a great deal of small crystals will be generated.[173,174] In contrast, if the freeze temperature is relatively high, ice crystals tend to grow and then nucleate, which means that large pores will be formed.

An empirical equation describes the relationship of the two neighboring lamellar crystals and ice growth rate (ν) [141]:

$$\lambda \propto \nu^{-n} \tag{10.6}$$

where n depends on particle size. According to Equation (10.6), the ice growth rate (ν) increases as λ decreases. In other words, the higher the freeze rate, the smaller are the pores that are formed in kinetics. Figure 10.15[175] shows SEM images of the distribution of YSZ scaffolds with various freeze temperatures (–50°C to –5°C). A lower freeze temperature accelerates the ice growth, resulting in the formation of small, well-oriented columnar pores (Figure 10.15a). Increasing the freeze temperature means that the YSZ slurry takes longer to solidify; thus, the growth direction of ice might be distorted and form a dendritic structure (Figure 10.15b). In addition, increasing the freeze temperature to the melting point generates pores that are more disturbed (Figure 10.15c). Furthermore, YSZ particles have no time to rearrange under low temperatures[96]; therefore, the ice crystals have more opportunity to embed the YSZ particle, and thus the porosity will be improved. In conclusion, the competitive relation between the rate of nucleation and growth for ice crystals affects the final pore morphologies.

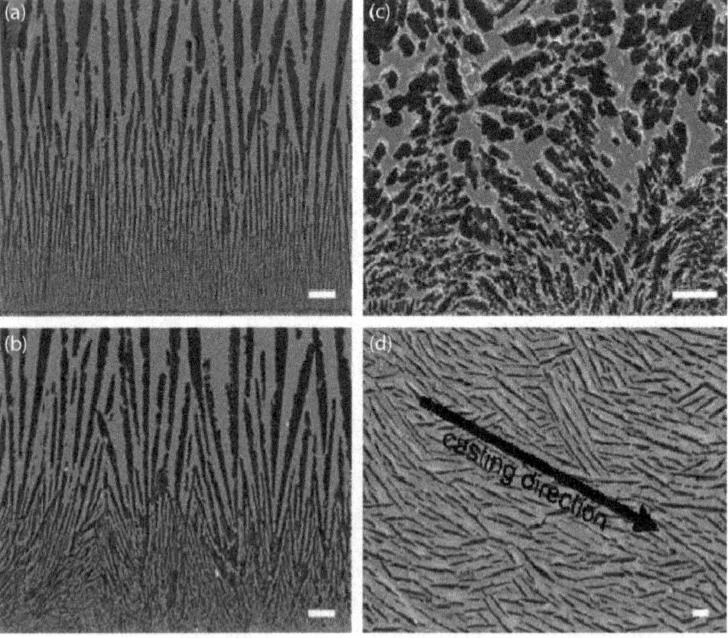

FIGURE 10.15 Cross-section SEM images of YSZ scaffolds frozen at different temperature: (a) frozen at –50°C, (b) frozen at –25°C, (c) frozen at –5°C, and (d) vertical-section SEM image of surface morphology (bar = 25 μm). (From Sofie, S.W., Fabrication of functionally graded and aligned porosity in thin ceramic substrates with the novel freeze–tape-casting process. *J. Am. Ceram. Soc.*, 2007, 90, 2024–2031. Copyright Wiley-VCH Verlag GmbH & Co KGaA. Reproduced with permission.)

10.2.2.2.4 Effect of Solids Loading

Solid loading in the slurry is also a critical factor that affects the pore morphology, size, porosity, and mechanical strength. Generally, the solid loading of an aqueous suspension ranges from 5 to 60 vol%.[176,177] A low rate of solid loading means high water content. In other words, the number of ices crystals increases. The ice crystals effectively reject the particles and are elongated along the temperature gradient and grow large. However, in the case of a too low rate of solid loading, the mechanical strength of the sample is inferior and has the potential to collapse during the sublimation process. In contrast, when the rate of solid loading is high, the ice crystals have difficulty rejecting the accumulated particles and the growth of ice crystals is hindered, finally forming a dense three-dimensional reticular structure. If the rate of solid loading is increased even more, more particles will be trapped and fail to form lamellar structures.

Yu Chen et al. [104] researched the effect of solid loading on pore morphology and distribution (Figure 10.16). They showed that when the solid loading of Ni-YSZ support is 15, 20, and 25 vol%, the porosity is 60.88%, 49.01%, and 37.57%, respectively. In the meantime, the average pore sizes are decreasing accordingly. Xue Wang et al.[93,178] prepared the YSZ electrode scaffold by freeze casting with various rates of solid loading (50–70 wt%, Figure 10.17).[93] When the solid loading is reduced to 50 wt%, the scaffolds form directional pores with high porosity. However, if the rate of solid loading is high (70 wt%), the scaffolds form a dense structure.

10.2.3 Summary and Perspectives

In this section of the chapter, the fundamental solidification features of the freeze casting method were discussed, and critical factors such as dispersion media, additives, solid loading, and freezing conditions were summarized. These critical factors affect pore morphology, size, and distribution. We found that it is easier to fabricate directional lamellar and columnar pores in water-based systems. Compared with aqueous systems, camphene systems show more desirable flexibility with a higher melting point, one that is near room temperature, that can be used to form dendritic pore morphology. Both glycerol and PVA additives can increase solution viscosity and hinder the diffusion of water moleculars, resulting in the formation of fine pores. A low freeze temperature (high freeze rate) is beneficial for ice crystals to grow rather than nucleate, leading to the generation of large pores. A slurry with a low rate of solid loading suffers less resistance from particles; therefore, ice crystals elongate along the a-axis (the direction of the temperature gradient) more easily and grow larger in size.

The freeze casting method has many advantages in fabricating YSZ scaffolds with complex shapes, such as remarkable controllability and flexibility. Therefore it is promising for application in the SOC field. Scaffolds with directional pores enable high porosity and large pore sizes; thus, the polarization resistance (R_p) caused by gas diffusion can be effectively reduced.[104,179] In addition, the freeze casting method has also been used in other fields, including medicine, biology, and chemical fields,[180–185] for instances, in preparing drugs, artificial bone, and catalyst carriers.[121–123,186] Furthermore, scaffolds synthesized by freeze casting exhibit both high porosity and desirable mechanical strength, both of which eliminate the limitation in porosity for traditional electrodes when considering mechanical strength. Compared

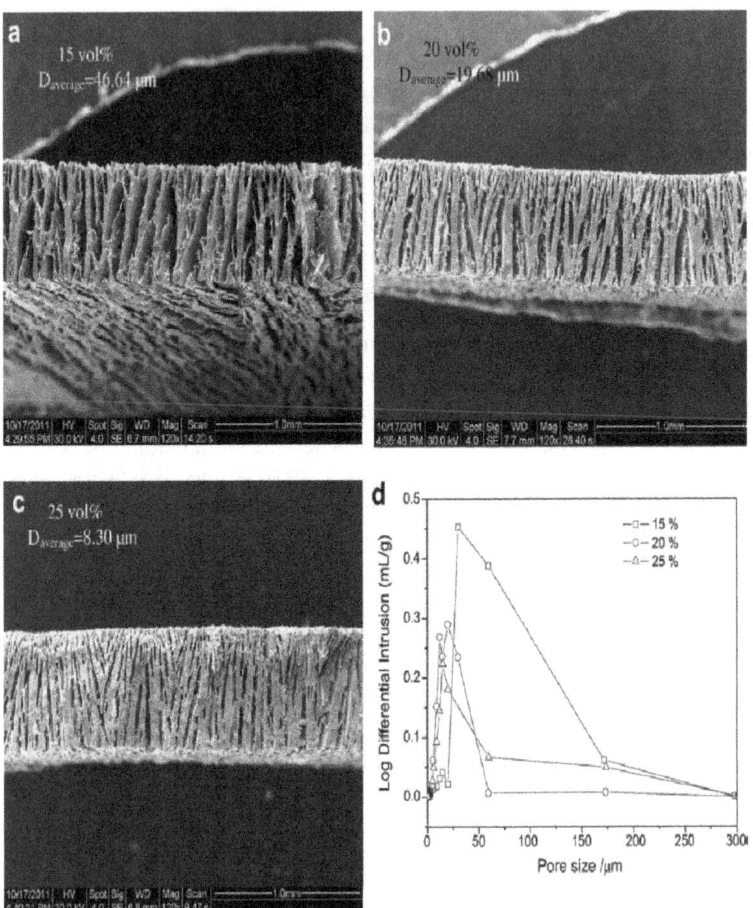

FIGURE 10.16 SEM images of Ni-YSZ scaffolds with various solid loading: (a) 15, (b) 20, (c) 25 vol%, and (d) pore size distribution, vertical section. (Reprinted from *J. Power Sources*, 213, Chen, Y. et al., Novel functionally graded acicular electrode for solid oxide cells fabricated by the freeze-tape-casting process, 93–99, Copyright 2012, with permission from Elsevier.)

with traditional fabrication methods (e.g., dry pressing and spray pyrolysis), scaffolds prepared by freeze casting have higher electrochemical activity and stability. Yu Chen et al.[104] tested the performance of cells with YSZ anode scaffolds prepared by freeze casting, and showed the power density (1.2 W cm^{-2} at 800°C) was higher than that of cells (0.65 W cm^{-2} at 800°C) whose scaffold was prepared by the dry-pressing method (Figure 10.18).[179]

However, several challenges still exist in freeze casting technology, such as the long production cycle and high cost. In addition, a novel antifreeze agent must be developed to inhibit the formation of large crystals.

In conclusion, although freeze casting technology has many problems to be solved, we believe that this method has bright prospects in the designing and synthesizing of scaffolds with well-oriented pores in the SOC field.

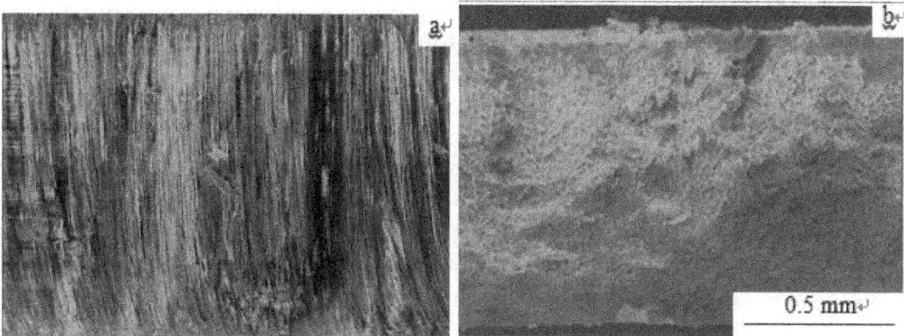

FIGURE 10.17 SEM images of scaffolds with various solid YSZ loading: (a) 50 wt%, (b) 70 wt%, vertical section. (From Wang, X. et al., A high performance solid oxide electrolysis cell support: CN201310279789.4 [P] 2013-07-04.)

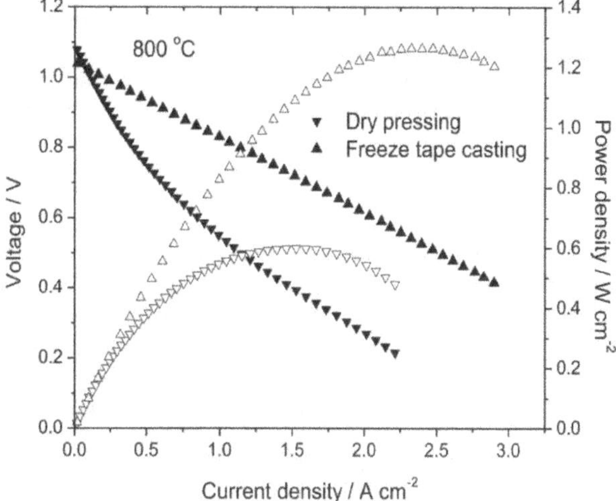

FIGURE 10.18 IP and IV curves of cells prepared by dry pressing and freeze tape casting methods (800°C).

REFERENCES

1. Brett, D. J. L., Atkinson, A., Brandon, N. P., Skinner, S. J. Intermediate temperature solid oxide fuel cells. *Chemical Society Reviews* **37**, 1568–1578 (2008).
2. Shao, Z., Zhou, W., Zhu, Z. Advanced synthesis of materials for intermediate-temperature solid oxide fuel cells. *Progress in Materials Science* **57**, 804–874 (2012).
3. Yamamoto, O. Solid oxide fuel cells: Fundamental aspects and prospects. *Electrochimica Acta* **45**, 2423–2435 (2000).
4. Steele, B. Material science and engineering: The enabling technology for the commercialisation of fuel cell systems. *Journal of Materials Science* **36**, 1053–1068 (2001).
5. Fleig, J. Solid oxide fuel cell cathodes: Polarization mechanisms and modeling of the electrochemical performance. *Annual Review of Materials Research* **33**, 361–382 (2003).
6. Adler, S. B. Factors governing oxygen reduction in solid oxide fuel cell cathodes. *Chemical Review* **104**, 4791–4843 (2004).

7. Sun, C., Hui, R., Roller, J. Cathode materials for solid oxide fuel cells: A review. *Journal of Solid State Electrochemistry* **14**, 1125–1144 (2010).
8. Tsipis, E. V., Kharton, V. V. Electrode materials and reaction mechanisms in solid oxide fuel cells: A brief review. *Journal of Solid State Electrochemistry* **12**, 1367–1391 (2008).
9. Schefold, J., Brisse, A. Long term testing of short stacks with solid oxide cells for water electrolysis. *ECS Transactions* **35**, 2915–2927 (2007).
10. Sharma, V. I., Yildiz, B. Degradation mechanism in $La_{0.8}Sr_{0.2}CoO_3$ as contact layer on the solid oxide electrolysis cell anode. *Journal of the Electrochemical Society* **157**, B441–B448 (2010).
11. Cai, Z., Kuru, Y., Han, J. W., Chen, Y., Yildiz, B. Surface electronic structure transitions at high temperature on perovskite oxides: the case of strained $La_{0.8}Sr_{0.2}CoO_3$ thin films. *Journal of the American Chemical Society* **133**, 17696–17704 (2011).
12. Wenqiang, Z., Bo, Y., Jingming, X. Investigation of single SOEC with BSCF anode and SDC barrier layer. *Fuel & Energy Abstracts* **37**, 837–842 (2012).
13. Petric, A., Huang, P., Tietz, F. Evaluation of La-Sr-Co-Fe-O perovskites for solid oxide fuel cells and gas separation membranes. *Solid State Ionics* **135**, 719–725 (2000).
14. Vohs, J. M., Gorte, R. J. High-performance SOFC cathodes prepared by infiltration. *Advanced Materials* **21**, 943–956 (2009).
15. Jiang, S. P. Nanoscale and nano-structured electrodes of solid oxide fuel cells by infiltration: Advances and challenges. *International Journal of Hydrogen Energy* **37**, 449–470 (2012).
16. Ding, D., Li, X., Lai, S. Y., Gerdes, K., Liu, M. Enhancing SOFC cathode performance by surface modification through infiltration. *Energy & Environmental Science* **7**, 552–575 (2014).
17. Vohs, J., Gorte, R. High-performance SOFC cathodes prepared by infiltration. *Advanced Materials* **21**, 943–956 (2009).
18. Fan, H., Keane, M., Li, N., Tang, D., Singh, P., Han, M. Electrochemical stability of $La_{0.6}Sr_{0.4}Co_{0.2}Fe_{0.8}O_{3-\delta}$-infiltrated YSZ oxygen electrode for reversible solid oxide fuel cells. *International Journal of Hydrogen Energy* **39**, 14071–14078 (2014).
19. Jiang, Z., Xia, C., Chen, F. Nano-structured composite cathodes for intermediate-temperature solid oxide fuel cells via an infiltration/impregnation technique. *Electrochimica Acta* **55**, 3595–3605 (2010).
20. Ding, D., Li, X., Lai, S. Y., Gerdes, K., Liu, M. Enhancing SOFC cathode performance by surface modification through infiltration. *Energy & Environmental Science* **7**, 552 (2014).
21. Sholklapper, T. Z., Kurokawa, H., Jacobson, C. P., Visco, S. J., De Jonghe, L. C. Nanostructured solid oxide fuel cell electrodes. *Nano Letters* **7**, 2136–2141 (2007).
22. Jiang, S. P. A review of wet impregnation—An alternative method for the fabrication of high performance and nano-structured electrodes of solid oxide fuel cells. *Materials Science and Engineering: A* **418**, 199–210 (2006).
23. Sholklapper, T. Z., Lu, C., Jacobson, C. P., Visco, S. J., De Jonghe, L. C. LSM-infiltrated solid oxide fuel cell cathodes. *Electrochemical and Solid-State Letters* **9**, A376–A378 (2006).
24. Vandillen, A., Terorde, R., Lensveld, D., Geus, J., Debjong, K. Synthesis of supported catalysts by impregnation and drying using aqueous chelated metal complexes. *Journal of Catalysis* **216**, 257–264 (2003).
25. Galarraga, C., Peluso, E., de Lasa, H. Eggshell catalysts for Fischer–Tropsch synthesis: Modeling catalyst impregnation. *Chemical Engineering Journal* **82**, 13–20 (2001).
26. Christos Agrafiotisa, A. T. Deposition of meso-porous g-alumina coatings on ceramic honeycombs by sol-gel methods. *Journal of the European Ceramic Society* **22**(4), 423–434 (2002).

27. Lekhal, A., Glasser, B. J., Khinast, J. G. Impact of drying on the catalyst profile in supported impregnation catalysts. *Chemical Engineering Science* **56**, 4473–4487 (2001).
28. Gardezi, S. A., Landrigan, L., Joseph, B., Wolan, J. T. Synthesis of tailored egg-shell cobalt catalysts for Fischer-Tropsch synthesis using wet chemistry techniques. *Industrial & Engineering Chemistry Research* **51**, 1703–1712 (2012).
29. Jiang, P. P., et al. Studying the fabrication and performance of three-way catalyst coating. *Journal of Inorganic Materials* **19**, 634–640 (2004).
30. Lee, S., Miller, N., Abernathy, H., Gerdes, K., Manivannan, A. Effect of Sr-doped $LaCoO_3$ and $LaZrO_3$ infiltration on the performance of SDC-LSCF cathode. *Journal of The Electrochemical Society* **158**, B735–B742 (2011).
31. Lee, S., et al. $Pr_{0.6}Sr_{0.4}CoO_{3-\delta}$ electrocatalyst for solid oxide fuel cell cathode introduced via infiltration. *Electrochimica Acta* **56**, 9904–9909 (2011).
32. Li, W., et al. Effect of adding urea on performance of Cu/CeO_2/yttria-stabilized zirconia anodes for solid oxide fuel cells prepared by impregnation method. *Electrochimica Acta* **56**, 2230–2236 (2011).
33. Lu, C., Sholklapper, T. Z., Jacobson, C. P., Visco, S. J., De Jonghe, L. C. LSM-YSZ cathodes with reaction-infiltrated nanoparticles. *Journal of the Electrochemical Society* **153**, A1115–A1119 (2006).
34. Knöfel, C., Wang, H., Thydén, K. T. S., Mogensen, M. Modifications of interface chemistry of LSM–YSZ composite by ceria nanoparticles. *Solid State Ionics* **195**, 36–42 (2011).
35. Hansen, K. K., Wandel, M., Liu, Y. L., Mogensen, M. Effect of impregnation of $La_{0.85}Sr_{0.15}MnO_3$/yttria stabilized zirconia solid oxide fuel cell cathodes with $La_{0.85}Sr_{0.15}MnO_3$ or Al_2O_3 nano-particles. *Electrochimica Acta* **55**, 4606–4609 (2010).
36. Nicholas, J. D., Barnett, S. A. Measurements and modeling of $Sm_{0.5}Sr_{0.5}CoO_{3-x}$-$Ce_{0.9}Gd_{0.1}O_{1.95}$ SOFC cathodes produced using infiltrate solution additives. *Journal of the Electrochemical Society* **157**, B536–B541 (2010).
37. Lou, X., et al. Controlling the morphology and uniformity of a catalyst-infiltrated cathode for solid oxide fuel cells by tuning wetting property. *Journal of Power Sources* **195**, 419–424 (2010).
38. Choi, J., Qin, W., Liu, M., Liu, M. Preparation and characterization of $(La_{0.8}Sr_{0.2})_{0.95}$ $MnO_{3-\delta}$ (LSM) thin films and LSM/LSCF interface for solid oxide fuel cells. *Journal of the American Ceramic Society* **94**, 3340–3345 (2011).
39. Jung, S. W., Vohs, J. M., Gorte, R. J. Preparation of SOFC anodes by electrodeposition. *Journal of the Electrochemical Society* **154**, B1270–B1275 (2007).
40. Yasuda, I., Ogasawaraa, K., Hishinuma, M., Kawada, T., Dokiyab, M. Oxygen tracer diffusion coefficient of (La, Sr)$MnO_{3\pm\delta}$. *Solid State Ionics* **86–88**, 1197–1201 (1996).
41. Ji, Y., Kilner, J., Carolan, M. Electrical properties and oxygen diffusion in yttria-stabilised zirconia (YSZ)–LaSrMnO (LSM) composites. *Solid State Ionics* **176**, 937–943 (2005).
42. Hojberg, J., Sogaard, M. Impregnation of LSM based cathodes for solid oxide fuel cells. *Electrochemical and Solid-State Letters* **14**, B77–B79 (2011).
43. Armstrong, T. J., Virkar, A. V. Measurement of O-2-N-2 effective diffusivity in porous media at high temperatures using an electrochemical cell. *The Electrochemical Society* **150**, A249–A256 (2003).
44. Sholklapper, T., Radmilovic, V., Jacobson, C., Visco, S., & Jonghe, L. Synthesis and stability of a nanoparticle-infiltrated solid oxide fuel cell electrode. *Electrochemical and Solid-State Letters* **10**, B74–B76 (2007).
45. Liang, F., et al. High performance solid oxide fuel cells with electrocatalytically enhanced (La, Sr)MnO3 cathodes. *Electrochemistry Communications* **11**, 1048–1051 (2009).
46. Petric, A., Huang, P., Tietz, F. Evaluation of La–Sr–Co–Fe–O perovskites for solid oxide fuel cells and gas separation membranes. *Solid State Ionics* **135**, 719–725 (2000).

47. Holc, J., Kuščer, D., Hrovat, M., Bernik, S., Kolar, D. Electrical and microstructural characterisation of $(La_{0.8}Sr_{0.2})(Fe_{1-x}Al_x)O_3$ and $(La_{0.8}Sr_{0.2})(Mn_{1-x}Al_x)O_3$ as possible SOFC cathode materials. *Solid State Ionics* **95**, 259–268 (1997).
48. Teraoka, Y., Zhang, H.M., Okamoto, K., Yamazoe, N. Mixed ionic-electronic conductivity of $La_{1-x}Sr_xCo_{1-y}Fe_yO_{3-\delta}$ perovskite-type oxides. *Materials Research Bulletin* **23**, 51–58 (1988).
49. G Dickemeier, M., Sasaki, K., Gauckler, L. J., Riess, I. Perovskite cathodes for solid oxide fuel cells based on ceria electrolytes. *Solid State Ionics* **86–88, Part 2**, 691–701 (1996).
50. Horita, T., et al. Oxygen reduction mechanism at porous $La_{1-x}Sr_xCoO_{3-\delta}$ cathodes/$La_{0.8}$ $Sr_{0.2}Ga_{0.8}Mg_{0.2}O_{2.8}$ electrolyte interface for solid oxide fuel cells. *Electrochimica Acta* **46**, 1837–1845 (2001).
51. Huang, Y. Y., Ahn, K., Vohs, J. M., Gorte, R. J. Characterization of Sr-doped $LaCoO_3$-YSZ composites prepared by impregnation methods. *Journal of the Electrochemical Society* **151**, A1592–A1597 (2004).
52. Samson, A., Sogaard, M., Knibbe, R., Bonanos, N. High performance cathodes for solid oxide fuel cells prepared by infiltration of $La_{0.6}Sr_{0.4}CoO_3$-delta into Gd-Doped Ceria. *Journal of the Electrochemical Society* **158**, B650–B659 (2011).
53. Armstrong, T. J., Rich, J. G. Anode-supported solid oxide fuel cells with $La_{0.6}Sr_{0.4}CoO_3$-lambda-$Zr_{0.84}Y_{0.16}O_2$-delta composite cathodes fabricated by an infiltration method. *Journal of the Electrochemical Society* **153**, A515–A520 (2006).
54. Samson, A. J., et al. Durability and performance of high performance infiltration cathodes. *Fuel Cells* **13**, 511–519 (2013).
55. Huang, Y. Y., Vohs, J. M., Gorte, R. J. An examination of LSM-LSCo mixtures for use in SOFC cathodes. *Journal of the Electrochemical Society* **153**, A951–A955 (2006).
56. Sase, M., et al. Interfacial reaction and electrochemical properties of dense (La, Sr) $CoO_{3-\delta}$ cathode on YSZ (1 0 0). *Journal of Physics and Chemistry of Solids* **66**, 343–348 (2005).
57. Ralph, J. M., Rossignol, C., Kumar, R. Cathode materials for reduced-temperature SOFCs. *Journal of the Electrochemical Society* **150**, A1518–A1522 (2003).
58. Wang, W., Gross, M. D., Vohs, J. M., Gorte, R. J. The stability of LSF-YSZ electrodes prepared by infiltration. *Journal of the Electrochemical Society* **154**, B439–B445 (2007).
59. Kong, J., Zhang, Y., Deng, C., Xu, J. Synthesis and electrochemical properties of LSM and LSF perovskites as anode materials for high temperature steam electrolysis. *Journal of Power Sources* **186**, 485–489 (2009).
60. Wang, W., Huang, Y., Jung, S., Vohs, J., Gorte, R. A comparison of LSM, LSF, and LSCo for solid oxide electrolyzer anodes. *Journal of the Electrochemical Society* **153**, A2066–A2070 (2006).
61. Meng, N., Leung, M., Leung, D. Technological development of hydrogen production by solid oxide electrolyzer cell (SOEC). *International Journal of Hydrogen Energy* **33**, 2337–2354 (2008).
62. Ciambelli, P., et al. AFeO(3) (A = La, Nd, Sm) and $LaFe_{1-x}Mg_xO_3$ perovskites as methane combustion and CO oxidation catalysts: structural, redox and catalytic properties. *Applied Catalysis B-Environmental* **29**, 239–250 (2001).
63. Xiaojing, Z., Wenjie, Huaju, L., Yong, L. Structural properties and catalytic activity of Sr-substituted $LaFeO_3$ perovskite. *Chinese Journal of Catalysis* **33**, 1109–1114 (2012).
64. Bidrawn, F., Lee, S., Vohs, J. M., Gorte, R. J. The effect of Ca, Sr, and Ba doping on the ionic conductivity and cathode performance of LaFeO(3). *Journal of the Electrochemical Society* **155**, B660–B665 (2008).
65. Huang, Y. Y., Vohs, J. M., Gorte, R. J. Fabrication of Sr-doped LaFeO(3)YSZ composite cathodes. *Journal of the Electrochemical Society* **151**, A3904–3909 (2004).

66. Zhou, Q., Zhang, Y., Shen, Y., He, T. Layered perovskite $GdBaCuCoO_{5+\delta}$ cathode material for intermediate-temperature solid oxide fuel cells. *Journal of the Electrochemical Society* **157**, B628–B632 (2010).
67. Simner, S. P., Anderson, M. D., Engelhard, M. H., Stevenson, J. W. Degradation mechanisms of La-Sr-Co-Fe-O_3SOFC cathodes. *Electrochemical and Solid-State Letters* **9**, A478–A481 (2006).
68. Lee, S., et al. Oxygen-permeating property of $LaSrBFeO_3$ (B=Co, Ga) perovskite membrane surface-modified by LaSrCoO3. *Solid State Ionics* **158**, 287–296 (2003).
69. Li, S., Jin, W., Huang, P., Xu, N., Shi, J. Comparison of oxygen permeability and stability of perovskite type $La_{0.2}A_{0.8}Co_{0.2}Fe_{0.8}O_{3-\delta}$ (A = Sr, Ba, Ca) membranes. *Industrial & Engineering Chemistry Research* **38**, 2963–2972 (1999).
70. Jiang, S. P. A comparison of O_2 reduction reactions on porous (La, Sr)MnO_3 and (La, Sr)(Co, Fe)O_3 electrodes. *Solid State Ionics* **146**, 1–22 (2002).
71. Esquirol, A., Brandon, N. P., Kilner, J. A., Mogensen, M. Electrochemical characterization of $La_{0.6}Sr_{0.4}Co_{0.2}Fe_{0.8}O_3$ cathodes for intermediate-temperature SOFCs. *Journal of the Electrochemical Society* **151**, A1847–A1855 (2004).
72. Liu Y. H., Chi B., Pu J., Li J. Performance degradation of impregnated $La_{0.6}Sr_{0.4}Co_{0.2}Fe_{0.8}O_3$+ Y_2O_3 stabilized ZrO_2 composite cathodes of intermediate temperature solid oxide fuel cells. *International Journal of Hydrogen Energy* **37**, 4388–4393 (2012).
73. Chen, J., et al. Nano-structured (La, Sr)(Co, Fe)O_3+YSZ composite cathodes for intermediate temperature solid oxide fuel cells. *Journal of Power Sources* **183**, 586–589 (2008).
74. Liu, Y., et al. A stability study of impregnated LSCF–GDC composite cathodes of solid oxide fuel cells. *Journal of Alloys and Compounds* **578**, 37–43 (2013).
75. Liu, Y., et al. Performance stability of impregnated $La_{0.6}Sr_{0.4}Co_{0.2}Fe_{0.8}O_{3-\delta}$–$Y_2O_3$ stabilized ZrO_2 cathodes of intermediate temperature solid oxide fuel cells. *International Journal of Hydrogen Energy* **39**, 3404–3411 (2014).
76. Liu, Y., et al. Performance stability and degradation mechanism of $La_{0.6}Sr_{0.4}Co_{0.2}Fe_{0.8}O_{3-\delta}$ cathodes under solid oxide fuel cells operation conditions. *International Journal of Hydrogen Energy* **39**, 15868–15876 (2014).
77. Tu, H. Y., Takeda, Y., Imanishi, N., Yamamoto, O. (Ln = Sm, Dy) for the electrode of solid oxide fuel cells. *Solid State Ionics* **100**, 283–288 (1997).
78. Fukunaga, H., Koyama, M., Takahashi, N., Wen, C., Yamada, K. Reaction model of dense $Sm_{0.5}Sr_{0.5}CoO_3$ as SOFC cathode. *Solid State Ionics* **132**, 279–285 (2000).
79. Koyama, M., et al. The mechanism of porous $Sm_{0.5}Sr_{0.5}CoO_3$ cathodes used in solid oxide fuel cells. *Journal of the Electrochemical Society* **148**, A795–A801 (2001).
80. Tu, H., Takeda, Y., Imanishi, N., Yamamoto, O. $Ln_{1-x}Sr_xCoO_3$(Ln = Sm, Dy) for the electrode of solid oxide fuel cells. *Solid State Ionics* **100**, 283–288 (1997).
81. Wang, F., Chen, D., Shao, Z. $Sm_{0.5}Sr_{0.5}CoO_3$-delta-infiltrated cathodes for solid oxide fuel cells with improved oxygen reduction activity and stability. *Journal of Power Sources* **216**, 208–215 (2012).
82. Han, D., et al. A micro-nano porous oxide hybrid for efficient oxygen reduction in reduced-temperature solid oxide fuel cells. *Scientific Reports* **2**, 462 (2012).
83. Suzuki, T., Awano, M., Jasinski, P., Petrovsky, V., Anderson, H. U. Composite (La, Sr)MnO_3–YSZ cathode for SOFC. *Solid State Ionics* **177**, 2071–2074 (2006).
84. Zhao, F., Zhang, L., Jiang, Z., Xia, C., Chen, F. A high performance intermediate-temperature solid oxide fuel cell using impregnated $La_{0.6}Sr_{0.4}CoO_{3-\delta}$ cathode. *Journal of Alloys and Compounds* **487**, 781–785 (2009).
85. Shah, M., Barnett, S. Solid oxide fuel cell cathodes by infiltration of $La_{0.6}Sr_{0.4}Co_{0.2}Fe_{0.8}O_{3-\delta}$ into Gd-Doped Ceria. *Solid State Ionics* **179**, 2059–2064 (2008).
86. Zhou, Y. C., Yuan, C., Liu, Y. D., Zhan, Z. L., Wang, S. R. Metal supported solid oxide fuel cells with infiltrated nanoelectrodes. *Material Research Innovations* **18**, 122–127 (2014).

87. Xia, C., Rauch, W., Chen, F. $Sm_{0.5}Sr_{0.5}CoO_3$ cathodes for low-temperature SOFCs. *Solid State Ionics, Diffusion & Reactions* **149**, 11–19 (2002).
88. Wang, F., Chen, D., Shao, Z. $Sm_{0.5}Sr_{0.5}CoO_{3-\delta}$-infiltrated cathodes for solid oxide fuel cells with improved oxygen reduction activity and stability. *Journal of Power Sources* **216**, 208–215 (2012).
89. Chen, J., et al. Performance of large-scale anode-supported solid oxide fuel cells with impregnated $La_{0.6}Sr_{0.4}Co_{0.2}Fe_{0.8}O_{3-\delta}$+$Y_2O_3$ stabilized ZrO_2 composite cathodes. *Journal of Power Sources* **195**, 5201–5205 (2010).
90. Jin, C., Yang, C., Chen, F. Effects on microstructure of NiO–YSZ anode support fabricated by phase-inversion method. *Journal of Membrane Science* **363**, 250–255 (2010).
91. Yang, C., Jin, C., Chen, F. Micro-tubular solid oxide fuel cells fabricated by phase-inversion method. *Electrochemistry Communications* **12**, 657–660 (2010).
92. Sofie, S. W. Fabrication of functionally graded and aligned porosity in thin ceramic substrates with the novel freeze tape-casting process. *Journal of the American Ceramic Society* **90**, 2024–2031 (2007).
93. Wang, X., Yu, B., Zhang, W. Q., Chen, J. A high performance solid oxide electrolysis cell support: CN201310279789.4 [P] 2013-07-04.
94. Cable, T. L., Sofie, S. W. A symmetrical, planar SOFC design for NASA's high specific power density requirements. *Journal of Power Sources* **174**, 221–227 (2007).
95. Deville, S. Freeze-casting of porous ceramics: A review of current achievements and issues. *Advanced Engineering Materials* **10**, 155–169 (2008).
96. Sofie, S. W., Dogan, F. Freeze casting of aqueous alumina slurries with glycerol. *Journal of the American Ceramic Society* **84**, 1459–1464 (2001).
97. Cable, T. L., Sofie, S. W. A symmetrical, planar SOFC design for NASA's high specific power density requirements. *Journal of Power Sources* **174**, 221–227 (2007).
98. Xin, X., Zhe, L., Zhu, Q. Fabrication of dense YSZ electrolyte membranes by a modified dry-pressing using nanocrystalline powders. *Journal of Materials Chemistry* **17**, 1627–1630 (2007).
99. Ma, Q. L., Ma, J. J., Zhou, S. A high-performance ammonia-fueled SOFC based on a YSZ thin-film electrolyte. *Journal of Power Sources* **164**, 86–89 (2007).
100. Zhou, S. T., Lin, X. P., Ai, D. S., Effect of composite pore former on the electrochemical performance of hydrogen electrode in SOEC. *Rare Metal Materials and Engineering* **42**, 700–703 (2012).
101. Yu, B., Zhang, W. Q., Liang, M. D. Effect of PMMA pore former on hydrogen production performance of solid oxide electrolysis cell. *Journal of Inorganic Materials* **26**, 807–812 (2011).
102. Zhao, C., Liu, R., Wang, S. Fabrication of a large area cathode-supported thin electrolyte film for solid oxide fuel cells via tape casting and co-sintering techniques. *Electrochemistry Communications* **11**, 842 (2009).
103. Kong, J., Sun, K., Zhou, D. Anode-supported IT-SOFC anode prepared by tape casting technique. *International Forum on Strategic Technology* **59**, 1169–1173 (2006).
104. Chen, Y., Bunch, J., Li, T. S., Mao, Z. P., Chen, F. L. Novel functionally graded acicular electrode for solid oxide cells fabricated by the freeze-tape-casting process. *Journal of Power Sources* **213**, 93–99 (2012).
105. Kim, S. D., Moon, H., Hyun, S.H. Performance and durability of Ni-coated YSZ anodes for intermediate temperature solid oxide fuel cells. *Solid State Ionics* **177**, 931–938 (2006).
106. Setoguchi, T., Okamoto, K., Eguchi, K. Effects of anode material and fuel on anodic reaction of solid oxide fuel cells. *Journal of the Electrochemical Society* **139**, 2875–2880 (1992).
107. Yu B., Liu M. Y., Zhang W. Q., et al. Polarization loss of single solid oxide electrolysis cells and microstructural optimization of the cathode. *Acta PhysicoiChimica Sinica.* **27**, 395–402 (2011).

108. Chroneos, A., Yildiz, B., Tarancón, A. Oxygen diffusion in solid oxide fuel cell cathode and electrolyte materials: mechanistic insights from atomistic simulations. *Energy & Environmental Science* **4**, 2774–2789 (2011).
109. Kuru, Y., Jalili, H., Cai, Z. Direct probing of nanodimensioned oxide multilayers with the aid of focused ion beam milling. *Advanced Materials* **23**, 4543–2548 (2011).
110. Chen, Z., Xiao, Q. G. Surface electronic structure transitions at high temperature on perovskite oxides: The case of strained $La_{0.8}Sr_{0.2}CoO_3$ thin films. *Journal of the American Chemical Society* **133**, 17696–17704 (2011).
111. Ullmann, H., Trofimenko, N., Tietz, F., St Ver, D., Ahmad-Khanlou, A. Correlation between thermal expansion and oxide ion transport in mixed conducting perovskite-type oxides for SOFC cathodes. *Solid State Ionics* **138**, 79–90 (2000).
112. Hideko, H., et al. Thermal expansion coefficient of yttria stabilized zirconia for various yttria contents. *Solid State Ionics* **176**, 613–619 (2005).
113. Szepes, A., Fehér, A., Szabó-Révész, P. Influence of freezing temperature on product parameters of solid dosage forms prepared via the freeze-casting technique. *Chemical Engineering & Technology* **30**, 511–516 (2007).
114. Fukasawa, T., Deng, Z. Y., Ando, M. Synthesis of porous silicon nitride with unidirectionally aligned channels using freeze-drying process. *Journal of the American Ceramic Society* **85**, 2151–2155 (2002).
115. Fukasawa, T., Deng, Z. Y., Ando, M., Ohji, T., Goto, Y. Pore structure of porous ceramics synthesized from water-based slurry by freeze-dry process. *Journal of Materials Science* **36**, 2523–2527 (2001).
116. Fukasawa, T., Ando M., Ohji T., Kanzaki, S. Synthesis of porous ceramics with complex pore structure by freeze-dry processing. *Journal of the American Ceramic Society* **84**, 230–232 (2001).
117. Szepes, A., Ulrich, J., Farkas, Z., Kovacs, J., Szabo-Revesz, P. Freeze-casting technique in the development of solid drug delivery systems. *Chemical Engineering and Processing* **238**, 230 (2007).
118. Laurie J., Bagnall C. M., Harris B. Colloidal suspensions for the preparation of ceramics by a freeze casting route. *Journal of Non-Crystalline Solids* **147**, 320–325 (1992).
119. Koch, D., Andresen, L., Schmedders, T. Evolution of porosity by freeze casting and sintering of sol-gel derived ceramics. *Journal of Sol-Gel Science and Technology* **26**, 149–152 (2003).
120. Jones, R. W. Near net shape ceramics by freeze casting. *Refractories and Industrial Ceramics* **20**, 117–120 (2000).
121. Kim, H. W., Knowles, J. C., Kim, H. E. Porous scaffolds of gelatin-hydroxyapatite nanocomposites obtained by biomimetic approach: Characterization and antibiotic drug release. *Journal of Biomedical Materials Research Part B Applied Biomaterials* **74**, 686–698 (2005).
122. Gong, Y., et al. Poly(lactic acid) scaffold fabricated by gelatin particle leaching has good biocompatibility for chondrogenesis. *Journal of Biomaterials Science Polymer Edition* **19**, 207–21 (2008).
123. Portero, A., Teijeiro-Osorio, D., Alonso, M. J., Remuñán-López, C. Development of chitosan sponges for buccal administration of insulin. *Carbohydrate Polymers* **68**, 617–625 (2007).
124. Cable, T. L., Setlock, J. A., Farmer, S. C. Regenerative performance of the NASA symmetrical solid oxide fuel cell design, 1–12, *International Journal of Applied Ceramic Technology* (2011).
125. Song, J. H., Koh, Y. H., Kim, H. E., Li, L. H., Bahn, H. J. Fabrication of a porous bioactive glass-ceramic using room-temperature freeze casting. *Journal of the American Ceramic Society* **89**, 2649–2653 (2006).

126. Koh, Y. H., Sun, J. J., Kim, H. E. Freeze casting of porous Ni–YSZ cermets. *Materials Letters* **61**, 1283–1287 (2007).
127. Han, J., Hong, C., Zhang, X. Highly porous ZrO_2 ceramics fabricated by a camphene-based freeze-casting route: Microstructure and properties. *Journal of the European Ceramic Society* **30**, 53–60 (2010).
128. Lu, K. Microstructural evolution of nanoparticle aqueous colloidal suspensions during freeze casting. *Journal of the American Ceramic Society* 90, 3753–3758 (2007).
129. Zhang, Y., et al. Effects of gelatin addition on the microstructure of freeze-cast porous hydroxyapatite cerarmics. *Ceramics International* **35**, 2151–2154 (2009).
130. Zhang, D., et al. Freeze gelcasting of aqueous alumina suspensions for porous ceramics. *Ceramics International* **38**, 6063–6066 (2012).
131. Zhang, Y., et al. Effects of rheological properties on ice-templated porous hydroxyapatite ceramics. *Materials Science and Engineering: C Materials for Biological Applications,* **33**, 340–346 (2013).
132. Deville, S., Saiz, E., Tomsia, A. P. Freeze casting of hydroxyapatite scaffolds for bone tissue engineering. *Biomaterials* **27**, 5480–5489 (2006).
133. Suciu, C., Tikkanen, H., Wærnhus, I. Water-based tape-casting of SOFC composite 3YSZ/8YSZ electrolytes and ionic conductivity of their pellets. *Ceramics International* **38**, 357–365 (2012).
134. Chen, R. F., Wang, C. A., Huang, Y., Ma, L .G., Lin, W. Y. Ceramics with special porous structures fabricated by freeze-gelcasting: Using tert-butyl alcohol as a template. *Journal of the American Ceramic Society* **90**, 3478–3484 (2007).
135. Hu L., Wang, C. A., Huang Y. Control of pore channel size during freeze casting of porous YSZ ceramics with unidirectionally aligned channels using different freezing temperatures. *Journal of the European Ceramic Society* **30**, 3389–3396 (2010).
136. Smorodin, V. Y. Mechanisms of heterogeneous ice nucleation onto mixed ice nuclei in the atmosphere. *Journal of Aerosol Science* **21**, S249–S253 (1990).
137. Bauerecker, S., Ulbig, P., Buch, V., Vrbka, L., Jungwirth, P. Monitoring ice nucleation in pure and salty water via high-speed imaging and computer simulations. *The Journal of Physical Chemistry C* **112**, 7631–7636 (2008).
138. Knight, C. A. Ice nucleation in the atmosphere advances in colloid and interface. *Science* **10**, 369–395 (1979).
139. Granasy, L., Oxtoby, D. W. Cahn–Hilliard theory with triple-parabolic free energy. II. Nucleation and growth in the presence of a metastable crystalline phase. *The Journal of the Chemical Physics* **112**, 2410–2419 (2000).
140. Levi, L. Structure of ice grown from droplet accretion and solidification process. *Journal of Crystal Growth* **22**, 303–310 (1974).
141. Deville, S., Saiz, E., Tomsia, A. P. Ice-templated porous alumina structures. *Acta Materialia* **55**, 1965–1974 (2007).
142. Körber, C., Rau, G., Cosman, M. D. Interaction of particles and a moving ice-liquid interface. *Journal of Crystal Growth* **72**, 649–662 (1985).
143. Chino, Y. & Dunand, D.C. Directionally freeze-cast titanium foam with aligned, elongated pores. *Acta Materialia* **56**, 105–113 (2008).
144. Casses, P., Azouni-Aidi, M. A. A general theoretical approach to the behavior of foreign particles at advancing solid-liquid interfaces. *Advances in Colloid and Interface Science* **50**, 103–120 (1994).
145. Ren L. L. Preparation of porous TiO_2 by a novel freeze casting. *Ceramics International* **35**, 1267–1270 (2009).
146. Mukai, S. R., Nishihara, H., Tamon, H. Porous properties of silica gels with controlled morphology synthesized by unidirectional freeze-gelation. *Microporous and Mesoporous Materials* **63**, 43–51 (2003).

147. Yu, J., Li, S., Lv, Y., Zhao, Y., Pei, Y. Preparation of silicon nitride–barium aluminum silicate composites by freeze gelation. *Materials Letters* **147**, 128–130 (2015).
148. Deville, S., et al. In situ X-Ray radiography and tomography observations of the solidification of aqueous alumina particle suspensions-Part I: Initial instants. *Journal of the American Ceramic Society* **92**, 2489–2496 (2009).
149. Deville, S., Maire, E., Lasalle, A. In situ X-ray radiography and tomography observations of the solidification of aqueous alumina particles suspensions. Part II: Steady state. *Journal of the American Ceramic Society* **92**, 2497–2503 (2009).
150. Landi, E., Valentini, F., Tampieri, A. Porous hydroxyapatite/gelatine scaffolds with ice-designed channel-like porosity for biomedical applications. *Acta Biomaterialia* **4**, 1620 (2008).
151. Kochs, M., Körber, C., Nunner, B. The influence of the freezing process on vapour transport during sublimation in vacuum-freeze-drying. *International Journal of Heat & Mass Transfer* **34**, 2395–2408 (1991).
152. Gannon, P., Sofie, S., Deibert, M. Thin film YSZ coatings on functionally graded freeze cast NiO/YSZ SOFC anode supports. *Journal of Applied Electrochemistry* **39**, 497–502 (2009).
153. Zuo, K. H., Zeng, Y. P., Jiang, D. L. Effect of polyvinyl alcohol additive on the pore structure and morphology of the freeze-cast hydroxyapatite ceramics. *Materials Science and Engineering: C* **30**, 283–287 (2010).
154. Lei, Q., Zhang, H. Controlled freezing and freeze drying: A versatile route for porous and micro-/nano-structured materials. *Journal of Chemical Technology & Biotechnology* **86**, 172–184 (2011).
155. Waschkies T., Oberacker, R., Hoffmann M. J. Control of lamellae spacing during freeze casting of ceramics using double-side cooling as a novel processing route. *Journal of the American Ceramic Society* **92**, S79–S84 (2009).
156. Cooney, M. J., Lau C., Windmeisser M. Design of chitosan gel pore structure: Towards enzyme catalyzed flow-through electrodes. *Journal of Materials Chemistry* **18**, 667–674 (2008).
157. Parks, W. M., Guo, Y. B. A casting based process to fabricate 3D alginate scaffolds and to investigate the influence of heat transfer on pore architecture during fabrication. *Materials Science & Engineering C* **28**, 1435–1440 (2008).
158. Macchetta, A., Turner, I. G., Bowen, C. R. Fabrication of HA/TCP scaffolds with a graded and porous structure using a camphene-based freeze-casting method. *Acta Biomater* **5**, 1319–1327 (2009).
159. Rubinstein, E. R., Glicksman, M. E. Dendritic growth kinetics and structure II. Camphene. *Journal of Crystal Growth* **112**, 97–110 (1991).
160. Hong, C. Q., Zhang, X. H., Han, J. C. Camphene-based freeze-cast ZrO_2, foam with high compressive strength. *Materials Chemistry & Physics* **119**, 359–362 (2010).
161. Çadırlı, E., Maraslı, N., Bayender, B. Dependency of the microstructure parameters on the solidification parameters for camphene. *Materials Research Bulletin* **35**, 985–995 (2000).
162. Araki, K., Halloran, J. W. Porous ceramic bodies with interconnected pore channels by a novel freeze casting technique. *Journal of the American Ceramic Society* **88**, 1108–1114 (2005).
163. Yoon, B. H., Lee, E. J., Kim, H. E., Koh, Y. H. Highly aligned porous silicon carbide ceramics by freezing polycarbosilane/camphene solution. *Journal of the American Ceramic Society* **90**, 1753–1759 (2007).
164. Fu, Q., Rahaman, M.N., Dogan, F. & Bal, B.S. Freeze casting of porous hydroxyapatite scaffolds. I. Processing and general microstructure. *Journal of Biomedical Materials Research Part B: Applied Biomaterials* **86B**, 125–135 (2008).
165. Zhang, Y. M., Hu, L. Y., Han, J. C., Jiang, Z. H. Freeze casting of aqueous alumina slurries with glycerol for porous ceramics. *Ceramics International* **36**, 617–621 (2010).

166. Rahaman, M. N., Fu, Q. Freeze-cast hydroxyapatite scaffolds for bone tissue engineering applications. *Journal of the American Ceramic Society* **91**, 4137–4140. (2008).
167. Lu, K. Microstructural evolution of nanoparticle aqueous colloidal suspensions during freeze casting. *Journal of the American Ceramic Society* **90**, 3753–3758 (2007).
168. Pachulski, N., Ulrich, J. New fields of application for sol-gel processes: Cold and vacuum-free 'compacting' of pharmaceutical materials to tablets. *Chemical Engineering Research and Design* **85**, 1013–1019 (2007).
169. Fu, Q., Rahaman, M. N., Dogan, F., Bal, B. S. Freeze-cast hydroxyapatite scaffolds for bone tissue engineering applications. *Biomedical Materials* **3**, 1–7 (2008).
170. Zuo, K. H., Zeng, Y. P., Jiang, D. L. Properties of microstructure-controllable porous yttria-stabilized ziroconia ceramics fabricated by freeze casting. *International Journal of Applied Ceramic Technology* **5**, 198–203 (2008).
171. Han, J.C., Hu, L.Y., Zhang, Y.M. & Zhou, Y.F. Fabrication of ceramics with complex porous structures by the impregnate-freeze-casting process. *Journal of the American Ceramic Society* **92**, 2165–2167 (2009).
172. Li, W. L., Lu, K., Walz, J. Y. Freeze casting of porous materials: Review of critical factors in microstructure evolution. *International Materials Reviews* **57**, 37–60 (2012).
173. McKee, C. T., Walz, J. Y. Effects of added clay on the properties of freeze-casted composites of silica nanoparticles. *Journal of the American Ceramic Society* **92**, 916–921 (2009).
174. Lu, K., Hammond, C., Qian, J. M. Surface patterning nanoparticle-based arrays. *Journal of Materials Science* **45**, 582–588 (2010).
175. Sofie, S. W. Fabrication of functionally graded and aligned porosity in thin ceramic substrates with the novel freeze–tape-casting process. *J. Am. Ceram. Soc.* **90**, 2024–2031 (2007).
176. Liu, G., Zhang, D., Meggs, C., Button, T. W. *Script Materialia* **62**, 466–468. (2010).
177. Munch, E., Franco, J., Deville, S. Porous ceramic scaffolds with complex architectures. *Journal of the Minerals, Metals and Materials Society* **60**, 54–58 (2008).
178. Wang X., Yu B., Zhang W. Q., et al. A support structure with gradient pores used for solid oxide cell. *Rare Mental Materials and Engineering* **42**, 717–719 (2013).
179. Zhang, L., Jiang, S. P., Wang, W., Zhang, Y. NiO/YSZ, anode-supported, thin-electrolyte, solid oxide fuel cells fabricated by gel casting. *Journal of Power Sources* **170**, 55–60 (2007).
180. Statham, M. J., Hammett, F., Harris, B. Net-shape manufacture of low-cost ceramic shapes by freeze-gelation. *Journal of Sol-Gel Science and Technology* **13**, 171–175 (1998).
181. Ho, M. H., et al. Preparation of porous scaffolds by using freeze-extraction and freeze-gelation methods. *Biomaterials* **25**, 129–138 (2004).
182. Mukai, S. R., Nishihara, H., Tamon, H. Porous properties of silica gels with controlled morphology synthesized by unidirectional freeze-gelation. *Microporous and Mesoporous Materials* **63**, 43–51 (2003).
183. Lu, K., Kessler, C. S., Davis, R. M. Optimization of a nanoparticle suspension for freeze casting. *Journal of the American Ceramic Society* **89**, 2459–2465 (2006).
184. Chino, Y., Dunand, D. C. Directionally freeze-cast titanium foam with aligned, elongated pores. *Acta Materialia* **56**, 105–113 (2008).
185. Nakagawa, K., Thongprachan, N., Charinpanitkul, T., Tanthapanichakoon, W. Ice crystal formation in the carbon nanotube suspension: A modelling approach. *Chemical Engineering Science* **65**, 1438–1451 (2010).
186. Minaberry, Y., Jobbágy, M. Macroporous bioglass scaffolds prepared by coupling sol–gel with freeze drying. *Chemistry of Materials* **23**, 2327–2332 (2011).

11 High-Temperature Electrochemical Process of CO_2 Conversion with SOCs 6

Advanced Structure (Heterostructure)

11.1 BRIEF INTRODUCTION FOR HETEROSTRUCTURE

Despite solid oxide cells (SOCs) having attracted much attentions and being developed at unprecedented rates in recent years, some technical hurdles must be overcome before their broad commercialization.[1,2] Existing SOC devices are typically operated at extremely high temperatures (800°C–1000°C); although such high operating temperatures accelerate the efficiency of SOCs significantly, they simultaneously lead to detrimental interactions between electrodes and electrolytes, which may damage cell compositions and material microstructures, leading to long-term durability loss and thus resulting in high cost of utilization.[3] For instance, the degradation rates of a five-cell SOC stack with 5000 hours of successful operation were approximately 15%/1000 hours (0–2000 hours) and 6%/1000 hours (2000–5000 hours) for two types of feed-gas compositions. In another case, 18% loss in H_2 production was observed after 1000 hours of operation performed by the Idaho National Laboratory with a 25-cell SOC stack.[4]

The key challenge to developing a more efficient, durable, and economically feasible SOCs system is to reduce operating temperatures to the intermediate range (500°C–800°C) and even lower (300°C–500°C) while maintaining high performance.[5,6] However, the reduced operation temperatures are usually found to cause other performance degradation issues, such as sluggish oxygen reduction reaction/oxygen evolution reaction (ORR/OER) kinetics on oxygen electrodes, low ionic conductivities in electrolytes, and increased polarization resistances throughout the whole SOC.[7] As a result, most of the existing cell components and structures may become inapplicable. Compared with the electrolyte or the fuel electrode, the energy loss comes mainly from the oxygen electrode. Thus, the sluggish ORR/OER kinetics on the oxygen electrode need to be improved accordingly.

In recent studies, two state-of-the-art strategies have been explored by researchers involving the development of more active materials and the tailoring of material structures.[8] For the former, it is urgent to find more active oxygen electrode materials, such as $Ba_{0.5}Sr_{0.5}Co_{0.8}Fe_{0.2}O_{3-\delta}$ (BSCF),[9] $SmNiO_3$ (SNO),[10] $REBaCo_2O_{5+\delta}$ (RE = Pr, Nd, and Gd),[11] and other materials with perovskite type or perovskite-related type structures that have shown promise in the acceleration of ORR/OER kinetics for intermediate temperature SOCs. And regarding the tailoring of material structures, the introduction of heterointerface possessing two phases can be an effective approach to improve the performance of electrodes in SOCs. This topic is the focus of this chapter.

The interface between different crystalline materials (heterointerface) can provide extraordinary local electronic or atomic structures and subsequently produce improved properties such as high ion/electronic conductivity,[12–14] high-performance thermoelectricity, high-efficiency catalytic activity,[15] super-wettability,[16,17] induced magnetization,[18–21] and so on. These properties are attributed to the physical interactions among spin, charge, lattice, and/or orbital, which can occur near the heterointerface.[14,22] Here the heterostructure can be regarded as a structure containing some heterointerfaces.

For instance, the enhancement of ionic conduction for the AgI/AgBr heterointerface was discovered in the 1980s,[23] attracting research attention to the two-phase conductor systems.[24–29] A heterointerface possessing CaF_2 and BaF_2 shows very fast ion conduction on the nanometer scale, which was caused from the redistribution of fluoride ions in the space-charge regions.[13] Another example, published in *Science*, for improving ion conduction is the heterointerface of YSZ (ZrO_2:Y_2O_3)/STO ($SrTiO_3$); the lateral ionic conductivity at the interface showed an amazing 10^8 times enhancement around room temperature, and the related mechanism was that the atomic reconstruction at the interface between highly dissimilar structures offered both a high-mobility plane and a large number of carriers. On the other hand, with respect to the increase of electronic conductivity, high-mobility electron gas at the heterointerface of $LaAlO_3$/$SrTiO_3$ was investigated by Ohtomo et al.[14,30] and Nakagawa et al.[31] They found that the heterointerface (electron-doped interface) between two insulators showed extremely high carrier mobility exceeding 10,000 $cm^2V^{-1}S^{-1}$, which was due to the built-in polarity discontinuity of the interface.

To sum up, the heterostructure can provide extraordinary structure and also show some distinct properties. Especially, the properties including ionic conductivities and electronic conductivities are closely related to ORR/OER activity. Thus the heterostructures are widely investigated in supercapacitors, solar cells, lithium batteries, fuel cells, etc. With respect to SOCs, several recent reports have shown that the heterostructure between ABO_3 and A_2BO_4 phases (A-site is rare earth or alkaline earth metal ions, B-site is transition metal ions, and O is the oxygen ions), typically the interface between $La_{1-x}Sr_xCoO_{3-\delta}$ and $(La_{1-x}Sr_x)_2CoO_{4\pm\delta}$, can exhibit remarkably accelerated ORR/OER kinetics.[5,32–43] More details are discussed below.

11.2 HETEROSTRUCTURE OF ABO_3/A_2BO_4 IN SOCs

In general, perovskite-structured oxides possess high electrical conductivities (approximately 500–2000 S cm^{-1} at 600°C)[44,45] but are limited by oxygen ion mobility, whereas A_2BO_4 phases exhibit very high oxygen incorporation and oxygen diffusion kinetics[42] but insufficient electrical conductivity under intermediate temperatures (approximately 50–100 S cm^{-1}).[45] Therefore, heterostructured oxides between ABO_3 and A_2BO_4 oxides have been studied extensively.[5,32,33,39,41,46,47] A prime case in point is the $(La, Sr)_2CoO_4/La_{0.6}Sr_{0.4}CoO_3$ heterostructure proposed by Sase et al.,[34,48] in which the oxygen exchange coefficient (k^q) of the heterostructure was enhanced by three orders of magnitude in comparison to that of single-phase $La_{0.6}Sr_{0.4}CoO_3$ at about 500°C. This surprising enhancement of oxygen exchange kinetics has been confirmed by various recent investigations under intermediate temperatures,[5,32–43] as summarized in Table 11.1.

Among the studies related to SOCs, heterostructured oxygen electrode materials can be roughly classified into six categories based on structure (as shown in Figure 11.1), including dense thin-films (Figure 11.1a),[39] dense multilayered (more than two layers) structures (Figure 11.1b),[42,51] and vertically aligned nanocomposite structures (Figure 11.1c)[5] prepared accurately through physical processes (e.g., pulsed laser deposition [PLD] or molecular beam epitaxy [MBE]); and the porous layered structures (Figure 11.1d),[36,54] porous composite structures (Figure 11.1e),[55] and porous infiltrated structures (Figure 11.1f)[40] prepared through chemical processes (e.g., one-pot method, glycine nitrate process). Of these electrode materials, electrodes prepared using physical methods at the nano-scale are more suitable for theoretical research but have limited practical applications, whereas porous heterostructured electrodes prepared using chemical processes provide larger triple phase boundaries (TPBs), $O_2/ABO_3/A_2BO_4$, areas and favorable electron-transport configurations for ORR/OER. In addition, porous heterostructured electrodes are inexpensive and less time consuming in their preparation. However, porous heterostructured ABO_3/A_2BO_4 electrodes have been rarely studied.[55]

TABLE 11.1
Summary of Literatures About Oxygen Electrodes with ABO_3/A_2BO_4 Interface from 2008 to 2018

Year	Material	Geometry	T (°C)[a]	Performance	Mechanism	References
2008	$La_{0.6}Sr_{0.4}CoO_{3-\delta}$/ $LaSrCoO_{4+\delta}$	Composite	500	Exchange coefficient k^a (cm s^{-1}) (from SIMS) enhanced 10^3 times	Fast oxygen-incorporation paths along the heterogeneous interface boundary	34
2009	$La_{0.6}Sr_{0.4}CoO_{3-\delta}$/ $LaSrCoO_{4+\delta}$	Composite	500	Exchange coefficient σ_e (S cm^{-2}) (from EIS) 10–$10^{1.5}$ times	Heterogeneous interface enhance the oxygen catalytic reaction	36
2010	$La_{0.8}Sr_{0.2}CoO_{3-\delta}$/ $LaSrCoO_{4+\delta}$	Thin film	550	ORR coefficient k^q (from EIS) enhance 10^3–10^4 times	Strain, space charge effects, increase in electronic structure or oxygen vacancy	39
2012	(La, Sr)$CoO_{3-\delta}$/(La, Sr)$_2CoO_{4+\delta}$	Computation	500	400 times faster oxygen incorporation kinetics	Anisotropy and strain	41
2012	$(La_{0.5}Sr_{0.5})_2CoO_{4\pm\delta}$/ $La_{0.6}Sr_{0.4}CoO_{3-\delta}$	Thin film	550	Approximately 10^4 times enhancement of surface oxygen kinetics	Heterogeneous interface	49
2012	$La_{0.8}Sr_{0.2}CoO_{3-\delta}$/ $(La_{0.5}Sr_{0.5})_2CoO_{4-\delta}$	Computation	500–600	Oxygen vacancy concentration enhanced 10^2–$10^{2.5}$ times	Sr segregation	50
2013	$La_{0.8}Sr_{0.2}CoO_3$/ $(La_{0.5}Sr_{0.5})_2CoO_4$	Multilayer	300	10^3 times increase in the rate of charge transfer to oxygen	Electronic structure	42
2014	$La_{0.8}Sr_{0.2}CoO_3$/ $(La_{0.5}Sr_{0.5})_2CoO_4$	Thin film	250	—	Electron donation and transfer of oxygen vacancies across heterointerface	51
2014	$(La_{1-y}Sr_y)_2CoO_{4\pm\delta}$/ $La_{1-x}Sr_xCoO_{3-\delta}$	Thin film	400–600	Anomalous Sr segregation	Cation diffusion	37
2015	$La_{0.8}Sr_{0.2}CoO_{3-\delta}$/ $LaSrCoO_{4+\delta}$	VAN[a]	320–400	ORR coefficient k^q (from EIS) enhanced about 10 times	Electronic activation and the more stable cation composition at interface	5

(Continued)

TABLE 11.1 (*Continued*)
Summary of Literatures About Oxygen Electrodes with ABO_3/A_2BO_4 Interface from 2008 to 2018

Year	Material	Geometry	T (°C)[a]	Performance	Mechanism	References
2015	$(La, Sr)_2CoO_{4\pm\delta}$/ $L_{a1-x}Sr_xMO_{3-\delta}$	Thin film	550	Surface exchange coefficients enhanced 10^2 times	Sr segregation	46
2015	$(La_{1-x}Sr_x)_2CoO_4$/ $SrTiO_3$ $(La_{1-x}Sr_x)_2CoO_4/LaSrAlO_3$	Thin film	400–500	20 times higher oxygen diffusion coefficient along ab-plane compared that along the c-axis	Cation diffusion	52
2017	$(La, Sr)_2FeO_{4-\delta}$/ $La_{0.8}Sr_{0.2}FeO_{3-\delta}$	Composite	650–800	The ORR activity enhanced 10 times	Lattice mismatch between LSF_{214} and LSF_{113}	40
2018	$La_{0.8}Sr_{0.2}CoO_{3-\delta}$/ $Nd_2NiO_{4+\delta}$	Thin film	300–415	The oxygen exchange kinetics is much slower	Charge transfer	53

[a] VAN = vertically aligned nanocomposite; T = temperature.

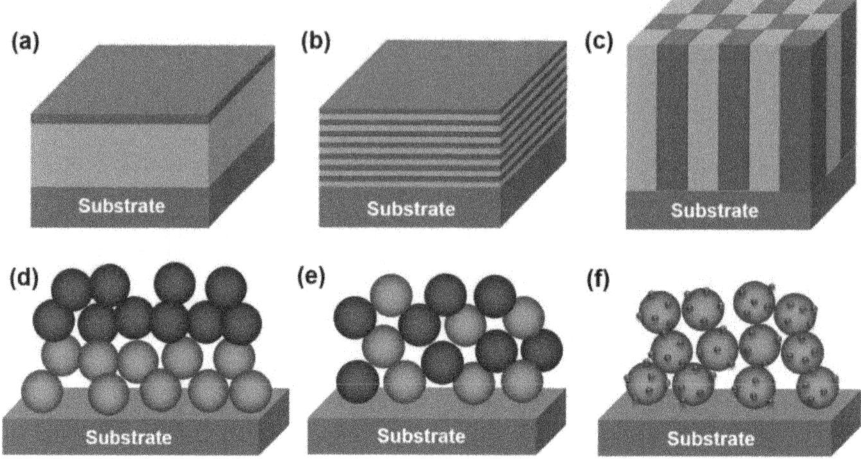

FIGURE 11.1 Schematic diagram of the heterostructures of ABO_3 and A_2BO_4. (a) Dense thin-film, (b) dense multilayered structure, (c) vertically aligned nanocomposite material, (d) porous layered structure, (e) porous composite structure, and (f) porous infiltrated structure. Here, yellow layers or spheres represent the ABO_3 phase, and green layers or spheres represent the A_2BO_4 phase. Note that the schematics are not to scale.

11.3 MECHANISM OF ORR/OER IN ABO_3/A_2BO_4 HETEROSTRUCTURE

To explain the mechanism about the accelerated ORR/OER kinetics of oxygen electrode in ABO_3/A_2BO_4 systems, some typical cases are investigated by researchers as shown in Table 11.2. Typically, the enhancement could be ascribed to interfacial properties, mainly including electronic structure,[5,42] anisotropy,[41] lattice strain[40,41,56] or mismatch in lattice parameter,[40] and cation inter-diffusion (e.g., Sr enrichment).[37,50,57]

11.3.1 Electronic Structure

Electron inter-transfer generally occurs between heterointerfaces, which can change the electronic structure and make the composite systems more superior for catalyzing an oxygen electrode process. As shown in Figure 11.2,[42] in a typical LSC_{113}/LSC_{214} heterostructure, excess electrons from LSC_{113} are spontaneously injected into LSC_{214}, raising the Fermi level in LSC_{214} and thus facilitating the charge transfer process at the $LSC_{214}(100)$ surface. Such electronic activation of LSC_{214}, concurrent with fast oxygen incorporation in the [100] direction in LSC_{214}, leads to stronger attraction of oxygen molecule on the $LSC_{214}(100)$ surface and thus faster ORR kinetics at the LSC_{113}/LSC_{214} heterostructure. This proposed mechanism was also confirmed by Ma et al. in vertically aligned nanocomposite LSC_{113}/LSC_{214} cathodes.[5]

TABLE 11.2
Mechanism of Oxygen Exchange Kinetics Enhancement of ABO_3/A_2BO_4 Heterostructure

No.	Heterostructure	Enhancement	Mechanism	Publications
1	$Nd_{0.5}Sr_{0.5}CoO_{3-\delta}/Nd_{0.8}Sr_{1.2}CoO_{4\pm\delta}$	10^2 to 10^3 times	Sr enrichment and decrease of cobalt valence states	*Nano Energy*[58]
2	$La_{0.8}Sr_{0.2}CoO_3/(La_{0.5}Sr_{0.5})_2CoO_4$	10^2 times	Electronic structure	*Adv. Energy Mater.*[42]
	$La_{0.8}Sr_{0.2}CoO_3/(La_{0.5}Sr_{0.5})_2CoO_4$	10 times		*J. Mater. Chem. A*[5]
3	$(La, Sr)CoO_{3-\delta}/(La, Sr)_2CoO_{4+\delta}$	10^2 times	Anisotropy	*Energ. Environ. Sci.*[41]
4	$(La, Sr)CoO_{3-\delta}/(La, Sr)_2CoO_{4+\delta}$	3×10^2 times	Lattice strain	*Energ. Environ. Sci.*[41]
	$La_{0.8}Sr_{0.2}FeO_{3-\delta}/La_{0.8}Sr_{1.2}FeO_{4-\delta}$	10 times	Mismatch in lattice parameter	*ACS Appl. Mater. Inter.*[40]
5	$La_{1-x}Sr_xCoO_{3-\delta}/(La_{1-y}Sr_y)_2CoO_{4\pm\delta}$ ($x = 0.25$ or 0.2, $y = 0.5$)	10^2 times	Cation inter-diffusion	*J. Phys. Chem. Lett.*[37]
	$La_{0.8}Sr_{0.2}CoO_{3-\delta}/(La_{0.5}Sr_{0.5})_2CoO_{4+\delta}$	—	Sr enrichment	*J. Phys. Chem. Lett.*[57]
	$La_{0.8}Sr_{0.2}CoO_{3-\delta}/(La_{0.5}Sr_{0.5})_2CoO_{4-\delta}$	10^2 to $10^{2.5}$ times	Sr interdiffusion	*Phys. Chem. Chem. Phys.*[50]

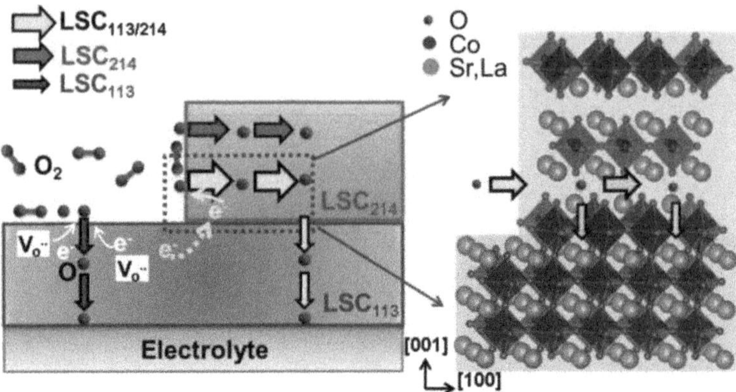

FIGURE 11.2 Mechanism that governs the enhancement of ORR activity at the interface of LSC_{113} and LSC_{214} layers. (From Chen, Y., et al., Electronic activation of cathode superlattices at elevated temperatures-source of markedly accelerated oxygen reduction kinetics. *Adv. Energy Mater.*, 2013, 3, 1221–1229. Copyright WILEY-VCH Verlag GmbH & Co. KGaA. Reprinted with permission.)

11.3.2 Anisotropy

Han et al.[41] found that the strongly anisotropic oxygen incorporation kinetics on the LSC_{214} was an important contributor for ORR enhancement. According to their density functional theory (DFT) simulation, for the anisotropic LSC_{214} system, the adsorption energies of oxygen molecules on the (100) and (001) surfaces were −2.02 and −0.20 eV, respectively. Such a difference represents the anisotropic adsorption strength that favors the $LSC_{214}(100)$ surface.

Regarding a specific oxygen electrode process, the oxygen incorporation path on the single-phase LSC_{113} is shown in Figure 11.3. O_2 adsorbs onto the $LSC_{113}(001)$ surface and then is incorporated and dissociated by the assistance of a migrating surface oxygen vacancy. By comparison, for LSC_{113}/LSC_{214} heterostructure, there are two parallel paths for oxygen incorporation, as shown in Figure 11.3b: path I (a–b, across the (100) surface of LSC_{214}) and path II (c-d across the (001) surface of LSC_{214}). Due to the anisotropic adsorption strength of $LSC_{214}(100)$ surface, which was discussed above, it exhibits a much faster oxygen diffusion rate through path I near the heterostructure, which finally contributes to more than 10^2 times acceleration of oxygen exchange.

11.3.3 Lattice Strain (or the Mismatch in Lattice Parameter)

Lattice strain has been demonstrated to alter the oxygen defect chemistry as well as the oxygen reaction and diffusion kinetics on perovskite oxides. Especially in a specific LSC system, increasing its planar tensile strain weakens the in-plane Co–O bonds by decreasing the Co d and lattice O p orbitals' hybridization, which

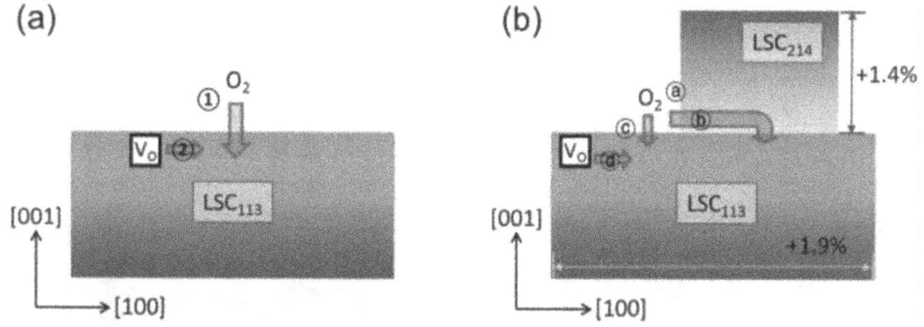

FIGURE 11.3 Illustration of the oxygen incorporation paths. (a) On the single phase LSC_{113} and (b) near the LSC_{113}/LSC_{214} heterointerface. The latter includes both path I (a,b) and path II (c,d). (Han, J. W., et al., *Energy Environ. Sci.*, 5, 8598–8607, 2012. Reprinted by permission of the Royal Society of Chemistry.)

consequently causes strengthening of the chemisorption of O_2 onto Co. An oxygen vacancy is more easily formed due to the weakening of the in-plane Co–O bonds upon tensile strain, thus resulting in the enhanced ORR kinetics.

Han et al. demonstrated that the LSC_{113}/LSC_{214} interface imposed a +1.9% planar strain in the (001) plane of LSC_{113} and a +1.4% planar strain in the (100) plane of LSC_{214}, as illustrated in Figure 11.4.[41] It has been proved that two such planes were the main paths for surface oxygen exchange (see path I in Figure 11.3). As a result, lattice strain in tandem with lattice anisotropy contributes to more than 10^3 times enhancement in electrochemical activities.

Aside from the LSC_{113}/LSC_{214} heterostructure, a similar mechanism was also proposed to explain the origin of ORR/OER kinetics enhancement in a $La_{0.8}Sr_{0.2}FeO_{3-\delta}/La_{0.8}Sr_{1.2}FeO_{4-\delta}$ heterostructured system.[40]

11.3.4 Cation Inter-diffusion

The large extent of cation inter-diffusion (e.g., Sr) and cation enrichment near heterointerface regions is expected to increase the number of oxygen vacancies,

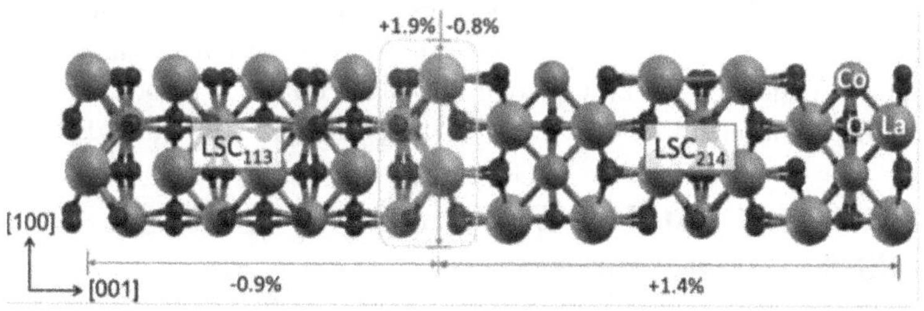

FIGURE 11.4 DFT model of the LSC_{113}/LSC_{214} structure and the theoretically estimated strain states induced near this interface. (Han, J. W., et al., *Energy Environ. Sci.*, 5, 8598–8607, 2012. Reprinted by permission of the Royal Society of Chemistry.)

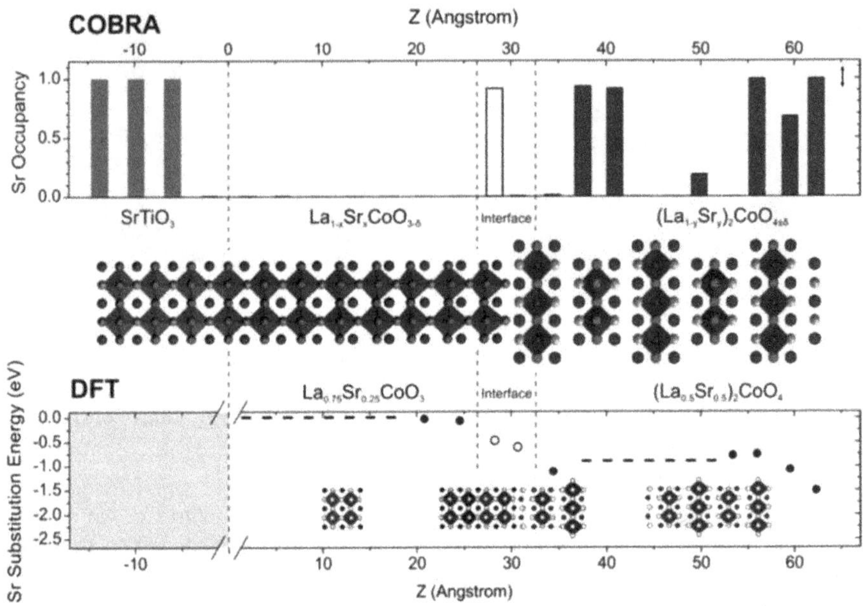

FIGURE 11.5 Top panel: layer-by-layer Sr concentration versus depth profile determined from differential COBRA for the as-deposited LSC thin film. Bottom panel: Ab initio Sr for La substitution energies (relative to that of bulk $La_{0.7}5Sr_{0.25}CoO_3$). (Reprinted from Feng, Z. X., et al., *J. Phys. Chem. Lett.*, 5, 1027–1034, 2014. Copyright 2014, American Chemical Society.)

which is also essential for fast oxygen adoption and incorporation. As shown in Figure 11.5, Sr was found to strongly segregate near the LSC_{214}/LSC_{113} interface, which was confirmed by Coherent Bragg Rod Analysis (COBRA) experiments and DFT simulation, the COBRA is a method that can provide elemental information through the energy dependence of atomic scattering cross sections, further allow for the determination of atomic concentrations of individual crystallographic sites. It is obvious that for a specific $La_{1-x}Sr_xCoO_3$ system, Sr enrichment (Sr occupancy on the A site of approximately 1) can arise the valence of B-site Co and shift the O 2p band center relative to the Fermi level. This can induce the generation of oxygen vacancies and promote oxygen transport kinetics.[37] A similar mechanism related cation inter-diffusion (e.g., Sr) was also proposed or confirmed by other researchers, although they mentioned inter-diffusion as Sr enrichment[57] or Sr interdiffusion.[50]

11.4 CURRENT CHALLENGES FOR ABO_3/A_2BO_4 HETEROSTRUCTURE

Heterostructured materials have attracted a lot of attention in recent years because of their unique properties and potential applications in SOC-based energy conversion devices. However, some challenges still need to be investigated for the further development and applications of ABO_3/A_2BO_4 in SOCs as introduced above.

11.4.1 Barriers to Accurately Detecting the Performance of a Heterointerface

A heterointerface is the interfacial region that exists between two layers of dissimilar solid-state materials. The surface can be easily detected by many typical methods, such as X-ray photoelectron spectroscopy (XPS), atomic force microscope (AFM), auger electron spectrometry (AES), scanning electron microscope (SEM), and so forth. The surface is very different from the interface, which is submerged below the surface and is extremely thin (as thin as several layers of atoms), which makes it challenging to probe its enhanced performance and detailed features accurately. Such technical barriers greatly hinder the investigations of the role of heterostructures and their impact on electrode performance. As a critical evaluation method, the measurement for electrochemical performance of thin films with heterostructure is of great significance for the research of SOCs. The challenges for detecting the performance of very thin film electrodes with heterointerfaces include realization of accurate microzone test, high temperature changing rate and wide temperature range, and quickly adjustable gas atmosphere range. However, the vast majority of the existing test platforms in laboratories for studying SOCs are specially designed for porous material half-battery testing, which is not satisfactory for thin-film electrode material testing.

11.4.2 Invisibility of Heterointerface and its Unclear Mechanism

Although several hypotheses have been proposed to explain the activation mechanisms of interfaces, including strain effects, changes in electronic structure, interfacial cation diffusion, and extension of TPBs, an overall theoretical model that can cover all these effects and direct evidence for those mechanisms are still lacking. Thus, the role of heterogeneous interfaces in SOC-based systems is not fully understood. Since the performance-enhancing level caused by interfaces may range from several times to several thousand times, once a breakthrough is made, the rational design of oxygen electrodes can be achieved and the electrochemical performance can be improved by many orders of magnitude.

To achieve this extended understanding, characterization for the detailed features of heterostructure is indispensable, while the position and size of heterostructures in SOCs materials are challenging to probe, as mentioned above. These barriers have hindered the investigations of the role of heterostructure. The development of advanced characterization techniques, such as secondary ion mass spectrometry (SIMS), focused ion beam (FIB), in situ scanning tunneling microscopy/scanning tunneling spectroscopy (STM/STS), and so on, makes it possible to detect local composition and the structure of heterointerfaces with very high resolution. However, the applications of these advanced methods for investigation of heterostructure, coupled with DFT calculation and prediction in SOCs, still face some challenges.

11.4.3 The Gap Between Theoretical Investigation and Practical Application

A heterostructure consists of a substrate and a cleanly lattice-matched abrupt epitaxial layer. It is typically manufactured by precisely controlled epitaxial growth technologies, including PLD, atomic laser deposition (ALD), MBE, chemical vapor deposition (CVD), and so on.[36,54] At the interfacial regions, electronic properties of materials depend on spatial properties, especially on small length scales across the interface. Because of the break in symmetry, heterointerfaces generally produce some unique structural characteristics compared to bulk counterparts, including dislocation, strain, and elemental non-stoichiometry, which can further induce various property changes in conductivity, ion transfer properties, catalytic activities, and so on. Nevertheless, some properties of thin-film electrodes with heterostructures in nanoscale may disappear when the heterostructures are applied in a practical system. For instance, the enhanced ORR/OER performance may weaken in practical porous heterostructured electrodes. Until now, porous heterostructured ABO_3/A_2BO_4 electrodes prepared using chemical methods have been rarely studied.[55] Therefore, developing porous heterostructured electrodes with practical and effective interfaces is a promising approach to address application problems.

Undoubtedly, all developments and breakthroughs concerning the challenges and possible directions discussed in this chapter rely largely on the development of material sciences and engineering, including the discovery and design of new materials and analytical thinking from chemistry, physics, and engineering perspectives to understand microscopic and macroscopic observations. Thus, future studies should highlight the application of heterostructures in various fields and reveal the origin of novel quantum states at the interfacial region. The combination of advanced atomic technologies with electrochemical testing methods can characterize electrode enhancements as functions of various structures and therefore provide guidelines for the development of electrode materials with better performance.

REFERENCES

1. Zheng, Y., Wang, J. C., Yu, B., et al. A review of high temperature co-electrolysis of H_2O and CO_2 to produce sustainable fuels using solid oxide electrolysis cells (SOECs): Advanced materials and technology. *Chemical Society Reviews* **46**, 1427–1463 (2017).
2. Zheng, Y., Zhang, W. Q., Li, Y. F., et al. Energy related CO_2 conversion and utilization advanced materials/nanomaterials, reaction mechanisms and technologies. *Nano Energy* **40**, 512–539 (2017).
3. Moçoteguy, P., Brisse, A. A review and comprehensive analysis of degradation mechanisms of solid oxide electrolysis cells. *International Journal of Hydrogen Energy* **38**, 15887–15902 (2013).
4. Sohal, M. S., O'Brien, J. E., Stoots, C. M., et al. Degradation issues in solid oxide cells during high temperature electrolysis. *Journal of Fuel Cell Science and Technology* **9**, 0110171 (2012).
5. Ma, W., Kim, J. J., Tsvetkov, N., et al. Vertically aligned nanocomposite $La_{0.8}Sr_{0.2}CoO_3$/$(La_{0.5}Sr_{0.5})_2CoO_4$ cathodes-electronic structure, surface chemistry and oxygen reduction kinetics. *Journal of Materials Chemistry A* **3**, 207–219 (2015).

6. Choi, S., Kucharczyk, C. J., Liang, Y. G., et al. Exceptional power density and stability at intermediate temperatures in protonic ceramic fuel cells. *Nature Energy* **3**, 202–210 (2018).
7. Lee, K. T., Manthiram, A. Comparison of $Ln_{0.6}Sr_{0.4}CoO_{3-\delta}$ (Ln = La, Pr, Nd, Sm, and Gd) as cathode materials for intermediate temperature solid oxide fuel cells. *Journal of the Electrochemical Society* **4**, A794–A798 (2006).
8. Irvine, J. T. S., Neagu, D., Verbraeken, M. C., et al. Evolution of the electrochemical interface in high-temperature fuel cells and electrolysers. *Nature Energy* **1**, 15014 (2016).
9. Shao, Z., Haile, S. M. A high-performance cathode for the next generation of solid oxide fuel cells. *Nature* **431**, 170–173 (2004).
10. Zhou, Y., Guan, X. F., Zhou, H., et al. Strongly correlated perovskite fuel cells. *Nature* **534**, 231–234 (2016).
11. Liu, S., Zhang, W., Li, Y., et al. $REBaCo_2O_{5+\delta}$ (RE ¼ Pr, Nd, and Gd) as promising oxygen electrodes for intermediate-temperature solid oxide electrolysis cells. *RSC Advances* **7**, 16332–16340 (2017).
12. Garcia-Barriocanal, J., Bruno, F. Y., Rivera-Calzada, A. "Charge Leakage" at $LaMnO_3$/$SrTiO_3$ interfaces. *Advanced Materials* **22**, 627–632 (2010).
13. Sata, N., Eberman, K., Eberl, K., et al. Mesoscopic fast ion conduction in nanometre-scale planar heterostructures. *Nature* **408**, 946–949 (2000).
14. Ohtomo, A., Hwang, H. Y. A high-mobility electron gas at the $LaAlO_3$/$SrTiO_3$ heterointerface. *Nature* **429**, 423–426 (2004).
15. Scanlon, D. O., Dunnill, C. W., Buckeridge, J., et al. Band alignment of rutile and anatase TiO_2. *Nature Materials* **12**, 798–801 (2013).
16. Feng, X., Feng, L., Jin, M. H., et al. Reversible super hydrophobicity to super hydrophilicity transition of aligned ZnO nanorod films. *Journal of the American Chemical Society* **126**, 62–63 (2004).
17. Lim, H. S., Kwak, D., Lee, D. Y., et al. UV-Driven reversible switching of a roselike vanadium oxide film between superhydrophobicity and superhydrophilicity. *Journal of the American Chemical Society* **129**, 4128–4129 (2007).
18. Singh, S., Haraldsen, J. T., Xiong, J., et al. Induced magnetization in $La_{0.7}Sr_{0.3}MnO_3$/$BiFeO_3$ superlattices. *Physical Review Letters* **113**, 047204 (2014).
19. Först, M., Caviglia, A. D., Scherwitzl R., et al. Spatially resolved ultrafast magnetic dynamics initiated at a complex oxide heterointerface. *Nature Materials* **14**, 883–888 (2015).
20. Li, M. D., Song, Q. C., Zhao, W. W., et al. Dirac-electron-mediated magnetic proximity effect in topological insulator/magnetic insulator heterostructures. *Physical Review B* **96**, 201301 (2017).
21. Liu, M., Jiang, L. Dialectics of nature in materials science: Binary cooperative complementary materials. *Science China Materials* **59**, 239–246 (2016).
22. Dagotto, E. Complexity in strongly correlated electronic systems. *Science* **309**, 257–262 (2005).
23. Joachim, M. Space charge regions in solid two-phase systems and their conduction contribution-I. Conductance enhancement in the system ionic conductor-'inert' phase and application on $AgCl:Al_2O_3$ and $AgCl:SiO_2$. *Journal of Physics and Chemistry of Solids* **46**, 309–320 (1985).
24. Lee, J. S., Adams, S., Maier, J. Transport and phase transition characteristics in $AgI:Al_2O_3$ composite electrolytes evidence for a highly conducting 7-layer AgI polytype. *Journal of the Electrochemical Society* **147**, 2407–2418 (2000).
25. Joachim, M. Ionic conduction in space charge regions. *Progress in Chemistry* **23**, 171–263 (1995).

26. Chen, C., Wang, Z., Kato, T., et al. Misfit accommodation mechanism at the heterointerface between diamond and cubic boron nitride. *Nature Communications* **6**, 6327 (2015).
27. Kang, B., Jang, M., Chung, Y., et al. Enhancing 2D growth of organic semiconductor thin films with macroporous structures via a small-molecule heterointerface. *Nature Communications* **5**, 4752 (2014).
28. Liu, J., Kargarian, M., Kareev, M., et al. Heterointerface engineered electronic and magnetic phases of $NdNiO_3$ thin films. *Nature Communications* **4**, 2714 (2013).
29. Chen, Y., Ojha, S., Tsvetkov, N., et al. Spinel/perovskite cobaltite nanocomposites synthesized by combinatorial pulsed laser deposition. *CrystEngcomm* **18**, 7745–7752 (2016).
30. Ohtomo, A., Muller, D. A., Grazul, J. L., et al. Artificial charge-modulation in atomic-scale perovskite titanate superlattices. *Nature* **419**, 378–380 (2002).
31. Nakagawa, N., Hwang, H. Y., Mullefr, D. A. Why some interfaces cannot be sharp. *Nature Materials* **5**, 204–209 (2006).
32. Stämmler, S., Merkle, R., Stuhlhofer, B., et al. Phase constitution, Sr distribution and morphology of self-assembled La-Sr-Co-O composite films prepared by PLD. *Solid State Ionics* **303**, 172–180 (2017).
33. Stämmler, S., Merkle, R., Maier, J. Oxygen electrocatalysis on epitaxial $La_{0.6}Sr_{0.4}CoO_{3-\delta}$ perovskite thin films for solid oxide fuel cells. *Journal of the Electrochemical Society* **164**, F454–F463 (2017).
34. Sase, M., Yashiro, K., Sato, K., et al. Enhancement of oxygen exchange at the hetero-interface of (La, Sr)CoO_3/(La, Sr)$_2CoO_4$ in composite ceramics. *Solid State Ionics* **178**, 1843–1852 (2008).
35. Sase, M., Hermes, F., Nakamura, T., et al. Promotion of oxygen surface reaction at the Hetero-Interface of (La, Sr)CoO_3/(La, Sr)$_2CoO_4$. *ECS Transactions* **7**, 1055–1060 (2007).
36. Yashiro, K., Nakamura, T., Sase, M., et al. Composite cathode of perovskite-related oxides, (La, Sr)$CoO_{3-\delta}$/(La, Sr)$_2CoO_{4-\delta}$, for solid oxide fuel cells. *Electrochemical and Solid-State Letters* **9**, B135–B137 (2009).
37. Feng, Z., Yacoby, Y., Gadre, M. J., et al. Anomalous interface and surface strontium segregation in $(La_{1-y}Sr_y)_2CoO_{4\pm\delta}/La_{1-x}Sr_xCoO_{3-\delta}$ heterostructured thin films. *The Journal of Physical Chemistry Letters* **5**, 1027–1034 (2014).
38. Lee, D., Lee, Y., Wang, X. R., et al. Enhancement of oxygen surface exchange on epitaxial $La_{0.6}Sr_{0.4}Co_{0.2}Fe_{0.8}O_3$ thin films using advanced heterostructured oxide interface engineering. *MRS Communications* **6**, 204–209 (2016).
39. Crumlin, E. J., Mutoro, E., Ahn, S., et al. Oxygen reduction kinetics enhancement on a heterostructured oxide surface for solid oxide fuel cells. *The Journal of Physical Chemistry Letters* **1**, 3149–3155 (2010).
40. Hong, T., Zhao, M., Brinkman, K., et al. Enhanced oxygen reduction activity on Ruddlesden–Popper phase decorated $La_{0.8}Sr_{0.2}FeO_{3-\delta}$ 3D heterostructured cathode for solid oxide fuel cells. *ACS Applied Materials & Interfaces* **9**, 8659–8668 (2017).
41. Han, J. W., Yildiz, B. Mechanism for enhanced oxygen reduction kinetics at the (La, Sr)CoO_3/(La, Sr)$_2CoO_{4+\delta}$ hetero-interface. *Energy & Environmental Science* **5**, 8598–8607 (2012).
42. Chen, Y., Cai, Z. H., Kuru, Y., et al. Electronic activation of cathode superlattices at elevated temperatures-source of markedly accelerated oxygen reduction kinetics. *Advanced Energy Materials* **3**, 1221–1229 (2013).
43. Zhao, C., Liu X. G., Zhang, W. Q., et al. Measurement of oxygen reduction-evolution kinetics enhanced (La, Sr)CoO_3/(La, Sr)$_2CoO_4$ hetero-structure oxygen electrode in operating temperature for SOCs. *International Journal of Hydrogen Energy* (2018). doi:10.1016/j.ijhydene.2018.04.128.
44. Li, F., Jiang, L., Zeng, R., et al. One-pot synthesized hetero-structured $Ca_3Co_2O_6$/$La_{0.6}Ca_{0.4}CoO_3$ dual-phase composite cathode materials for solid-oxide fuel cells. *International Journal of Hydrogen Energy* **40**, 12750–12760 (2015).

45. Gao, Z., Mogni, L. V., Miller, E. C., et al. A perspective on low-temperature solid oxide fuel cells. *Energy & Environmental Science* **9**, 1602–1644 (2016).
46. Lee, D., Lee, Y., Hong, W. T., et al. Oxygen surface exchange kinetics and stability of (La, Sr)$_2$CoO$_{4\pm\delta}$/La$_{1-x}$Sr$_x$MO$_{3-\delta}$ (M = Co and Fe) hetero-interfaces at intermediate temperatures. *Journal of Materials Chemistry A* **3**, 2144–2157 (2015).
47. Lee D., Lee, Y., Grimaud, A., et al. Enhanced oxygen surface exchange kinetics and stability on epitaxial La$_{0.8}$Sr$_{0.2}$CoO$_{3-\delta}$ thin films by La$_{0.8}$Sr$_{0.2}$MnO$_{3-\delta}$ decoration. *The Journal of Physical Chemistry* **118**, 14326–14334 (2014).
48. Sase, M., Hermes, F., Yashiro, K., et al. Enhancement of oxygen surface exchange at the hetero-interface of (La, Sr)CoO$_3$/(La, Sr)$_2$CoO$_4$ with PLD-layered films. *Journal of the Electrochemical Society* **155**, B793–B797 (2008).
49. Crumlin, E.J., Ahn, S.J., Lee, D., et al. Oxygen electrocatalysis on epitaxial La$_{0.6}$Sr$_{0.4}$CoO$_{3-\delta}$ perovskite thin films for solid oxide fuel cells. *Journal of the Electrochemical Society* **159**, F219–F225 (2012).
50. Gadre, M. J., Lee, Y. L., Morgan, D. Cation interdiffusion model for enhanced oxygen kinetics at oxide heterostructure interfaces. *Physical Chemistry Chemical Physics* **14**, 2606–2616 (2012).
51. Tsvetkov, N., Chen, Y., Yildiz, B. Reducibility of Co at the La$_{0.8}$Sr$_{0.2}$CoO$_3$/(La$_{0.5}$Sr$_{0.5}$)$_2$CoO$_4$ hetero-interface at elevated temperatures. *Journal of Materials Chemistry A* **2**, 14690 (2014).
52. Chen, Y., Téllez, H., Burriel, M., et al. Segregated chemistry and structure on (001) and (100) surfaces of (La$_{1-x}$Sr$_x$)$_2$CoO$_4$ override the crystal anisotropy in oxygen exchange kinetics. *Chemistry of Materials* **27**, 5436–5450 (2015).
53. Lenser, C., Lu, Q. Y., Crumlin, E. J., et al. Charge transfer across oxide interface probed by in situ X-ray photoemission and Absorption spectroscopy. *The Journal of Physical Chemistry C* **122**, 4841–4848 (2018).
54. K. Yashiro, Nakamura, T., Sase, M., et al. Electrode performance at hetero-interface of perovskite-related oxides, (La, Sr)CoO$_{3-\delta}$/(La, Sr)$_2$CoO$_{4-\delta}$. *ECS Transactions* **7**, 1287–1292 (2007).
55. Li, F., Jiang, L., Zeng, R., et al. LaSrCoO$_{4\pm d}$ cathode with high electro-catalytic activity for solid-oxide fuel cells. *International Journal of Hydrogen Energy* **42**, 29463–29471 (2017).
56. Kubicek, M., Cai, Z. H., Ma, W., et al. Tensile lattice strain accelerates oxygen surface exchange and diffusion in La$_{1-x}$Sr$_x$CoO$_{3-\delta}$ thin films. *ACS Nano* **7**, 3276–3286 (2013).
57. Feng, Z., Crumlin, E. J., Hong, W. T., et al. In situ studies of the temperature-dependent surface structure and chemistry of single-crystalline (001)-oriented La$_{0.8}$Sr$_{0.2}$CoO$_{3-\delta}$ perovskite thin films. *The Journal of Physical Chemistry Letters* **4**, 1512–1518 (2013).
58. Zheng, Y., Li, Y. F., Wu, T., et al. Oxygen reduction kinetic enhancements of intermediate-temperature SOFC cathodes with novel Nd$_{0.5}$Sr$_{0.5}$CoO$_{3-\delta}$/Nd$_{0.8}$Sr$_{1.2}$CoO$_{4\pm\delta}$ hetero-interfaces. *Nano Energy* **51**, 711–720 (2018).

12 High-Temperature Electrochemical Process of CO_2 Conversion with SOCs 7

The Significant Phenomenon of Cation Segregation

12.1 INTRODUCTION OF CATION SEGREGATION IN PEROVSKITE-BASED SOC ELECTRODES

Electrodes are the core components of solid oxide cells (SOCs) because they directly determine the overall performance and durability of the device. In particular, the surface chemistry of the electrodes plays a crucial role in determining their electrical properties and electrocatalytic activities.[1–3] Perovskite-based oxides have become the most widely used electrode materials for SOCs, and the surface process on perovskite electrodes has been deeply investigated.[4,5] It has been widely accepted that the chemical environment of cations at the surface of perovskite-based oxides can be quite different from that within the lattice. As a commonly observed phenomenon, both A- and B-site cations in perovskite-based oxides could enrich at the surface region and may separate out from the initial oxides, leading to the formation of a surface secondary phase on the surface, in a process known as cation segregation. Typically, oxygen electrode materials run at high temperatures and strong oxidizing atmosphere, where A-site cations normally segregate out in the form of oxides. For instance, many researchers demonstrated the surface enrichment of A-site substituent cations in perovskite-based materials, including Pb^{2+} segregation in (La, Pb)$MnO_{3-\delta}$, as well as Sr^{2+} segregation in $Sm_{1-x}Sr_xCoO_{3-\delta}$, $La_{1-x}Sr_xMnO_{3-\delta}$, and $La_{0.6}Sr_{0.4}Co_yFe_{1-y}O_{3-\delta}$. In contrast, B-site segregation seems to be more desirable at a reduced environment, and its segregated component usually involves metallic particles.[6–8]

Even though the thickness of the A-site cations enriched layer is normally quite thin [for example, Sr-rich layer in (LaSr)$CoO_{3-\delta}$ is found to be no deeper than 20 nm],[1] and the secondary phase induced by segregation typically accounts for small areas at the surface, the rate of the oxygen exchange reaction can be reduced by several orders of magnitude.[9,10] On the other hand, the exsolution of B-site metallic particles

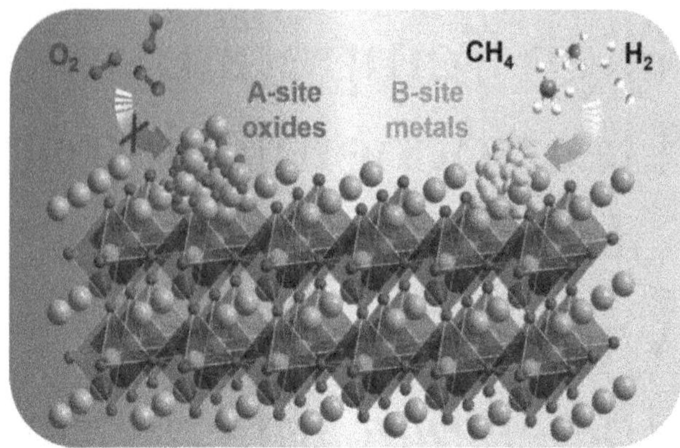

FIGURE 12.1 Illustration of A- and B-site segregation in a perovskite-based system and its influences on surface chemistry and catalytic properties. (Li, Y. et al., *Chem. Soc. Rev.*, 46, 6345–6378, 2017. Reproduced by permission of The Royal Society of Chemistry.)

under reducing conditions was reported to significantly enhance catalyst activity and stability for the electrodes.[6] In fact, in situ exsolution of highly active B nanoparticles has been proven to be a promising method for enhancing electrocatalytic activities of fuel electrodes.[6,11–13] Figure 12.1 illustrates several possible effects of A- and B-site cation segregation on surface chemistry and catalytic properties.

In this chapter, both conventional and several state-of-the-art surface characterization technologies used in detecting surface segregation will be listed and discussed. The main driving forces of cation segregation will be analyzed, and the influence of cation segregation on electrochemical performance will be reviewed. Based on this, several representative works enhancing SOC activity by surface engineering will be introduced. Finally, we will summarize the key points reviewed in this chapter and provide some critical insights into future research in this field. The methodology and mechanistic understanding of the surface processes are applicable to other materials systems in a wide range of applications, including thermochemical photo-assisted splitting of H_2O/CO_2 and even metal-air batteries.

12.2 CHARACTERIZATION OF SURFACE SEGREGATION

In this section, we will present some common techniques for probing and detecting surface cation segregation in perovskite-based oxide materials (Table 12.1). In addition, Figure 12.2 shows some typical probing depths and the information accessible by these techniques when applied to a perovskite-based system. For each technique, we will briefly introduce the principle of operation, unique capabilities, and specific applications in investigating surface segregation. Specifically, in Section 12.2.8, recent advancements in characterizing surface segregation under near SOC operation conditions will be highlighted.

TABLE 12.1
Representative Characterization Technologies to Investigate the Structure and Composition of Cation Segregation

Information Categories	Technologies and Applications	Representative Works
Cations composition	LEIS (elemental composition at the outmost layer)	[1,3,15–29]
	XPS (elemental composition as well as chemical environment near surface)	[11,15,30–41]
	AES and SEM (local elemental inhomogeneities with lateral resolution at the surface)	[7,8,12,15,30,42–49]
	SIMS (surface elemental distribution and compositional evolution from the outmost surface to bulk)	[9,50–60]
	STEM and EDS (local elemental distribution with lateral resolution at the surface)	[7,8,11,12,15,43,61–67]
Surface segregation and phase separation state	AFM and STM/STS (surface morphology and electronic structure)	[15,41,44,45,48,68–71]
	XRD (crystal structure and lattice parameters, as well as the generation of secondary phases)	[8,11,44,62,63,72–75]

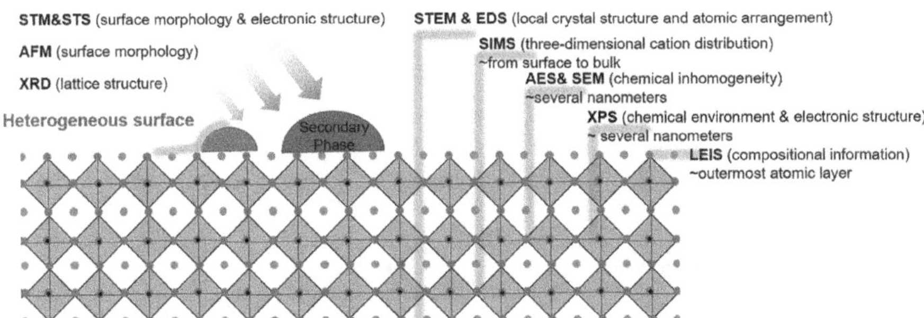

FIGURE 12.2 Several surface sensitive technologies with different depth resolutions that can be used to characterize surface segregation by probing cations composition or surface phase separation state in a perovskite model system. (Li, Y. et al., *Chem. Soc. Rev.*, 46, 6345–6378, 2017. Reproduced by permission of The Royal Society of Chemistry.)

12.2.1 Low-Energy Ion Scattering

Low-energy ion scattering (LEIS) spectroscopy, sometimes referred to as ion scattering spectroscopy (ISS), is an analytical technique that can be used to probe the structure and composition of the outermost surface. With combined LEIS and ion sputtering (a technique that removes the atoms or ions from the surface), the compositional information at near-surface region can be obtained layer-by-layer. This approach is quite different from some conventional detection methods that can only implement the average measure of composition over several atomic layers. LEIS impinges the outermost surface by a primary ion beam (the most widely applied cations include He^+, Ne^+, and Ar^+, and sometimes K^+ and Na^+) with certain energy of E_0 (generally 20 eV–10 keV) and records the amount of the reflected primary ion beam with energy of E_1, as illustrated in Figure 12.3a. By measuring the energy difference of E_0 and E_1, the category and stoichiometry of surface atoms and other structural information like the lattice arrangement can be precisely obtained, but specific chemical information like the oxidation state or binding environment cannot be detected.[21] In addition, due to the relatively high probability of ionic neutralization at the surface, the information depth of LEIS is typically only one or two layers of the surface, which makes LEIS one of the most sensitive and effective surface analysis methods.[23]

One representative work that used LEIS to study the segregation of three types of perovskite-based oxides (perovskites, Ruddlesden-Popper [RP] phases, and double perovskites) was carried out by Kilner et al.[1] Figure 12.3b shows the results characterizing the perovskite $La_{0.6}Sr_{0.4}Co_{0.2}Fe_{0.8}O_{3-\delta}$ (LSCF-113). Seen in the top

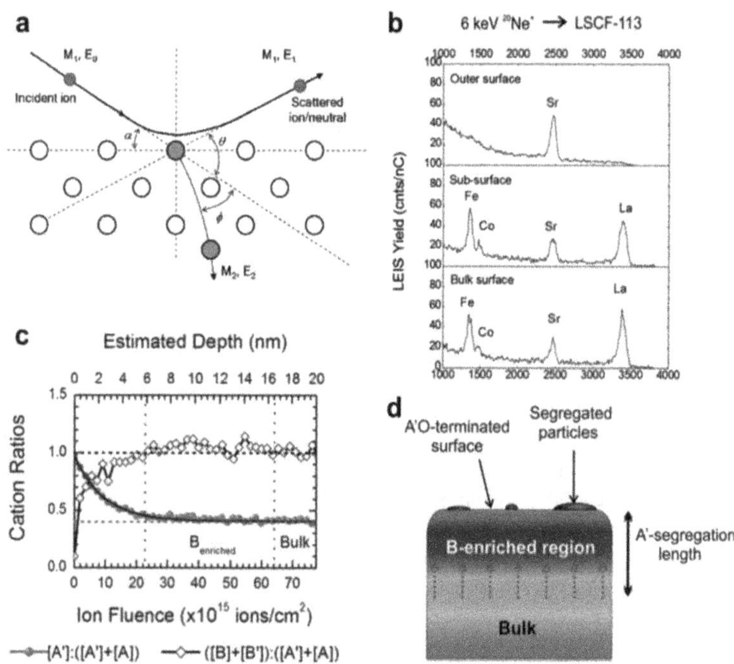

FIGURE 12.3 (a) The principle of LEIS; (b) the composition of several layers at different depths in LSCF-113, characterized by LEIS; (c) chemical composition in near-surface region of LSCF-113 with perovskite structure, characterized by LEIS depth profiles; (d) distribution of elements in near-surface region of these perovskite oxides. (Druce, J. et al., *Energ. Environ. Sci.*, 7, 3593–3599, 2014. Reproduced by permission of The Royal Society of Chemistry.)

column, the compositions of the topmost atomic layer are completely Sr-dominated, confirming an A'O-terminated surface intuitively. The full coverage of Sr dopant with no visible signal of other cations (La, Fe, or Co) at the outermost surface suggests that the surface chemical composition is completely different from that in the bulk due to Sr segregation. Figure 12.3c also shows LEIS depth profile results of the doping cations ratio A'/(A' + A) and the B-site transition metal cations ratio (B + B')/(A' + A) in the longitudinal direction with high depth resolution (less than 1 nm), reflecting the compositional evolution along the [001] direction.

Using LEIS, Kilner's group found that the tendency of Sr segregation in the LSCF system can increase in the presence of $Ce_{1-x}Gd_xO_{2-0.5x}$.[26] They did systematic work on applying LEIS to study the surface chemistry of perovskite oxide-based materials including $La_{0.6}Sr_{0.4}CoO_{3-\delta}$, confirming that the removal of the water-soluble Sr-rich surface layer can significantly reduce the polarization resistance.[3,19] By LEIS investigation on various $LnBaCo_2O_{5+\delta}$ systems, they declared that surface Ba segregation can take place fast, even at low annealing temperatures.[16,17] As for $PrLaNiO_{4+\delta}$, although it still has an A-rich surface, neither of two A-site cations were found to be preferential at the surface.[25,29] Based on their investigation, the distribution of elements in the near-surface region of these perovskite oxides can be described as shown Figure 12.3d.[1] According to the proposed model, the outermost surface of these three materials is normally governed by A-site cations, while the subsurface is dominated by transition metal ions. Such restructuring is believed to be driven by the segregation of the A-site cations to the surface, leaving a subsurface region lacking A-site cations.

One could further induce the formation of oxide (AO), hydroxide (AOH), or carbonate (ACO_3) secondary phases at the surface and lead to obvious changes in the surface topography. In addition, Kilner's group studied surface reconstruction of $NdGaO_3$ in (110) orientation at various atmospheres by LEIS,[23] as well as the surface chemistry of anisotropic $La_2NiO_{4+\delta}$ in different crystal orientations.[20,28] Chen et al. carried out LEIS depth profiling in $(La_{1-x}Sr_x)_2CoO_4$ systems from [001] and [100] orientation.[15] Using LEIS, Kucernak et al. demonstrated surface La enrichment in rhombohedral and orthorhombic $LaMnO_3$ powder synthesized by glycine combustion.[24] McIntosh et al. carried out LEIS detection on a series of Ruddlesden-Popper oxides (including n = 1, 2, and 3). They found that all the surfaces showed significant Sr segregation, with the Sr/La ratios of RPn2 > RPn3 > RPn1.[27] These results show that LEIS is a quantitative characterizing technology that is of great value for detecting cation segregation in various perovskite systems.

12.2.2 Auger Electron Spectroscopy and Scanning Electron Microscope

Auger electron spectroscopy (AES) is a surface science and materials science analysis technology that applies high-energy electron beam as the excitation source. For AES detecting, the incident electron beam first interacts with the inner electrons and stimulates them to a higher energy level. These electrons then return to the inner orbit accompanied by a release of energy. The energy may be emitted in the form of X-rays, or excite others to be free electrons (which are Auger electrons). These electrons are then directed into an electron multiplier for analysis.

By combining AES with depth profiling, Langell et al. characterized the composition of La_2NiMnO_6 double perovskite with vertical resolution.[42] Note that the result from AES is not as precise as that given by LEIS, as AES can merely probe an average of several atomic layers.[21]

Scanning electron microscope (SEM) is an approach that produces electronic images of a sample by scanning it with a focused electrons beam. It can apply a variety of physical signals on the sample for comprehensive analysis, and it can directly observe the larger sample wide range of magnification and depth of field characteristics. Utilizing SEM, Malzbender et al. found the increase of grain sizes and surface roughness in $La_{0.58}Sr_{0.4}Co_{0.2}Fe_{0.8}O_{3-\delta}$ thin films, which are caused by Sr segregation.[43] In addition, SEM is also widely applied in probing the exsolved metallic nanoparticles at the surface of perovskites in reducing atmospheres.[7,8,12]

The combination of AES and SEM provides information of the elemental distribution accompanied with surface topography in the exact region, which is extremely practical for the study of cation segregation at the surface. A representative work involving surface AES and SEM characterization of surface inhomogeneities on $La_{0.6}Sr_{0.4}CoO_{3-\delta}$ (LSC) was carried out by Yildiz's group.[31] Figure 12.4 shows the SEM results and (a) the composition from Auger electron spectroscopy for (b) Sr, (c) La, and (d) Co on LSC films after annealing. It is clear that Sr mainly concentrates on the surface particles, while La is enriched in the remaining surface regions that are particle-free, which is in agreement with the model (Figure 12.3d) proposed by Kilner et al.[1]

Using AES and SEM, Yildiz's group did systematic investigations on perovskite systems such as (La, Sr)MnO_3, (La, Sr)MnO_3, (La, Ba)MnO_3,[45,76] and RP-structured $(La_{1-x}Sr_x)_2CoO_4$,[15] as well as the heterointerfaces between $(La_{1-x}Sr_x)CoO_3$ (LSC_{113}) and $(La_{1-x}Sr_x)_2CoO_4$ (LSC_{214}).[77] In addition to their group, Shao-Horn's group conducted AES and SEM analysis on $La_{0.8}Sr_{0.2}CoO_{3-\delta}$, $La_{0.8}Sr_{0.2}MnO_{3-\delta}$ decorated $La_{0.8}Sr_{0.2}CoO_{3-\delta}$, and $(La_{0.5}Sr_{0.5})_2CoO_{4+\delta}$ decorated $La_{0.8}Sr_{0.2}CoO_{3-\delta}$ thin films to distinguish the compositional difference between surface particles and particle-free region.[47] With temperature-dependent AES, Salvador et al. observed the evolution of the surface cation ratio at elevated temperatures on $La_{0.7}Sr_{0.3}MnO_3$ thin-film.[48] In summary, AES and SEM is also a commonly used and intuitive technology for detecting surface elemental distribution with lateral resolution, which is particularly practical in characterizing cation enrichment and phase separation at the surface.

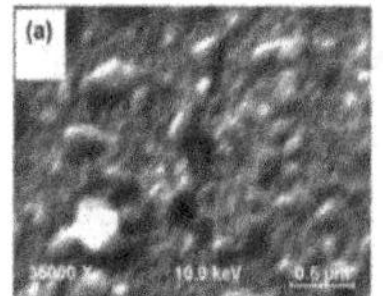

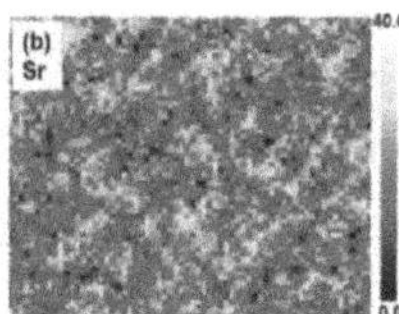

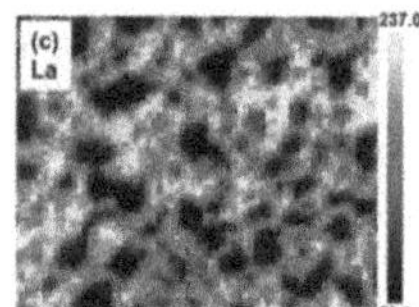

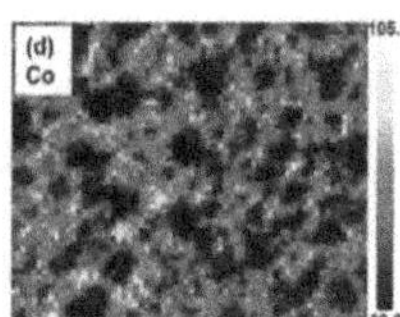

FIGURE 12.4 (a) SEM image and the composition from Auger electron spectroscopy for (b) Sr, (c) La, and (d) Co for LSC films after annealing. (Reprinted with permission from Cai, Z. et al., *Chem. Mater.*, 24, 1116–1127, 2012. Copyright 2012 American Chemical Society.)

12.2.3 X-Ray Photoelectron Spectroscopy

X-ray photoelectron spectroscopy (XPS), based on photoelectric effect, is a substantially nondestructive approach. When X-ray with a certain amount of energy irradiates the surface of a sample, the electrons in the atoms of the substance are released from the atoms into free electrons. Accordingly, different characteristic absorption peaks in the spectrum reflect different types of elements. The information depth of XPS is several nanometers,[78] depending on the categories of the test elements. By changing photo energy or the photoemission angle, XPS can also provide composition information with depth resolution in nondestructive ways. For instance, the evolution of A'/A or A/B ratios were probed in perovskite $La_{0.65}Sr_{0.35}MnO_3$/$La_{0.65}Pb_{0.35}MnO_3$ and RP-structured $(La_{1-x}Sr_x)_2CoO_4$, respectively.[15,31,32] In addition to compositional change, XPS has the capacity to determine the evolution of the chemical environment in the vertical direction according to the shift of binding energy. Representative work was done on $La_{1-x}Sr_xCoO_3$ (LSC_{113}) systems by Yildiz's and Shao-Horn's group,[31,37] as shown in Figure 12.5.

In LSC systems, the peak position of $Sr_{surface}$ (or surface-bound Sr, which includes the SrO-terminated surface or some SrO-like components) is slightly higher than that of $Sr_{lattice}$ (or lattice-bound Sr, which is in the perovskite lattice), indicating the slightly different 3d features for two forms of Sr. Note that there are double peaks ($3d_{3/2}$ and $3d_{5/2}$) for both $Sr_{surface}$ and $Sr_{lattice}$ 3d states, which is caused by the coupling of the orbital motion and spin motion. Figure 12.5 shows that the ratio of $Sr_{lattice}/Sr_{surface}$ decreases from surface to bulk by both approaches, which is circumstantial evidence for the formation of the secondary phase at the outermost surface. The chemical environment of surface-bound O and lattice-bound O can also be distinguished by analyzing O 1s center.[37] In addition, the valence change

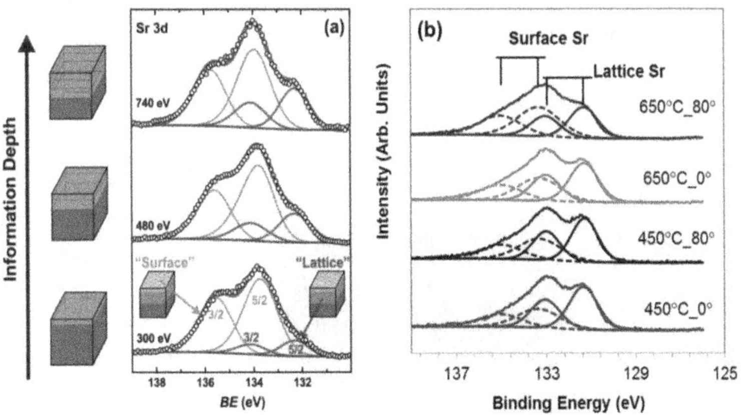

FIGURE 12.5 XPS results at different photon energies or emission angles provide information from the surface to varying depths into the LSC films: (a) Spectra as a function of incident X-ray energy, (b) Sr 3d region of the photoelectron spectra on 450°C and 650°C at emission angles of 0° and 80°. (Reprinted with permission from Cai, Z. et al., *Chem. Mater.*, 24, 1116–1127, 2012. Copyright 2012 American Chemical Society; Crumlin, E.J. et al., *Energ. Environ. Sci.*, 5, 6081–6088, 2012. Reproduced by permission of The Royal Society of Chemistry.)

of B-site cations is also detectable by XPS.[39] For example, the oxidation state of Co can be determined by its L_3-edge position or the presence of satellite peaks in the spectrum.[30,41] For B-site segregation, the exsolution of transition metal substance could be also observed by XPS.[11]

In conclusion, XPS is a very common means for probing surface elemental composition as well as the chemical environment. Nevertheless, conventional XPS detection is typically implemented at room temperature in a vacuum, which is far from the operating conditions for SOCs. In Section 12.2.8, we will further review the recent developments in the use of in situ XPS to characterize surface chemistry at elevated temperatures and/or air conditions, or under electrochemical polarization, where the origin of cation segregation in electrode materials can be revealed more precisely.

12.2.4 Scanning Transmission Electron Microscope and Energy-Dispersive X-Ray Spectroscopy

Scanning transmission electron microscope (STEM) is a type of transmission electron microscope that provides intuitive information of surface morphology and chemical composition. Energy-dispersive X-ray spectroscopy (EDS/EDX) is an analytical technique used for the determination of the relative content of elements. Combining these two detection techniques gives vertical distribution of the relative content of different ions. With STEM and EDS, Chen et al. characterized the structure and chemistry of the Sr-rich particles at the segregated surfaces of RP-related $(La_{1-x}Sr_x)_2CoO_4$ with atomic resolution.[15] The combined results of STEM and EDS are given in Figure 12.6, which shows the distribution of elements in the longitudinal direction.

EDS and STEM can also be carried out in the lateral direction to provide an elemental distribution mapping, which is a little similar to probing by AES and SEM. Thus, it is also an effective approach for measuring surface inhomogeneities caused by cation segregation.[10,44] Using this technology, Jiang et al. confirmed that surface A-site segregation is accompanied by chromium deposition and poisoning in (La, Sr)(Co, Fe)O_3 and $Ba_{0.5}Sr_{0.5}Co_{0.8}Fe_{0.2}O_{3-\delta}$ systems.[62–65] Specifically, this approach also

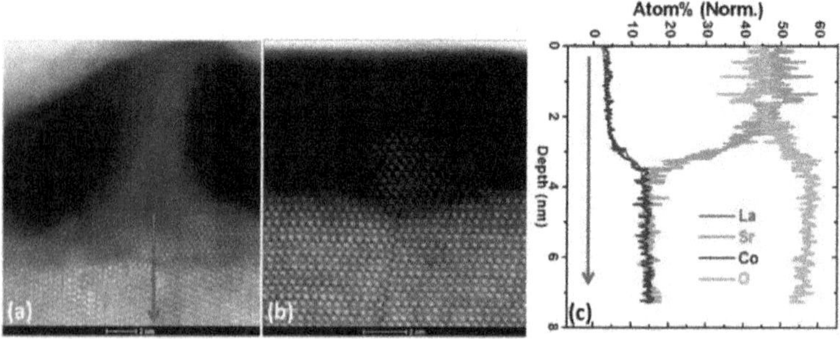

FIGURE 12.6 The results of STEM along (a) [100] and (b) [001] orientations, as well as EDS composition profile along (c) [100] in $(La_{1-x}Sr_x)_2CoO_4$. (Reprinted with permission from Chen, Y. et al. *Chem. Mater.*, 27, 5436–5450, 2015. Copyright 2015 American Chemical Society.)

shows significant advantages in characterizing the island-distributed transition metal substance that is in situ exsolved from the initial perovskite-based lattice.[7,8,11,12]

12.2.5 Secondary Ion Mass Spectroscopy

Secondary ion mass spectroscopy (SIMS) is a highly sensitive surface analysis technology with high vertical resolution (several atomic layers).[51,53,54] For SIMS static analysis, a focused primary ion beam is applied on the sample for a stable bombardment, which can transfer its energy to the lattice atoms, excite them to move to the surface, and lead to surface ionization. The ionized particles (including sputtered atoms, molecules, and radicals) are then separated by mass-to-charge ratio. Collecting the secondary ions separated by mass spectrometry, the element composition and distribution of the sample at various depth can be detected.

With SIMS, Fleig et al. probed the surface composition of as-prepared/annealed $La_{0.6}Sr_{0.4}CoO_{3-\delta}$ deposited by pulsed laser deposition (PLD) at various temperatures (Figure 12.7),[9] and Lacorre et al. detected the vertical distribution of cations as well as segregation behaviors at the interface of $La_2Mo_2O_9$ (electrolyte) and $La_{0.8}Sr_{0.2}MnO_{3-\delta}$ (electrode).[60] Both indicated visible Sr segregation at the surface or interface. It could also be inferred that the segregated components can be removed by acid treatments.

In addition, SIMS has the specific sensibility for distinguishing isotopes. Using SIMS, Fleig et al. investigated Sr diffusion behaviors in $La_{0.6}Sr_{0.4}CoO_{3-\delta}$ (LSC) thin films by isotope tracer.[58] They found that Sr migration intensifies significantly as the temperature increases. Combining ^{18}O isotope exchange depth profiling with ToF-SIMS has become an effective approach to measure the surface exchange coefficient (k) and the tracer diffusion coefficient (D) for SOC electrode materials.[52,55–57]

In summary, SIMS, together with depth profiling, provides a measure that can detect the detailed composition of a material layer-by-layer. This is a little similar to LEIS, but SIMS is also able to give a much broader compositional survey over much larger depth ranges than are available from LEIS.[55] Besides depth profiling, SIMS also has the capacity to provide lateral mapping of individual elements with micron-level resolution.[53,54]

12.2.6 Atomic Force Microscope and Scanning Tunneling Microscope

Atomic-force microscopy (AFM) is a type of scanning probe microscopy used typically for detecting the surface morphologies of materials. It detects heterogeneous surface with a resolution on the nanoscale. AFM has no special requirements for the sample, making it a practical technology to characterize cation segregation and separation of surface particles.[15,44,45,68] Using AFM, Yildiz et al. obtained the amplitude images in Ba-, Sr-, and Ca-doped $LaMnO_3$ systems at various temperatures.[45] The volume of surface particles was then calculated as a semi-quantitative indicator to evaluate surface segregation and phase separation behaviors (Figure 12.8). The results suggested the highest dopant segregation level in the $La_{1-x}Ba_xO_{3-\delta}$ system at elevated temperatures.

Scanning tunneling microscopy (STM), based on the tunneling effect of quantum mechanics, is the predecessor of AFM. In contrast to AFM, better conductivity is

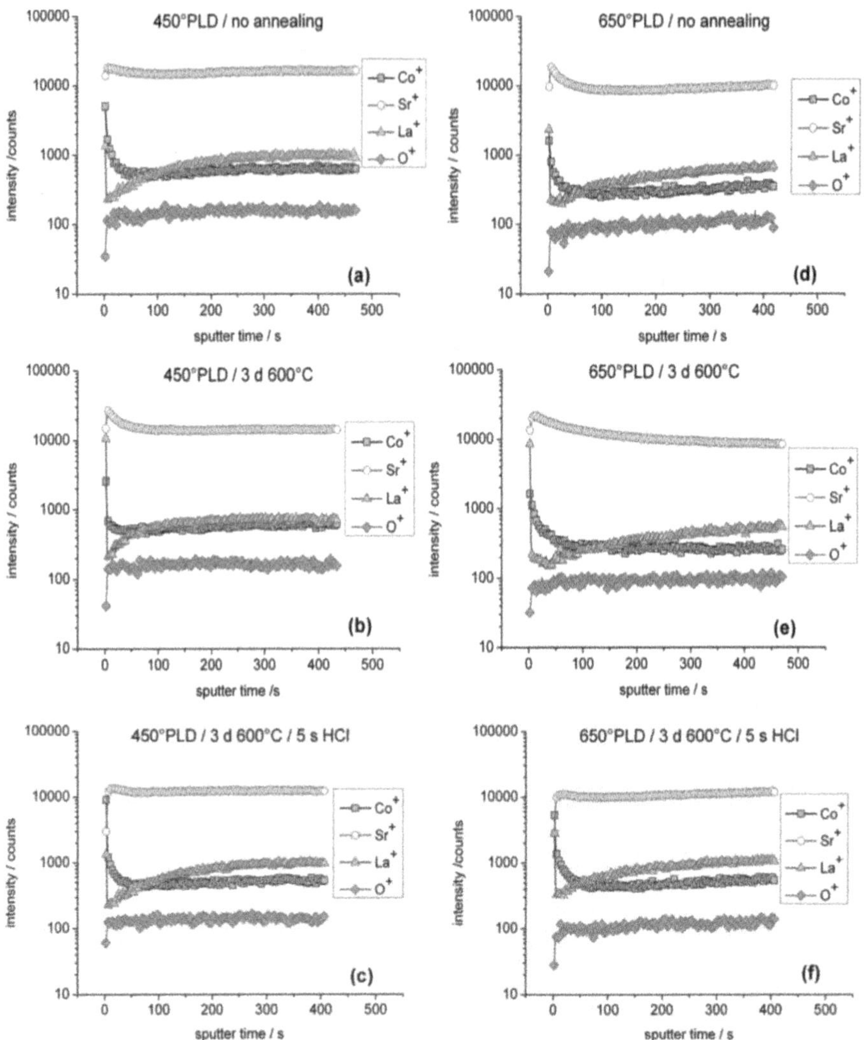

FIGURE 12.7 Positive secondary ion signals of different LSC films prepared at 450°C (left side) and 650°C (right side), respectively, (a, d) before annealing (b, e), after three days of annealing, and (c, f) as well as after such annealing and subsequent chemical etching. (Reprinted with permission from Kubicek, M. et al., *J Electrochem. Soc.*, 6, B727–B734, 2011. With permission from ECS, Copyright 2008.)

required for the detected samples to implement STM test. However, it can provide both surface microscopy and local electronic structure information by linking with scanning tunneling spectroscopy (STS). These features exactly meet the requirements for investigating the electrode materials for SOCs. Using STM/STS, Bonnell et al. studied the surface structures and chemistry of the annealed $SrTiO_3$ (001).[70] They identified the particles and particle-free regions of annealed $SrTiO_3$ to be Sr-rich islands and $SrTiO_3$ surface, respectively, based on their different electronic structures (Figure 12.9).

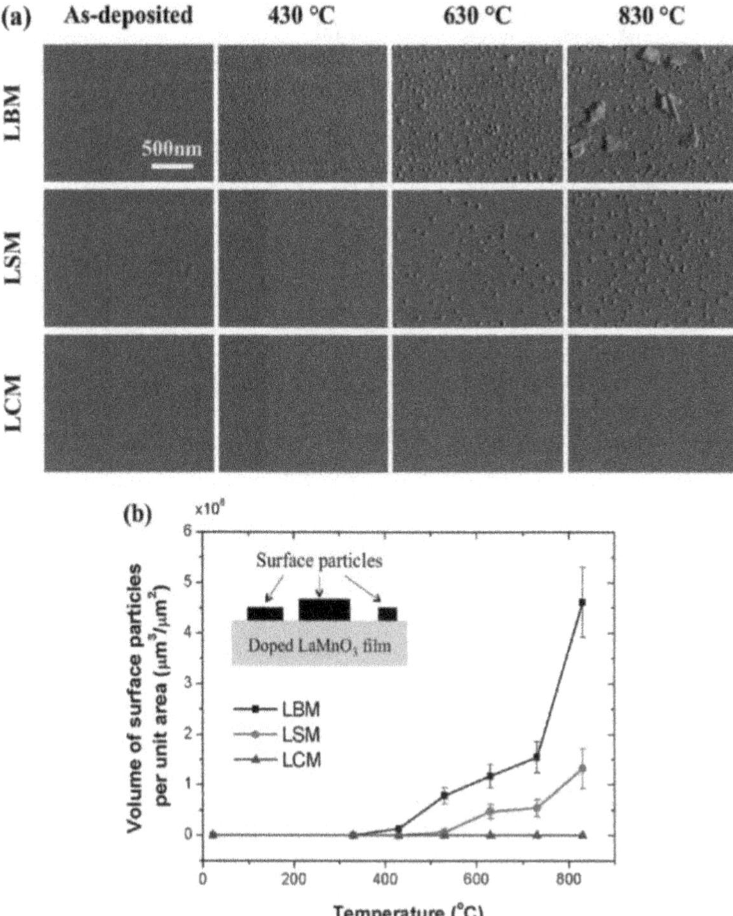

FIGURE 12.8 (a) AFM amplitude images of the Ba-, Sr-, and Ca-doped $LaMnO_3$ thin film surfaces as a function of annealing temperature, (b) the volume per unit area of the surface particles as a function of annealing temperature for doped $LaMnO_3$. (Reprinted with permission from Lee, W. et al., *J. Am. Chem. Soc.*, 135, 7909–7925, 2013. Copyright 2013 American Chemical Society.)

For the test of SOC electrode materials, only the results observed at elevated temperatures can effectively reflect operating conditions of the actual situation because the electronic structure changes reversibly with temperature. A lot of information will be missed if the tested is done at room temperature. Many researchers have done excellent work on perovskite-based materials using STM/STS. For example, Salvador et al. observed a decrease in tunneling conductance in STS on the Sr-rich surface of $La_{1-x}Sr_xMnO_3$ (LSM), suggesting that the surface secondary phase was more inactive for electron transfer compared to the perovskite component.[48] Similarly, Yildiz et al. reported that the surface chemical heterogeneities on dropped LSM were

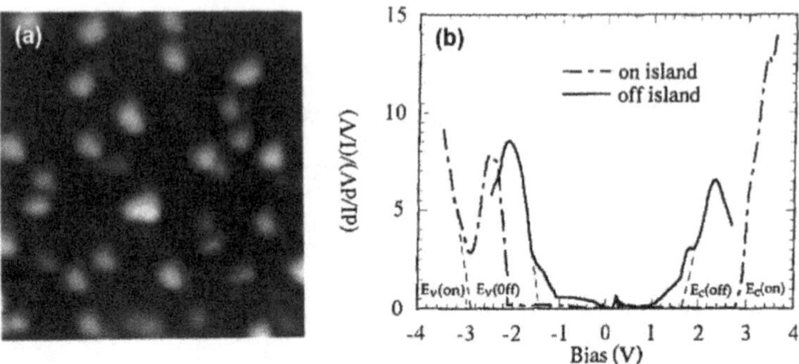

FIGURE 12.9 (a) Surface morphologies of $SrTiO_3$ films, (b) the electronic structure difference between the areas on island and areas off island. (Reprinted from *Surf. Sci.*, 310, Liang, Y. and Bonnell, D.A., Structures and chemistry of the annealed $SrTiO_3(001)$ surface, 128–134, Copyright 1994, with permission from Elsevier.)

electronically insulating.[45] They also investigated the influence of tensile or compressive strain on several perovskite-based $La_{0.7}Sr_{0.3}MnO_3$ and $La_{0.8}Sr_{0.2}CoO_3$ systems.[41,69] They found that the tensile strained surfaces coupling to less Sr segregation exhibit a more enhanced electronic density of states (DOSs) near the Fermi level following this transition, indicating a more highly active surface for electron transfer in oxygen reduction. In addition, by systematic research in $SrTi_{1-x}Fe_xO_3$ systems, they proposed that surface Sr segregation was accompanied by a downshift of the O 2p state relative to the Fermi level as well as a decreased amount of available oxygen vacancies, indicating blocked activities caused by segregation.[71]

12.2.7 X-Ray Diffraction

X-ray diffraction (XRD) is one of the primary technologies used to examine the physicochemical makeup and to detect lattice parameters of unknown crystals. With respect to segregation, XRD has the capacity to characterize the formation of several secondary phases generating at the surface. For instance, Bae et al. observed SrO and/or BaO segregation along with the generation of $SrCrO_4$ at the surface of $Ba_{0.5}Sr_{0.5}Co_{0.8}Fe_{0.2}O_{3-\delta}$ (BSCF),[62] Lacorre et al. also detected the formation of $Sr_{1-x}La_{2x/3}MoO_4$ scheelite-type phases at the surface induced by Sr segregation of $La_2Mo_2O_9/La_{0.8}Sr_{0.2}MnO_{3-\delta}$ hetero systems.[59]

In addition, XRD can probe for the presence of surface metal nanoparticles. For example, Han et al. observed surface nano-sized Co-Fe alloy exsolution on the $Pr_{0.8}Sr_{1.2}(Co, Fe)_{0.8}Nb_{0.2}O_{4+\delta}$ matrix;[72] Świerczek et al. confirmed the formation of Fe_3Ni alloy at the surface of $Sr_2FeMo_{0.65}Ni_{0.35}O_{6-\delta}$ after reduction at 850°C in pure H_2;[11] and Luo et al. demonstrated that, after exposure to an oxidizing atmosphere, the Ni nanoparticles on Ni-doped $(La_{0.7}Sr_{0.3})CrO_3$ perovskite could transform into perovskite again without generating impurities, indicating the reversible segregation of transition metals during redox cycling.[8]

12.2.8 State-of-the-Art Characterization Methods for Surface Segregation

To investigate the origin and impact of surface cation segregation (and some related surface processes) at near operating conditions, some state-of-the-art characterization methods under operando conditions have been developed. In this section, we will introduce the principle and capacity of several in situ methods, as well as their applications in characterization surface cation segregation.

12.2.8.1 In Situ XPS

Generally, when a sample excited by an electron beam, such as during LEIS or AES detection, it is necessary to use an ultrahigh vacuum to prevent the formation of carbon deposits on the sample and mask the surface to be measured. However, the softer feature of X-rays makes it possible to observe the surface for several hours at moderate vacuum levels without affecting the test results. Significant progress for XPS is the development of in situ XPS, which can provide real-time information in the observation of dynamics over a wide range of time scales and higher spatial resolution.[78] Such capability can facilitate the study of how surfaces evolve under the working condition of SOC and other energy devices. Figure 12.10 shows the ambient pressure XPS (a category of in situ XPS).[35]

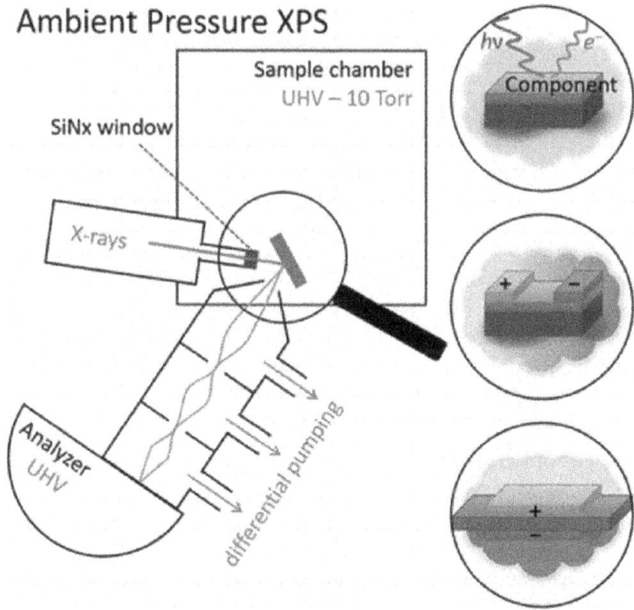

FIGURE 12.10 Schematic of the ambient pressure X-ray photoelectron spectroscopy. (Reprinted with permission from Stoerzinger, K.A. et al., *Acc. Chem. Res.*, 48, 2976–2983, 2015. Copyright 2015 American Chemical Society.)

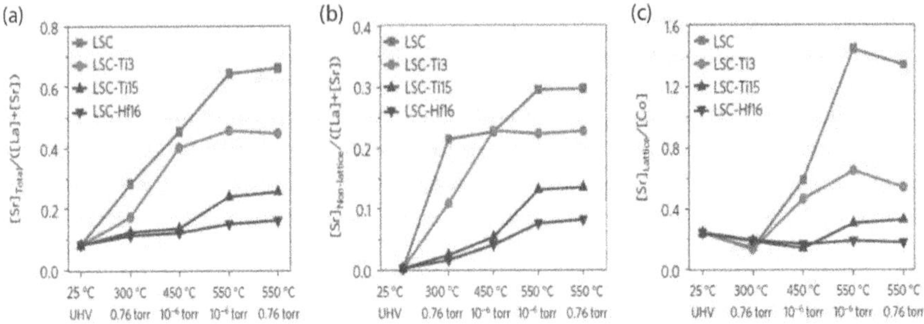

FIGURE 12.11 Surface chemical stability on LSC dense thin films. Concentration ratios $[Sr]_{Total}/([La]+[Sr])$ (a), $[Sr]_{Non\text{-}lattice}/([La]+[Sr])$ (b), and $[Sr]_{Lattice}/[Co]$ (c) at the surface of LSC and three modified LSC thin films were measured at different temperatures and oxygen partial pressures. (Reprinted by permission from Macmillan Publishers Ltd. *Nat. Mater.*, Tsvetkov, N. et al., 2016, Copyright 2016.)

In situ XPS can be applied to characterize the nature of a surface at elevated temperatures and oxygen partial pressures, which is closer to the actual operating conditions for SOCs. Using this technology, Yildiz et al. detected surface chemistry in several modified LSC films.[44] Figure 12.11 shows four films that exhibit a higher Sr ratio or $Sr_{surface}$ ratio as the temperature or oxygen partial pressure increases, suggesting that cation segregation can be more serious in the near operating conditions for SOCs. Shao-horn et al. detected the alternation of surface Sr 3d feature (128 eV to 140 eV in the spectrum) in $La_{1-x}Sr_xCoO_{3-\delta}$ perovskite thin films in several heating and cooling cycles, suggesting that the evolution of the surface structure and chemistry is a function of temperature.[35,47] They also observed the presence of absorbed oxygen on $La_{0.8}Sr_{0.2}CoO_{3-\delta}$ films, whose binding energy (approximately 538 eV and 539.5 eV) corresponds to molecular O_2 in the air.[37] With the use of in situ XPS, the evolution of $Sr_{surface}$ (approximately 133.9 eV), $Sr_{lattice}$ (approximately 138 eV), $O_{surface}$ (approximately 531.9 eV), and $O_{lattice}$ (approximately 529.9 eV) on $Sr_2Co_2O_5$ thin films were probed, respectively, in the range of 300°C to 600°C.[36] In addition, they carried out their in situ detections under extrinsic electrical potentials (0 to 800 mV) at high temperatures (approximately 500°C to 800°C), and they observed shifts of the top of the valence band binding energy and changes of the Sr 3d and O 1s spectral components under applied bias.[34] Using the same approach, Fleig et al. investigated oxidation state changes of transition metals as well as water-splitting kinetics under extrinsic electrochemical polarization.[33,40]

12.2.8.2 Environmental Transmission Electron Microscopy

Recent development of environmental transmission electron microscopy (ETEM) allowed researchers to probe the dynamics of structure change for catalysis at elevated

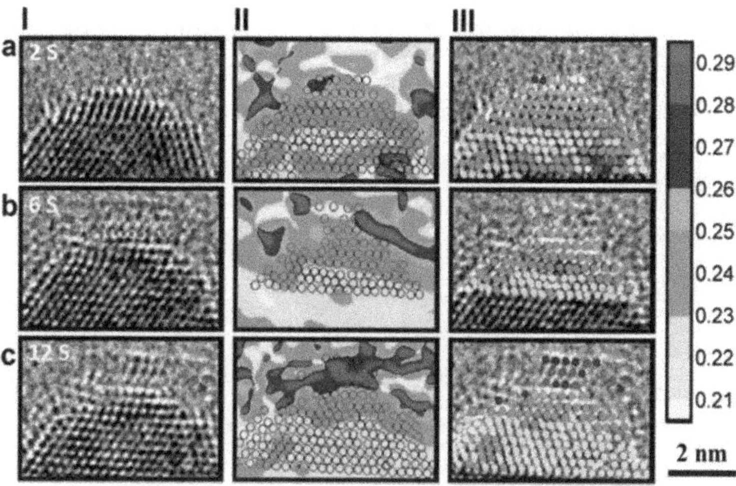

FIGURE 12.12 Tracking of Co segregation in $Pt_{0.5}Co_{0.5}$ nanoparticles under an oxidizing environment, characterized by ETEM in 0.1 mbar O_2 at 250°C. A series of images acquired at (a) 2 s, (b) 6 s, (c) 12 s in situ during oxidation. The atomic displacement maps of the nanoparticle (I), the atomic displacement maps of the nanoparticle with color scale (II and III). (Reprinted with permission from Xin, H.L. et al., *Nano Lett.*, 14, 3203–3207, 2014. Copyright 2014 American Chemical Society.)

temperatures and in a gas enviroment.[79] Xin et al. observed the fast segregation of Co in Pt-Co bimetallic nanoparticles in oxidizing environments (see Figure 12.12).[80] Han et al. observed nanoscale structural oscillations during an oxygen evolution reaction (OER) process within BSCF particles using ETEM.[81]

12.2.8.3 In Situ SIMS

Huber et al. reported that $La_{0.75}Sr_{0.25}Cr_{0.5}Mn_{0.5}O_{3\pm\delta}$ (LSCrM) exhibited pronounced performance improvement after cathodic polarization.[50,60] To understand the origin of the enhanced activity, they studied the surfaces of operating LSCrM electrodes with in situ SIMS after applying different electrical potentials in the chamber. They found that, after annealing, the electrode surfaces were significantly rich in Sr. Nevertheless, subsequent cathodic polarization decreases surface Sr concentration, while anodic polarization increases Sr accumulation at the electrode surface.

According to Figure 12.13, the LSCrM surface becomes depleted in Sr and Mn, and both elements diffuse into the bulk of the electrode and onto the electrolyte surface. The diffusion of the different species under cathodic bias is accompanied by an improvement of the electrocatalytic activity towards the oxygen reduction reaction by impedance spectroscopy. This in situ study addresses the mechanism of this activation process as well as reveals the relationship between surface cation segregation and electrode performance.

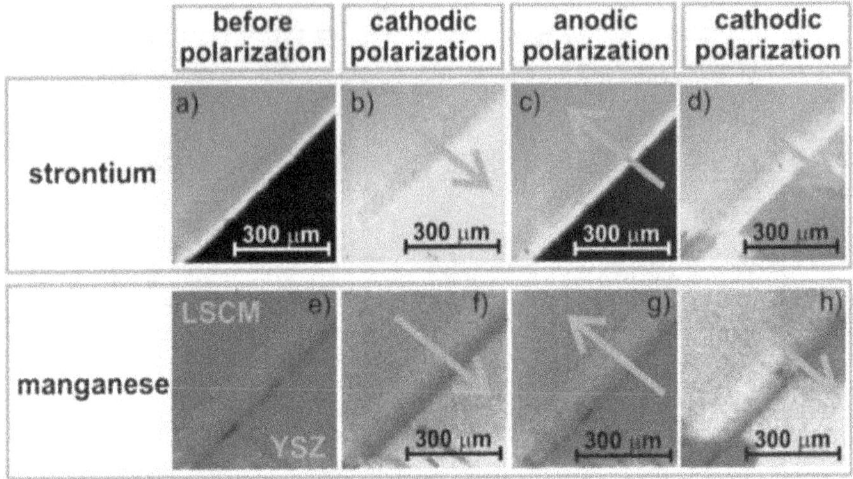

FIGURE 12.13 *In situ* SIMS images of the interface between LSCrM and YSZ at different applied bias measured at 10^{-6} mbar oxygen partial pressure and 500°C. The signal of strontium: before polarization (a), under cathodic polarization (b, d) or anodic bias (c); the signal of manganese: before polarization (e), under cathodic polarization (f, h) or anodic bias. (Reprinted with permission from Huber, A. et al., *Phys. Chem. Chem. Phys.*, 14, 751–758, 2012. Copyright Owner Societies 2012.)

12.2.8.4 Resonant Soft X-Ray Reflectivity and Resonant Anomalous X-Ray Reflectivity

Resonant soft x-ray reflectivity (RXR) is an element-specific and nondestructive way to study electronic properties and phenomena at surfaces and buried interfaces. (The technique was proposed by Geck et al.[82]) Similarly, resonant anomalous X-ray reflectivity (RAXR) is a powerful technique for measuring element-specific distribution profiles across surfaces and buried interfaces.[83] From the measured RAXR spectra, Perret et al. determined the Sr profile through a 3.5 nm-thick $La_{0.6}Sr_{0.4}Co_{0.2}Fe_{0.8}O_{3-\delta}$ film at 973 K and at an oxygen partial pressure of 150 Torr (Figure 12.14a).[84] According to the measured scatters, information of element distribution at each layer could be fitted (Figure 12.14b), which could be further revised by crystal truncation rod (Figure 12.14c). The results imply that temperature-dependent RAXR measurements can provide atomic scale insights into temperature- and environment-dependent elemental segregation processes.

12.2.8.5 Raman Spectroscopy and In Situ Raman Spectroscopy

Raman spectroscopy is a technique ideally suited for gathering surface information due to its sensitivity to a vast variety of chemical species. Using this method, Jiang et al. investigated the relationship between the surface Sr segregation, Cr deposition, and poisoning in $La_{0.6}Sr_{0.4}Co_{0.2}Fe_{0.8}O_{3-\delta}$ electrodes.[85]

In situ raman spectroscopy can be mainly used to characterize the contamination of impurities for SOC electrodes, such as SiO_2 or carbon deposition.[86,87] The results can reflect the long-term durability for the electrode of SOCs. Using in situ raman

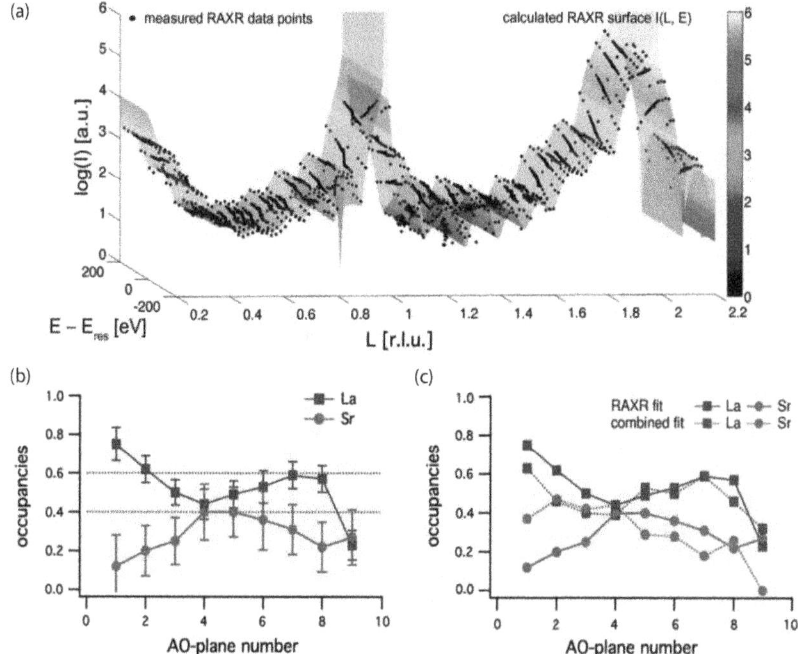

FIGURE 12.14 (a) Fit of all measured RAXR data points at 973 K and pO_2 = 150 Torr, (b) Sr and La occupancies retrieved through fits of RAXR data, (c) a comparison of RAXR fitted occupancies (solid lines) and results from the combined fit of a non-resonant and a resonant crystal truncation rod (dotted lines). (Reprinted with permission from Perret, E. et al., *J. Appl. Cryst.*, 46, 76–87, 2013. Copyright International Union of Crystallography 2013.)

spectroscopy, Michael et al. found that the majority products of butane pyrolysissmethane, ethylene, and propylenes form a variety of carbon deposits that affect Ni/YSZ anode performance in various ways.[88] They also found that the initial stages of carbon formation on Ni/YSZ anodes depend on the chemical nature of the incident fuel and that both the type of carbon formed (ordered or disordered) and the amount of carbon deposited are important when considering the impact of graphite formation on solid oxide fuel cell (SOFC) electrochemical performance (Figure 12.15).

12.2.8.6 In Situ X-Ray Fluorescence

In-situ X-ray fluorescence (XRF) is a surface-sensitive, nondestructive technique used for elemental analysis on the surface of materials.[89] Using in situ X-ray fluorescence, Fister et al. found evidence of reversible surface segregation in (001)-oriented $La_{0.7}Sr_{0.3}MnO_3$ thin films over a wide range of temperatures (25°C–900°C) and oxygen partial pressures (pO_2 = 0.15–150 Torr), which are closer to the real working environment of SOCs (Figure 12.16).[90] The Sr surface concentration is observed to increase with decreasing pO_2, suggesting that the surface oxygen vacancy concentration plays a significant role in controlling the degree of segregation. The decrease in segregation with temperature and increased pO_2 may imply that Sr segregation is driven by a buildup of oxygen vacancy-Sr interactions near the surface.

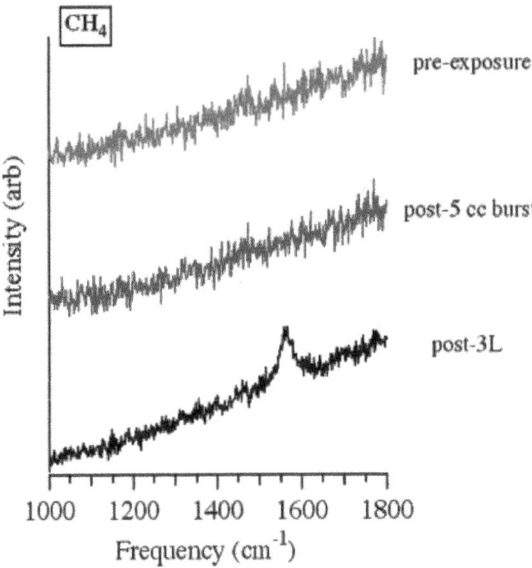

FIGURE 12.15 Raman spectra acquired from a Ni/YSZ porous anode prior to exposure to CH_4 (top), after exposure to 5 cm^3 of CH_4 (middle), and after exposure to 3 L of CH_4 (bottom). All measurements were made at OCV and at a temperature of 715°C. (Reprinted with permission from Pomfret, M.B. et al., *J. Phys. Chem. C*, 112, 5232–5240, 2008. Copyright 2014 American Chemical Society.)

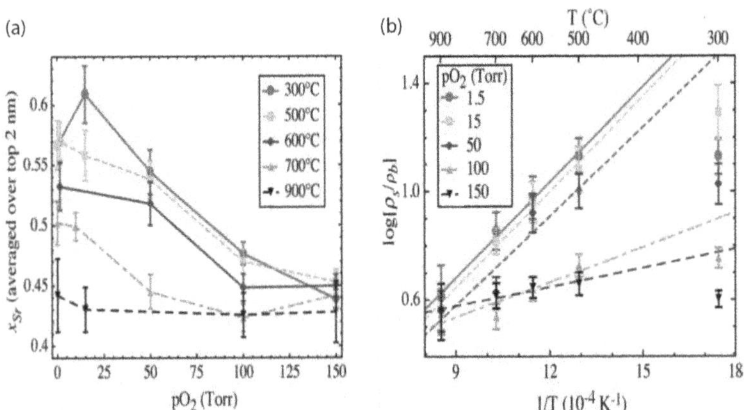

FIGURE 12.16 (a) Sr surface region concentration and oxygen partial pressure at T = 300°C–900°C, (b) Sr surface enrichment as a function inverse temperature for a variety of oxygen partial pressures. (Reprinted with permission from Fister, T.T. et al., *Appl. Phys. Lett.*, 93, 151904, 2008. Copyright 2008 by the American Institute of Physics.)

12.3 FACTORS INFLUENCING SEGREGATION LEVEL IN PEROVSKITE-BASED OXIDES

Cation segregation is a heterogeneous process that is affected by a combination of many factors, including the nature of the materials (e.g., cation non-stoichiometry, dopant species, crystallinity, lattice strain, etc.) and environmental conditions (e.g., gas environment, temperature and thermal history, electrical polarization, etc.), as summarized in Figure 12.17 and Table 12.2. In this section, we will emphasize the specific factors that may influence the degree of cation segregation in perovskite-based oxides and try to provide a systematic mechanism of cation segregation.

12.3.1 The Effect of Cation Non-Stoichiometry

For perovskite-based materials, cation deficiency has profound influences on electrode performances as well as segregation behaviors.[122] A segregation process

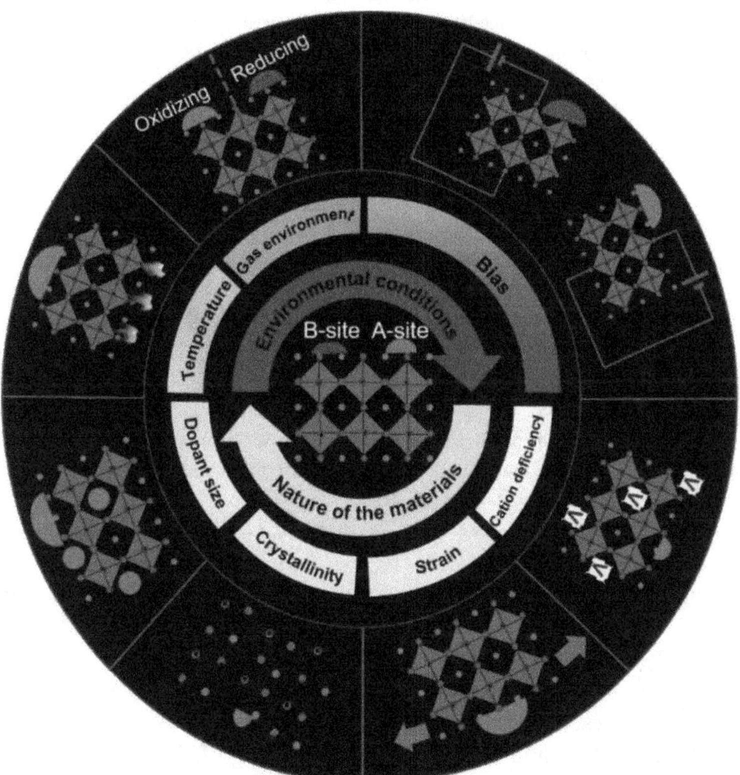

FIGURE 12.17 Factors influencing the degree of cation segregation in perovskite oxides: cation deficiency, dopant size, crystallinity, lattice strain, gas environment, temperature, and electrical polarization. The green balls represent A-site cations, the red balls represent B-site cations, and the size of segregated species reflects the tendency of cation segregation. (Li, Y., et al., *Chem. Soc. Rev.*, 46, 6345–6378, 2017. Reproduced by permission of The Royal Society of Chemistry.)

TABLE 12.2
Factors Influencing Cation Segregation Behavior in Perovskite-Based Systems

Influencing Factors	Impacts on Cation Segregation and Possible Reasons	Related Studies
Cation non-stoichiometry	A-site deficiency suppresses A-site segregation by allowing for more free space in the lattice and reducing elastic energy, promoting B-site segregation.	68,91–95
	B-site deficiency typically promotes A-site segregation.	93
Cation species	Cations with larger radii tend to segregate to the surface to reduce elastic energy in perovskite systems.	45,92,96–100
Crystallinity	High crystallinity degree leads to a more orderly arrangement and less free space in the lattice, driving the larger cations to the surface.	9,30
Lattice strain	Materials with more tensile surface (relative to bulk) can accommodate more cations with larger radii, thus promoting surface cation segregation.	43,69,101–105
Temperature and thermal history	Cation segregation is significantly favored at elevated temperatures and depends on annealing time.	6,17,41,44,47,48,58,102,106–108
Gas environment	Reducing atmosphere: transition metal ions exhibit low oxidation states with larger sizes, leading to B-site segregation.	6,7,11,12,92,109
	Oxidizing atmospheres: B-site cations exhibit high oxidation states with smaller sizes, driving A-site cations with fewer positive charges and greater radii towards the surface.	45,90,102,110
	Reactive gas environment typically reacts with segregated cations, driving more A-site cations to the surface.	44,85,98,111–113
Electrical polarization	Cathodic polarization generally suppresses A-site surface concentrations, while anodic polarization facilitates A-site accumulation.	13,33,40,50,60,114–121

involves cation migration towards the surface, so both A- or B-site deficiency can be expected to suppress the segregation tendency of the corresponding cations, respectively. By investigating cation segregation behaviors in several $(La_{0.8}Sr_{0.2})_xMnO_3$ model systems (x = 0.95, 1, and 1.05, corresponding to A-deficient, non-deficient, and B-deficient states, respectively), Lu et al. confirmed that surface Sr enrichment was decreased as a function of x after annealing in 500°C for 5 h.[93] Smoother surface morphology, lower Sr/(Sr+La) ratio, and lower $Sr_{lattice}/Sr_{surface}$ ratio were also observed in A-deficient $(La_{0.8}Sr_{0.2})_{0.95}MnO_{3-\delta}$ system, according to Yildiz et al., Gerdes et al., and Ludwig et al.[68,94,95] Thus, it can be assumed that A-site deficiency in these perovskites would suppress A-site segregation by allowing for more free space in the lattice and reducing elastic energy.

To systematically investigate the relationship between B-site non-stoichiometry and its segregation behaviors, Irvine et al. synthesized perovskite-structured

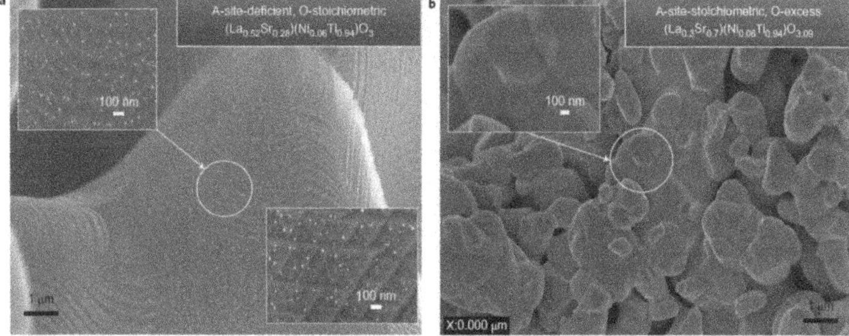

FIGURE 12.18 Comparison of (a) B-site segregation behaviors in initial A-deficient $La_{0.52}Sr_{0.28}Ni_{0.06}Ti_{0.94}O_3$ and (b) A-stoichiometric $La_{0.3}Sr_{0.7}Ni_{0.06}Ti_{0.94}O_{3.09}$ sample after reducing at 930°C in 5% H_2/Ar for 20 h. (Reprinted by permission from Macmillan Publishers Ltd. *Nat. Chem.*, Neagu, D. et al., 2013, Copyright 2013.)

$(La_{1-x}Sr_x)_yNi_{0.06}Ti_{0.94}O_{3+\delta}$ materials with or without B-site enrichment, which showed different surface chemistry after annealing at high temperatures in reducing atmospheres.[91] As shown in Figure 12.18, metal nanoparticles were presented on the surface of A-deficient (and B-rich) $La_{0.52}Sr_{0.28}Ni_{0.06}Ti_{0.94}O_3$, and no obvious surface particles were found on the surface of A-stoichiometric $La_{0.3}Sr_{0.7}Ni_{0.06}Ti_{0.94}O_{3.09}$. In addition, they also observed the exsolution of transition metal substance in a B-rich $La_{0.4}Sr_{0.4}Ni_{0.06}Ti_{0.94}O_{2.97}$ system.[92] These results demonstrated that a cation ratio above stoichiometry is an important driving force for surface segregation.

12.3.2 The Effect of the Cation Species

Perovskite-based oxides used as SOC electrodes are usually doped with various cations. In these modified perovskite systems, both host cations and dopant cations may segregate towards the surface, depending on their intrinsic nature and environments. It seems that cations with less positive charges and larger radii are more likely to segregate out, such as the widely observed Sr^{2+} segregation in $Sm_{1-x}Sr_xCoO_{3-\delta}$, $La_{1-x}Sr_xMnO_{3-\delta}$, $La_{0.6}Sr_{0.4}Co_yFe_{1-y}O_{3-\delta}$, and Pb^{2+} segregation in $(La, Pb)MnO_{3-\delta}$ (mainly electrostatic dependent), as well as La^{3+} segregation in $La_{2/3}Ca_{1/3}MnO_3$ thin films (mainly elastic dependent).

To investigate the segregation behaviors of various doped perovskites, Ramprasad et al. carried out first-principle-based calculations to study segregation behaviors of A-doped $LaMnO_3$ systems.[98,99] They confirmed that three categories of dopants (Ca, Sr, and Ba) in (La, D)MnO_3 have a tendency to segregate to the surface, while the tendency for dopant cation segregation becomes stronger as a function of increased dopant ionic size (the radius of Ca, Sr, and Ba is 1.34 Å, 1.44 Å, and 1.61 Å, respectively).

Based on AFM, XPS, and AES analysis on several Ca^{2+}, Sr^{2+}, and Ba^{2+} doped $LaMnO_3$ systems (referred to as LCM, LSM, and LBM, respectively), Lee et al.

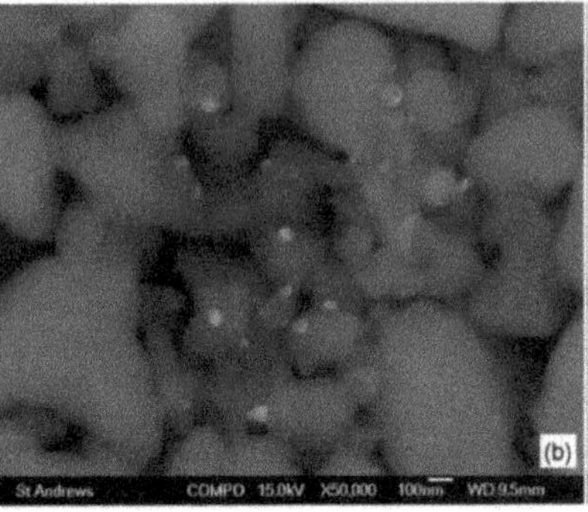

FIGURE 12.19 SEM images of (a) $La_{0.4}Sr_{0.4}Fe_{0.06}Ti_{0.94}O_{2.97}$ and (b) $La_{0.4}Sr_{0.4}Ni_{0.06}Ti_{0.94}O_{2.94}$ cathodes. It is obvious that lager amounts of surface nanoscale particles segregate out in the Ni-doped material. (Tsekouras, G. et al., *Energ. Environ. Sci.*, 6, 256–266, 2013. Reproduced by permission of The Royal Society of Chemistry.)

summarized that a smaller size mismatch between the host and dopant cations could reduce the segregation level of the dopant and suppress the formation of the secondary phase particles at the surfaces,[45] which was consistent with the calculations of Ramprasad et al.

It could be summarized from the discussion above that the lattice mismatch and the electrostatic interaction are the main origins of cation segregation in perovskite-based oxides. When replacing the trivalent A^{3+} with the bivalent D^{2+} in perovskites (which generates negative D'_A defects), more surface oxygen vacancies ($V_O^{\cdot\cdot}$ with positive formal charges) would be formed to reach the electrical neutralization. The electrostatic interaction between these two defects would therefore drive the substitutional D^{2+} cations towards the surface. This is how electrostatic interaction affects cation segregation. The lattice mismatch is caused by the larger ionic radius of D^{2+}, however, than that of A^{3+} ions, pushing the dopant D^{2+} cations to the surface and reducing elastic energy in perovskite systems.

This principle is also effective in the case of B-site segregation. For example, in Ni^{2+} or Fe^{3+}-doped $La_{0.4}Sr_{0.4}D_{0.06}Ti_{0.94}O_{2.97}$ systems (D represents the doped transition metals), Ni^{2+} with larger ion size (the radius of Ni^{2+} is 0.69 Å, while that of Fe^{3+} is 0.645 Å) has a greater tendency to segregate to the surface after annealing at 900°C in wet 5% H_2/Ar atmospheres.[92] Figure 12.19 shows that the Ni^0 surface particles exsolved on the surface of $La_{0.4}Sr_{0.4}Ni_{0.06}Ti_{0.94}O_{2.94}$ were obviously larger (approximately 60 to 90 nm) and more intensive compared to the Fe^0 particles (approximately 30 to 60 nm) exsolved on the surface of $La_{0.4}Sr_{0.4}Fe_{0.06}Ti_{0.94}O_{2.97}$.

12.3.3 The Effect of Crystallinity

Besides stoichiometry and chemical composition factors, high crystallinity also leads to relatively high cation segregation levels. For instance, Kelires et al. reported a significant suppression of phase separation in amorphous Si-Ge alloys.[123] This principle can be the same as that in perovskite-based materials. For doped $La_{0.6}Sr_{0.4}CoO_{3-\delta}$ thin-film electrodes prepared by PLD, the reduced deposition temperature accompanied with low film crystallinity could lead to a strong increase of the electrochemical oxygen exchange rate, as reported by Fleig et al.[9] To develop Fleig et al.'s results further, Cai et al. proposed that the high crystallinity degree could lead to a more orderly arrangement and less free space in the lattice, and finally drive the dopants to the surface.[30] AFM and XPS results also demonstrated that $La_{0.6}Sr_{0.4}CoO_{3-\delta}$ with poor crystallinity exhibited a smaller amount of surface particles and weaker signal of $Sr_{surface}$, suggesting a lower Sr segregation degree. Compared to the fully crystalline perovskite, there are more spaces in the more disordered lattices, which could accommodate a larger amount of Sr^{2+} cations within bulk.

12.3.4 The Effect of Lattice Strain

Strain effect also plays an important role in determining cation segregation level. According to Jalili et al., $La_{0.7}Sr_{0.3}MnO_3$ deposited on a $SrTiO_3$ substrate with tensile strain seems to appear higher Sr segregation levels compared to those deposited on $LaAlO_3$ substrate with compressive strain.[103] It might be attributed to surface expansion that can accommodate more Sr cations with a relatively large radius compared to the bulk. Szot et al. also found massive Sr segregation in $SrTiO_3$ subjected to mechanically tensile strain.[102] In contrast, on a compressive $La_{0.6}Sr_{0.4}Co_{0.2}Fe_{0.8}O_{3-\delta}$ surface, the segregation of much smaller and fewer particles was observed at the grain boundaries compared to the stress-free surface.[43,101]

Lattice strain can also be introduced by doping cations. For instance, partial substitution of A-site La^{3+} in $La_{0.7}Sr_{0.3}MnO_3$ with smaller Gd^{3+} (122 pm for Gd^{3+}, and 136 pm for La^{3+}) was implemented by Park et al.[105] The lattice parameter therefore decreased with increasing Gd content, resulting in less free space for accommodating larger Sr^{2+} in bulk. Therefore, the elastic interactions would drive it to free surfaces to minimize the elastic energy. Using energy resolved XPS, they did confirm that the $Sr_{surface}/Sr_{lattice}$ ratio increased as a function of Gd content. Based on the discussion above, one can see that applying compressive strain by surface coating, mechanical compressing, or suitable cations doping can be an effective way to suppress surface segregation.

12.3.5 The Effect of Temperature and Thermal History

Cation segregation is thought to be influenced by temperature in both kinetics and thermodynamics. Because cation diffusion lengths could be estimated from the expression $\sqrt{Dt}$ (D = diffusivity coefficient, t = time) at a certain temperature, the cation segregation process is therefore temperature- and time-dependent. Using the isotope labeled method, Fleig et al. estimated Sr diffusion kinetics in $La_{0.6}Sr_{0.4}CoO_{3-\delta}$ thin films at a temperatures of 625°C–800°C.[58] Kilner et al. applied LEIS and AFM to investigate the evolution of the surface composition and morphology of several double-perovskite type materials, showing the dynamic chemical environment of these surfaces as a function of annealing time.[16,17] In addition, they found that the outermost surface of $LnBaCo_2O_{5+\delta}$ changed significantly when the samples were subjected to 400°C for a relatively short time. The (Ba+Ln) surface peak increased to an extremely high level in the first 15 min, and then increased relative slowly in the subsequent 8 h. It could be inferred that the increasing concentration of Ba and Ln at the surface suppressed A-site segregation. Siebert et al. found that, in $La_{0.8}Sr_{0.2}MnO_3$ powders, the surface Sr-rich particles could be removed, and surface stoichiometry could be recovered after annealing at 800°C in air.[107] The surface particles reappeared at temperatures up to 1200°C. In other research, surface (Sr+La)/Mn ratio in $La_{0.7}Sr_{0.3}MnO_3$ thin film was found to obviously increase until the temperature exceeded 400°C,[48] and the surface roughness in $La_{1-x}Sr_xMnO_3$ thin film also increased from 973 to 1023 K,[108] suggesting that cation segregation is much more likely at elevated temperatures.

Feng et al. investigated the evolution of the surface structure and chemistry of $La_{0.8}Sr_{0.2}CoO_{3-\delta}$ perovskite thin films during several thermodynamic cycles.[47] During the first thermal cycling, the $Sr_{surface}/Sr_{lattice}$ ratio was found to decrease with increased temperature and to increase with decreased temperature (Figure 12.20c). Nevertheless, during the subsequent thermal cycling, this ratio exhibited the opposite tendency (Figure 12.20d). To explain these facts, they proposed that, during the first thermal cycling, a RP structured $(La, Sr)_2CoO_{4+\delta}$ (LSC_{214}) might be formed by adding the surface segregated SrO particles into perovskite unit cells. Because

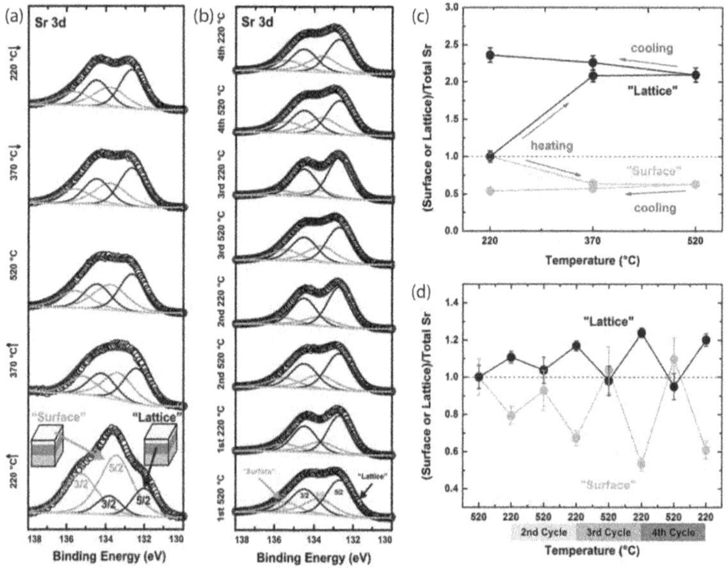

FIGURE 12.20 The evolution of the surface structure and chemistry of LSC_{113} during several heating and cooling cycles. (a) Sr 3d XPS spectra during the first heating, (b) Sr 3d XPS spectra during subsequent temperature cycling, (c) Sr 3d intensities of "lattice" and "surface" components during the first heating, (d) Sr 3d intensities of "lattice" and "surface" components during subsequent temperature cycling. (Reprinted with permission from Feng, Z. et al., *J. Phys. Chem. Lett.*, 4, 1512–1518, 2013. Copyright 2013 American Chemical Society.)

LSC_{214} has similar Sr 3d features to LSC_{113}, the Sr 3d peaks in the two compounds were indistinguishable (Figure 12.20a and b). Based on this assumption, the increase of $Sr_{lattice}$ peak might be attributed to the formation of LSC_{214}-like particles for LSC_{113} while heating. The formation of LSC_{214} at surface could enhance its thermal stability. Consequently, during the subsequent thermal cycling, $Sr_{lattice}/Sr_{surface}$ ratio continued increasing, indicating the accumulation of LSC_{214} on the surface of LSC_{113} (Figure 12.20d).

12.3.6 The Effect of Atmosphere

The environmental atmosphere also has complex influences on cation segregation. For instance, the redox features of external atmosphere can determine the category of segregated cations. According to existing research, B-site segregation is typically reported to occur at reducing environments, while A-site segregation generally occurs at neutral or oxidizing atmospheres. This could be explained by a combination of electrostatic interaction and lattice strain. Because the B-site transition metals have variable oxidation states, they can change their valence as well as ionic radii depending on redox conditions. In oxidizing atmospheres, B-site cations typically exhibit relatively high oxidation states with smaller size, therefore driving A-site cations with fewer positive formal charges and greater radii to the surface. In contrast, these transition metal ions can appear at relatively low oxidation states with larger size in reducing atmospheres, which eventually leads to B-site segregation.

For A-site cation in oxidizing atmospheres, the increasing oxygen pressure has opposite effects on its segregation behaviors. On the one hand, higher oxygen pressure is beneficial for oxygen incorporation into the lattice and oxidation of the transition metal cations, thus increasing the oxidation state of the transition metal cations and shrinking its radius, as well as causing the contraction of the lattice.[124,125] The narrowed lattice then leads to a more obvious size mismatch to the substitutional ions, which might finally drive them towards the surface. On the other hand, a large amount of oxygen ions incorporating into the lattice decreases the concentration of oxygen vacancies ($V_O^{\bullet\bullet}$) at the surface, which have opposite formal charge from the divalent substitutional ions' defects (D_A). The lack of oxygen vacancies would therefore suppress the segregation of dopants according to electrostatic interaction.

To estimate the effects of oxygen partial pressure on the surface chemistry, three doped perovskite materials, $La_{0.8}Ca_{0.2}MnO_3$ (LCM), $La_{0.8}Sr_{0.2}MnO_3$ (LSM), and $La_{0.8}Ba_{0.2}MnO_3$ (LBM), were annealed in 760 Torr, 1×10^{-6} Torr, and 1×10^{-9} Torr O_2 by Lee et al.[45] They found that the higher oxygen pressure drove cations towards the surface with segregation and secondary phase formation. After annealing in 1×10^{-6} Torr oxygen, LBM films exhibited visible changes in the chemistry and structure at the surface. In contrast, LSM films did not show any evident changes below 1×10^{-6} Torr, but they generated secondary phase in 760 Torr oxygen. These results indicated that A-site segregation could be accelerated at oxidizing environments. In contrast, using in situ TXRF, Salvador et al. found a decrease tendency for Sr segregation with increased oxygen pressure.[95] The segregation heat varied from −9.5 to −2.0 kJ/mol, with oxygen pressure increasing from 0 to 150 Torr. These results reflect that the increase of oxygen pressure can have opposite effects on A-site segregation.

In addition to the redox conditions, some small molecules and impurities in the surrounding environment can also influence A-site segregation level. For example, Bucher et al. found that after a treatment to $La_{0.58}Sr_{0.4}Co_{0.2}Fe_{0.8}O_{3-\delta}$ films under 600°C for 1000 h in a dry 1% O_2-Ar atmosphere, there was only a moderate enrichment of La and Sr within the first 30–35 nm at the surface.[112] However, after an additional treatment for 1000 h in a humid atmosphere (30% relative humidity under equilibrium conditions at 25°C), Sr was strongly enriched within about 5 nm depth, with a depletion of La at the surface, indicating an obvious promotion. Besides, CO_2 could also affect Sr segregation behaviors in $(La_{0.8}Sr_{0.2})_{0.98}MnO_{3\pm\delta}$ systems.[111]

Jiang et al. found that the presence of Cr components (such as Cr_2O_3) could also significantly promote Sr segregation levels and form $SrCrO_4$ phase in systems.[85,113] Similarly, Lu et al. reported that in a $(La_{0.8}Sr_{0.2})_{0.95}MnO_3$ system, surface Sr segregation was accelerated by Cr deposition.[93] Although there were no convincing explanations for these phenomena, they might be attributed to the reaction between small molecules/impurities and segregated Sr-contained particles at the surface, which transfers surface Sr into new forms (such as $Sr(OH)_2$, $SrCO_3$, $SrCrO_4$, etc.), therefore reducing its chemical potential at the surface and driving more lattice Sr out.

12.3.7 The Effect of Electrical Polarization

SOC electrode materials usually perform under external polarization, which is reported to impact surface segregation as well. It was proven by both in situ SIMS and XPS that in $La_{1-x}Sr_xMnO_3$ electrode systems, cathodic polarization decreased Sr surface concentration, while anodic polarization (in both SOFC mode and solid oxide electrolysis cell [SOEC] mode) caused strontium accumulation at the electrode surface (Figure 12.21). In other Co contained electrode systems, such as $La_{0.6}Sr_{0.4}Co_{0.2}Fe_{0.8}O_{3-\delta}$ and $PrBaCo_2O_{5+\delta}$ systems, anodic current passage also resulted in substantial inactive SrO/BaO segregation on the electrode surface.[116–118] Such compositional and structural alternations are preferable under cathodic polarization conditions because the electron injection driven by the electric field is basically equivalent to a chemical reduction process. Although the mechanism is still unclear, these results can offer potential strategies for rationally controlling surface chemistry.

According to the discussions in this section, one can find that cation segregation is not just determined by a simple driving force; instead, it is a synthesized process determined by various materials and environmental conditions. So far, there is no unified theory for explaining all the segregation behaviors in these systems. However, the development of in situ characterization methods is expected to be crucial for advancing our insights into the origin of cation segregation in various perovskite-based systems, as well as determining the main driving forces for cation

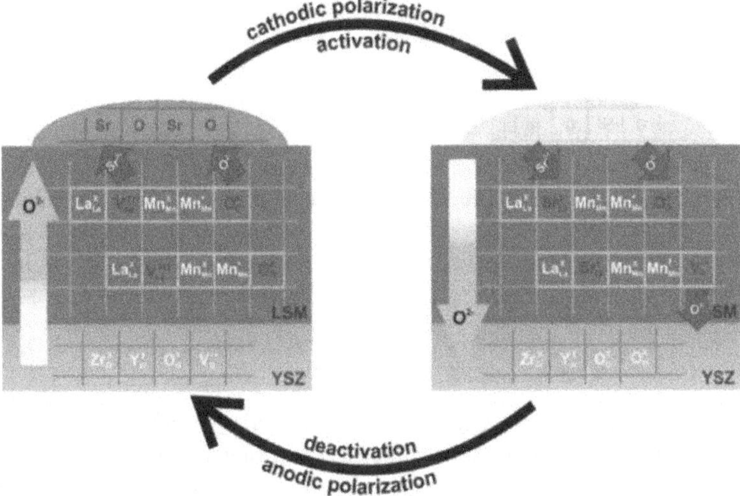

FIGURE 12.21 Defect chemistry of the $(La_{0.8}Sr_{0.2})_xMnO_3$ phase during the electrochemical polarization experiments. (Reprinted from *J. Catal.*, 294, Huber, A. et al., In situ study of activation and de-activation of LSM fuel cell cathodes: Electrochemistry and surface analysis of thin-film electrodes, 79–88, Copyright 2012, with permission from Elsevier.)

segregation under various conditions. By taking advantage of these understandings, it is possible to control cation segregation behaviors in perovskite-based materials to create new functionalities.

12.4 INFLUENCES OF CATION SEGREGATION ON ELECTROCHEMICAL ACTIVITY OF SOC ELECTRODES

Due to the relatively slow reaction kinetic, oxygen reduction reaction (ORR) and oxygen evolution reaction (OER) activity on electrodes surface is considered to dominates the performance of SOCs. Cation segregation has been reported to greatly impact the surface reaction steps of ORR and OER, such as surface absorption/exchange, electron transfer, molecular dissociation, and ion incorporation/migration, and so on, hence strongly influencing the activity of electrodes. The descriptors widely used for representing electrode ORR/OER activity in the literature include surface polarization resistance, surface exchange coefficient, and conductivity. Correspondingly, electrochemical impedance spectroscopy (EIS),[9,30,44,126,127] four-point dc-measurements,[61,112,128] and ^{18}O isotope exchange depth profiling with ToF-SIMS[56,129] are some of the most commonly used approaches for determining the activity for SOC electrodes.

This section will discuss how cation segregation influences surface activity. The effects of both A-site and B-site segregation on ORR/OER activity in perovskite-based materials will be discussed in detail. Generally, B-site segregation improves surface conductivity and enhances catalysis activity, while the influence of A-site segregation is more complex. As one of the most common A-site dopants, Sr segregation generally occurs on the electrode surface of SOCs after synthesis and in an oxidizing atmosphere. Therefore, for A-site cations, the most thoroughly investigated Sr segregation on ORR/OER activity in Sr-doped materials will be introduced in detail as a case study. Because Sr segregation generally occurs in a strong oxidizing atmosphere, we mainly focus on how cation segregation influences oxygen electrode materials for SOCs. Specifically, the separation of surface secondary phase is thought to be harmful to electrode performance; however, in certain cases, surface Sr enrichment may enhance oxygen surface kinetics. In addition, the influence of cation segregation in Sr-free electrode materials will be also reviewed.

12.4.1 Influence of Sr Segregation

Many investigations demonstrated that Sr segregation and secondary phase separation would increase surface resistance and have detrimental effects for ORR or OER activity of oxygen electrodes. A representative work focusing on how Sr segregation affects the surface polarization resistance of SOC electrode materials (LSC) was carried out by Markus et al. and Cai et al.[9,30] Because of the increasing of Sr segregation on the surface, which led to the formation of secondary phase (Figure 12.22), the surface polarization resistance of $La_{0.6}Sr_{0.4}CoO_{3-\delta}$ films deposited at 650°C (LSC_650°C) and 450°C (LSC_450°C) were found to increase with annealing time, which indicated performance degeneration. As shown in Figure 12.22, LSC_650°C with greater surface roughness and higher Sr segregation level exhibits an order greater resistance and a faster degeneration rate during annealing compared to LSC_450°C with less Sr segregation. In addition, HCl etching (0.14 mol/L for 10 s) of the 72 h annealed LSC_650°C and LSC_450°C removed the surface particles or Sr-enrichment layer and left a uniform surface with nearly stoichiometric element ratios on both samples. In addition to the removal of the Sr segregation layer, they

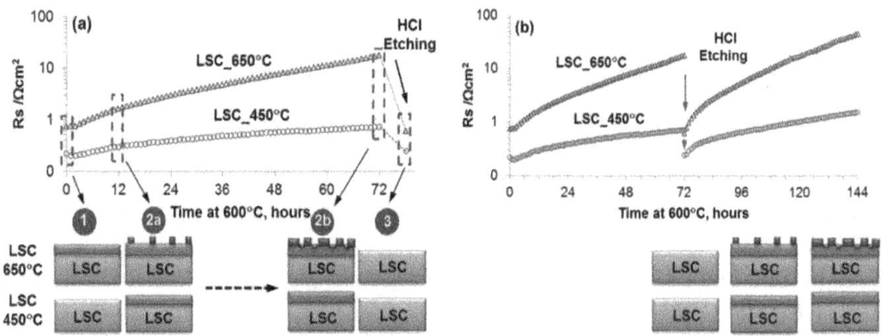

FIGURE 12.22 (a) Surface polarization resistance, Rs, of LSC grown at 650°C measured by impedance spectroscopy at 600°C in air and the likely mechanisms that govern the differences between the activity and stability of these cathode films, (b) further evolution of Rs of LSC after HCL etching measured at 600°C. (Reprinted with permission from Cai, Z. et al., *Chem. Mater.*, 24, 1116–1127, 2012. Copyright 2012 American Chemical Society.)

observed significant enhancement of electrode performance, as shown in the red box in Figure 12.22a. Although both films appear to have a stoichiometric surface, further annealing after HCl etching again leads to Sr segregation and degradation of LSC electrode performance. Note that the LSC_650°C shows a much faster degradation rate compared to LSC_450°C (Figure 12.22b).

In agreement with these results, some researchers confirmed that the separation of the SrO-based secondary phase would cause performance degeneration for ORR/OER, while removal of the components could eliminate this passivation. Pan et al. proposed that the resistance variation of $La_{0.6}Sr_{0.4}Co_{0.2}Fe_{0.8}O_{3-\delta}$ half-cell might be attributed to surface segregation of Sr-rich species.[130] Jiang et al. demonstrated that the surface exchange coefficients decrease in $La_{0.6}Sr_{0.4}Co_{0.2}Fe_{0.8}O_{3-\delta}$ electrode was accompanied by Sr segregation and formation of $SrCrO_4$ phase.[85,137] To eliminate the negative effects by Sr segregation, Tuller et al. treated the surface of $SrTi_{1-x}Fe_xO_{3-\delta}$ with a 10% diluted buffered HF solution for 20 s, and they found an obvious decrease in resistance after the removal of Sr-contained species.[132] Using in situ SIMS, Huber et al. demonstrated that in $(La_{0.8}Sr_{0.2})_{0.92}MnO_3$ or $La_{0.75}Sr_{0.25}Cr_{0.5}Mn_{0.5}O_{3-\delta}$ thin films prepared by PLD, surface segregation could be recovered by applying cathodic polarization, which is accompanied with significant performance activation.[50,60] Furthermore, Tsvetkov et al. also found that surface modification by several typical B-site cations can significantly reduce segregation level as well as enhance oxygen surface exchange kinetics.[44,76]

Although many researchers regarded Sr segregation as a detrimental process, in certain cases, the positive effects of Sr enrichment on electrode performance were also reported in the literature. Baumann et al. observed strong electrochemical activation of $La_{0.6}Sr_{0.4}Co_{0.8}Fe_{0.2}O_3$ electrode accompanied by surface Sr enrichment after cathodic polarization by a repetition of short bias pulses,[118] Shao-Horn et al. carried out cathodic polarization on $La_{0.8}Sr_{0.2}MnO_3$ thin-films, which was found to increase Sr and Mn concentration on the interface between the electrode and the electrolyte and to increase electrode performance.[133] By decorating $SrTiO_3$ single crystal substrate with alkaline earth (Ca, Sr, Ba) oxide coating, Wagner et al. found strong enhancement of the surface oxygen exchange rate.[134] Mutoro et al. also demonstrated that decorating $La_{0.8}Sr_{0.2}CoO_{3-\delta}$ thin films with SrO nanoparticles could enhance its surface exchange coefficient by an order of magnitude (Figure 12.23).[135]

Sr segregation may have significant influences on the electrochemical performance of SOC oxygen electrodes. It is of great importance to study the mechanism behind Sr segregation's influence on electrode performance. Based on understanding and controlling of surface chemistry related to properties, it might be possible to enhance the activity of Sr-contained electrodes by adjusting Sr segregation behaviors.

12.4.1.1 Mechanism 1: Blocking Effects

Combining EIS measurement with surface analysis, researchers attributed the performance degradation of SOC electrodes to the blocking effects of charge transfer process due to surface-bound Sr (mainly surface secondary phases).[30,37,126,130,136] This is due to the insulating nature of those SrO-like segregation phase. For example, the energy gap of SrO is as high as 6 eV,[71] which is much larger than that of perovskite-based electrodes. The presence of such an insulating layer can block the charge exchange process between surface and abosorbed oxygen.

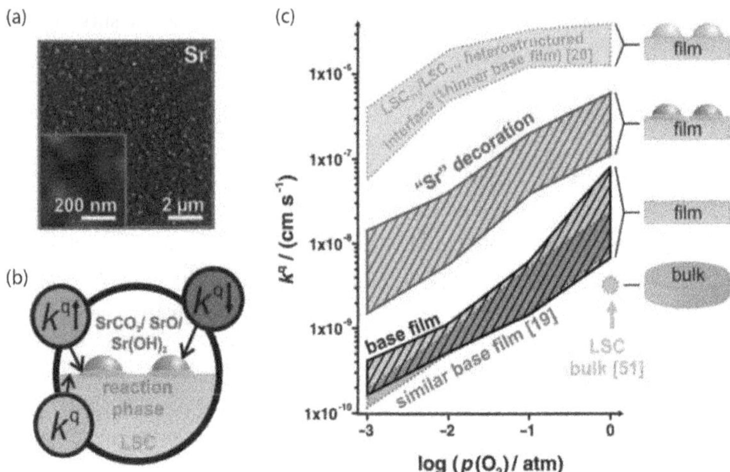

FIGURE 12.23 (a) Surface morphology of SrO decorated $La_{0.8}Sr_{0.2}CoO_{3-\delta}$ films after EIS testing, (b) possible mechanism of electrochemical activity enhancement by surface "Sr" decoration, (c) comparison of surface exchange values for different $La_{0.8}Sr_{0.2}CoO_{3-\delta}$ based samples in the range of 510°C to 550°C at various oxygen pressures. (Mutoro, E. et al., *Energ. Environ. Sci.*, 4, 3689–3696, 2011. Reproduced by permission of The Royal Society of Chemistry.)

The representative works focusing on the effect of Sr segregation on ORR activity were carried out in $SrTi_{1-x}Fe_xO_3$ (STFx) system by Tuller et al. and Chen et al.[71,132] By XPS analyses for various STFx systems, they found that increasing Fe content in the bulk led to the increase of surface Sr content as well as the segregation level (Figure 12.24a). Due to the insulating nature of this Sr-enriched surface phase, the increase of Sr segregation with Fe content led to the increase in the band gap. The surface electronic structure as a function of Fe content is contrary to that in the bulk (Figure 12.24b).[137] By removing the Sr-enriched phase using chemical etching, the polarization of ORR quantified by EIS decreased significantly (Figure 12.24c and d), indicating a recovery of surface activity. In an ORR process, several successive reaction steps involving electron transfer were likely to take place during oxygen exchange on the surface:

$$O_{2\_ads} \rightarrow 2O_{ads};\ 2O_{ads} + 2e^- \rightarrow 2O^-_{ads} \text{ or}$$

$$2O_{ads} + 2e^- \rightarrow O^{2-}_{2\_ads} \rightarrow 2O^-_{ads}$$

$$O^-_{ads} + V_O^{\bullet\bullet} + e^- \rightarrow O_O^{\times}$$

On the basis of surface chemistry deduced from AR-XPS analysis and the large band gap values measured by STS, Chen et al. conclude that the Sr segregation layer is more likely to be in the form of SrO_x on the surface of STF films. The electronic

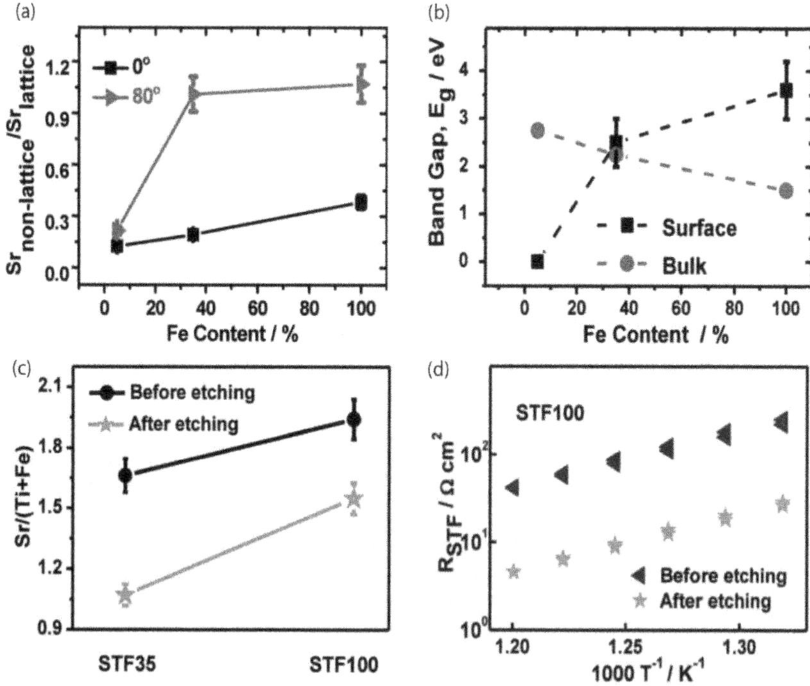

FIGURE 12.24 (a and b) The alternation of surface elemental composition and band gap as a function of Fe doping content in STFx systems, (c and d) the evolution of surface composition and polarization resistance before and after etching by a 10% diluted buffered HF solution for 20 s. (Chen, Y. et al., *Energ. Environ. Sci.*, 5, 7979–7988, 2012. Reproduced by permission of The Royal Society of Chemistry; Jung, W.C. and Tuller, H.L., *Energ. Environ. Sci.*, 5, 5370–5378, 2012. Reproduced by permission of The Royal Society of Chemistry.)

structure results demonstrate that such a SrO_x segregation layer on STF degrades the oxygen reduction kinetics by blocking electron transfer from bulk to surface (Figure 12.25a and b).

Similar to Jung et al. and Chen et al.'s work,[71,132] Oh et al. proposed that the separated SrO-like phase on the surface acts as an insulation layer and leads to an increase in ohmic resistance of $La_{0.6}Sr_{0.4}CoO_{3-\delta}$ thin film electrodes.[68] Luo et al. also proposed that the SrO coverage at the surface can block the contact between oxygen and activated B-site metals in perovskite materials.[6,138,139] Note that, in Chen et al.'s work, the downshift of the O 2p band relative to the Fermi level at the surface was observed in STM/S and XPS measurement. According to Shao-Horn et al.,[140,141] such a downshift caused the increase of oxygen vacancies formation energy, hence leading to a decrease in oxygen vacancy content on the surface as well as degradation of surface activity (Figure 12.25c and d).

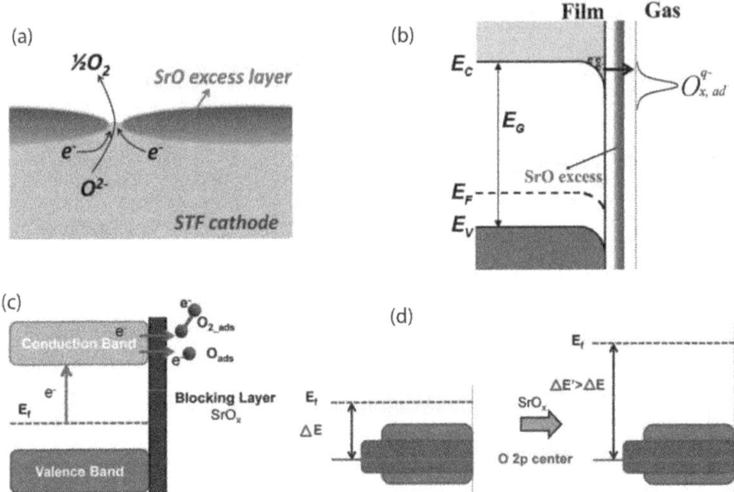

FIGURE 12.25 (a–c) Surface SrO segregation layer blocks the contact and electron transfer between absorbed oxygen and activated sites in STFx systems, (d) the downshift of the O 2p state relative to Fermi level caused by the Sr segregation layer coverage. All these factors contribute to the degeneration of electrochemical performance. (Chen, Y. et al., *Energ. Environ. Sci.*, 5, 7979–7988, 2012. Reproduced by permission of The Royal Society of Chemistry; Jung, W.C. and Tuller, H.L., *Energ. Environ. Sci.*, 5, 5370–5378, 2012. Reproduced by permission of The Royal Society of Chemistry.)

12.4.1.2 Mechanism 2: Inducing Detrimental Side Reactions

Several investigations indicate that Sr segregation and the formation of SrO particles at the surface may be related to chromium deposition and poisoning, which would have combined effects on performance degeneration. By in situ XRD, Bae et al. indicated that SrO and/or BaO segregation on the surface of BSCF cathode could intensify the deposition of chromium.[62] By XRD and XPS analyses, the deposition of $SrCrO_4$, $CrO_{2.5}$ and Cr_2O_3 phases was observed by Wei et al. in LSCF anode under SOEC conditions.[131] In another study of LSC and LSM oxygen electrodes at SOFC working conditions, Jiang et al. pointed out that chromium deposition occurs preferentially on the segregated SrO particles rather than on transition metal oxides (Figure 12.26).[63,85]

In addition to volatile Cr-contained species, small molecules such as H_2O or CO_2 are also thought to be attracted by surface SrO-like segregated particles, further leading to the generation of an insulation phase (such as $Sr(OH)_2$ and $SrCO_3$) and performance degradation. Bucher et al. investigated the long-term stability of $La_{0.58}Sr_{0.4}Co_{0.2}Fe_{0.8}O_{3-\delta}$ and $La_{0.6}Sr_{0.4}CoO_{3-\delta}$ in dry and humid atmospheres at 600°C and 800°C, respectively.[61,112] By surface elemental analysis (e.g., SEM, TEM, XPS, and EDXS/WDXS) and electrochemical testing (dc-conductivity relaxation measurements) for thousands of hours, they concluded that electrodes in humid environments could not only exhibit much severe Sr segregation but also arouse Si and Cr poison at Si- and Cr-sources. By XRD, Hu et al. found that in $(La_{0.8}Sr_{0.2})_{0.98}MnO_3$ electrodes,

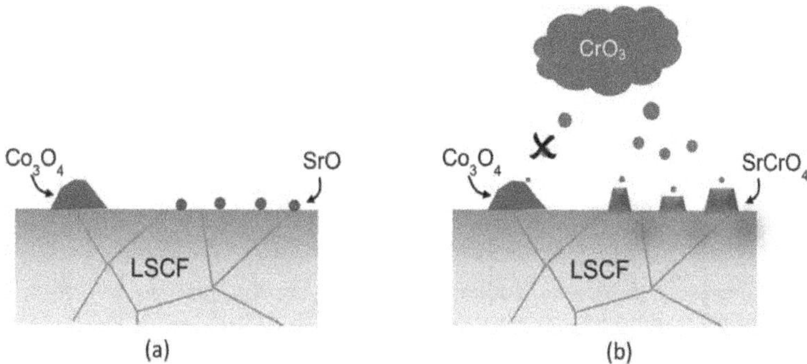

FIGURE 12.26 Scheme of (a) segregation and formation of Co_3O_4 and SrO and (b) the selective and preferential deposition and reaction of Cr species with segregated SrO on the surface of the LSCF electrode material. (Zhao, L. et al., *J. Mater. Chem. A*, 2, 11114–11123, 2014. Reproduced by permission of The Royal Society of Chemistry.)

surface segregated SrO could react with H_2O when cooling down in humidified air.[142] Sharma et al. proposed that (La, A)MnO_3 (A represents alkaline earth metal) systems, the surface terminal oxygen could attract H_2O in the atmosphere via H-bond.[98] This interaction would enhance the tendency for cation segregation and cause the catalytic activity degradation, thus leading to a conductivity drop. In a study performed by Darvish et al., CO_2 was found to have similar effects (forming $SrCO_3$ and reducing electrical conductivity) on $(La_{0.8}Sr_{0.2})_{0.98}MnO_{3\pm\delta}$ system.[111] These side reactions typically accelerate Sr segregation, which may synergistically lead to an increased degradation rate in perovskite-based electrodes.

12.4.1.3 Mechanism 3: Generating Active Phases

Although many researchers have reported the detrimental effects of Sr segregation on ORR/OER activity of SOCs electrode, in certain cases Sr enrichment was shown to improve the electrode performance.[118,134,135] Such improvement was thought to be related to the generation of an activated but nonequilibrium surface, as well as the formation of a high-performance heterointerface.[118,135]

Shao-Horn et al. investigated cation enrichment behaviors in $La_{1-x}Sr_xMnO_3$ thin films.[133] They found that Sr/La exchange might occur at the presence of cathodic polarization, which caused the formation of new composition at the near surface region. According to AES and XPS analysis, Sr and Mn enrichment together with La depletion was observed on the surface (compared to their average contents at the edge regions of electrode), suggesting that its composition could be approximately referred to as $Sr_xMn_yO_z$. Similarly, in a study on $La_{0.6}Sr_{0.4}Co_{0.8}Fe_{0.2}O_{3-\delta}$ thin film electrodes, Maier et al. confirmed the enrichment of Sr and Co after applying cathodic polarization.[118]

By XRD and in situ XPS analysis for $La_{0.8}Sr_{0.2}CoO_{3-\delta}$ systems, Mutoro et al. proposed that an RP-phased $(La, Sr)_2CoO_{4+\delta}$ might be generated on the LSC surface,[135] and Sr-enriched particles. Feng et al. also demonstrated the accumulation of $(La, Sr)_2CoO_{4+\delta}$ at the surface during several thermodynamic cycles, seemly because the LSC_{214} is more stable than LSC_{113} at elevated temperatures.[47] According to their

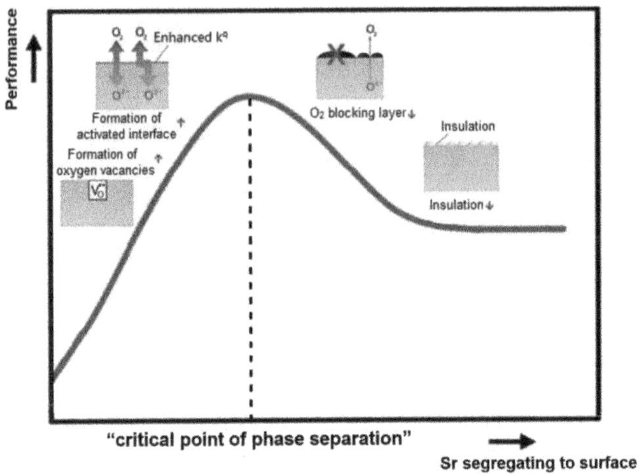

FIGURE 12.27 Schematic diagram of "critical point of phase separation" hypothesis about the correlation between Sr segregation and electrode performance for Sr doped $LaCoO_3$. (Li, Y. et al., *Chem. Soc. Rev.*, 46, 6345–6378, 2017. Reproduced by permission of The Royal Society of Chemistry.)

findings, the formation of this interfacial region with high Sr content may contribute to the enhancement surface exchange kinetic as well as decreasing polarization resistance.[133,143,144]

To summarize the influence of Sr segregation on the performance of (La, Sr) CoO_3 and other Sr-doped materials with similar characteristics, a "critical point of phase separation" hypothesis is proposed (Figure 12.27). As Sr initially segregates to the surface, the impact of the Sr segregation appears to be similar to increasing the Sr^{2+} dopant level in bulk $LaCoO_3$. At this stage, the electrode properties increase with the surface Sr content due to the increase in oxygen vacancies and the change in electronic structure. As the Sr content at the surface further increases, surface Sr concentrations will eventually reach a critical point (the so-called critical point of phase separation). As a result, inert SrO-like particles separate out to block the reaction between oxygen and electrode surfaces or form insulting phases, leading to the decrease of electrode performance. Therefore, Sr segregation is beneficial to electrode performances before the critical point and harmful after that.

12.4.2 A-Site Segregation on Sr-Free Electrode Materials

In addition to the commonly used Sr-contained electrode materials, Sr-free composite for oxygen electrodes was also widely studied. Chen et al. introduced several such materials with promising electrochemical properties, such as $LaNiO_3$, La_2NiO_4, $CoFe_2O_4$, and their derivatives.[5] Despite the absence of Sr, cation segregation is still generally observed in these systems.

$LaNi_{1-x}Fe_xO_3$ (LNF) and its derivatives are a series of state-of-the-art perovskites that possess good catalytic activity and high electrical conductivity.[145–150] Nevertheless, their performance was reported to suffer from Cr poisoning, which was related to the segregation of B-site Ni.[151,152] The separated NiO phase could further react with Cr-species to generate a Ni-substituted secondary phase.[153] Using XRD, Komatsu et al. found that in LNF systems, the residual NiO at the surface could further lead to the formation of a $NiCr_2O_4$ phase at the presence of Cr_2O_3 powder.[152]

Another typical Sr-free material with promise for SOC oxygen electrodes is RP-structured La_2NiO_4.[154–156] By LEIS depth profiling, Kilner et al. found its surface to be dominated by A-site La.[1] In another study on $PrLaNiO_{4+\delta}$, a derivative of La_2NiO_4, they also observed surface enrichment of both Pr and La with no obvious priority.[25] With respect to these materials, the properties seem to be less affected by cation segregation compared to that of Sr-contained materials. Pederson et al. compared the performance degradation rate in several common electrode materials,[157] and their results are shown in Figure 12.28. Although the as-prepared BSCF and LSCF electrodes exhibit promising properties, serious degeneration is presented in the process of continuous operation due to Sr segregation under electrochemical polarization. In contrast, the Pr_2NiO_4 electrode appears to have much more stability during the operation for 1000 hours. They proposed that the degradation mechanism in Sr-contained electrodes might be related to Sr segregation accompanied by the demixing of cations under electrochemical potential.[157]

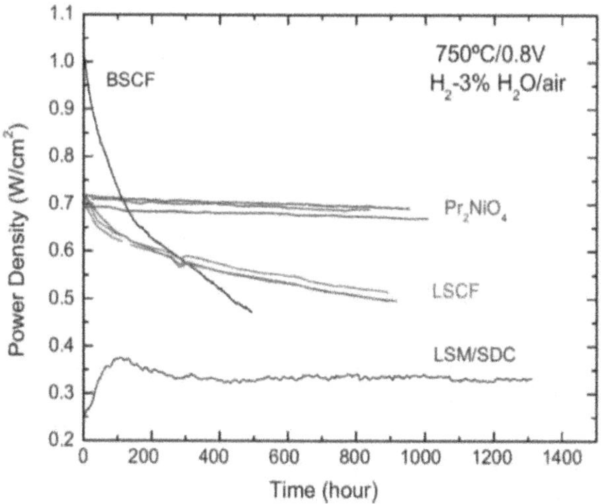

FIGURE 12.28 Electrochemical performance of cells with LSM, LSCF, Pr_2NiO_4, or BSCF cathodes as a function of time at 750°C and 0.8 V. (Reprinted from *Electrochim. Acta*, 71, Zhou, X.D. et al., Electrochemical performance and stability of the cathode for solid oxide fuel cells: V. high performance and stable Pr_2NiO_4 as the cathode for solid oxide fuel cells, 44–49, Copyright 2012, with permission from Elsevier.)

12.4.3 Influence of B-Site Segregation

The segregation of B-site transition metals under reduction conditions is also referred to as exsolution.[158–160] Since these exsolved metallic particles are thermally stable against reuniting at operating conditions, exsolution is proposed to be a promising approach that can be applied in SOCs. The most common metals that exsolved from perovskite-based materials include Fe, Co, and Ni. To generate metallic particles coverage, both reducing agent and cathodic polarization are available approaches.

As reviewed by Irvin et al., B-site exsolution is an effective approach to improve the electrochemical performance of electrode materials.[6] Wu et al. found that with the exsolved Ni nanocrystals on the surface of $(La_{0.2}Sr_{0.8})(Ti_{0.9}Mn_{0.1})O_{3-\delta}$ (LSTM), it exhibited enhanced catalytic activity for CO_2 electrolysis[161] Shao et al. proposed that the exsolved NiO particles could significantly reduce the area-specific resistance of the $SrFe_{0.85}Ti_{0.1}Ni_{0.05}O_{3-\delta}$ (STFNi) electrode.[7] These results are shown in Figure 12.29.

In addition to improving electrochemical activity, B-site exsolution can also enhance the stability for hydrogen electrodes at both SOFC and SOEC modes. For instance, Neagu et al. reported that the surface morphology of Ni-$La_{0.52}Sr_{0.28}Ni_{0.06}Ti_{0.94}O_3$ remained unchanged after exposure to 20% CH_4/H_2 at 800°C for 4 h, showing a good structural stability against coking.[109] In another study of Co/Fe exsolved $(Pr_{0.4}Sr_{0.6})_3(Fe_{0.85}Mo_{0.15})_2O_7$ systems, Luo et al. found no graphite generated during polarization for 10 h, indicating improved anti-coking properties.[138] The results are shown in Figure 12.30. In addition, many researches have confirmed that the surface composition and electrochemical performance of these exsolved systems exhibit no significant decline after several redox cycles, indicating considerable operational stability.[13,162–165]

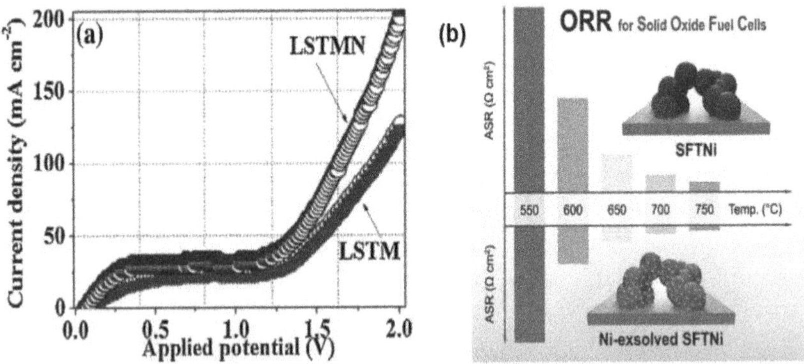

FIGURE 12.29 (a) Current-voltage curves of SOEC with cathodes based on $(La_{0.2}Sr_{0.8})(Ti_{0.9}Mn_{0.1})O_{3-\delta}$ (LSTM) and $(La_{0.2}Sr_{0.8})_{0.9}(Ti_{0.9}Mn_{0.1})_{0.9}Ni_{0.1}O_{3-\delta}$ (LSTMN) with exsolved Ni at 800°C, (b) area-specific resistance (ASR) of $SrFe_{0.85}Ti_{0.1}Ni_{0.05}O_{3-\delta}$ (STFNi) cathode and that decorated by NiO. (Reprinted from *Electrochim. Acta*, 153, Li, Y. et al., Efficient carbon dioxide electrolysis based on perovskite cathode enhanced with nickel nanocatalyst, 325–333, Copyright 2015, with permission from Elsevier; Reprinted with permission from Yang, G. et al., *ACS Appl. Mater. Interfaces*, 8, 35308–35314, 2016. Copyright 2016 American Chemical Society.)

(a)

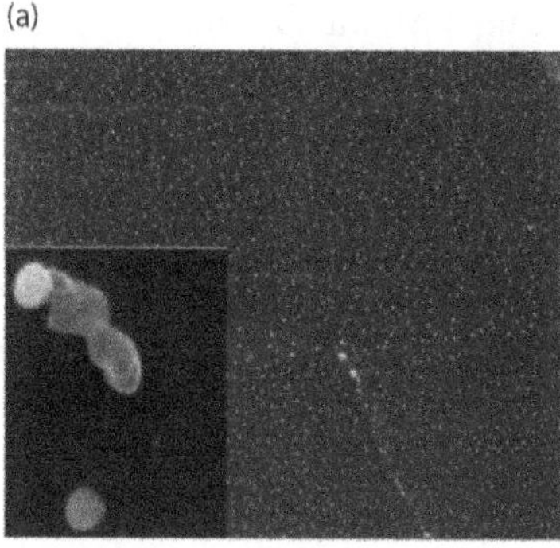

(b)

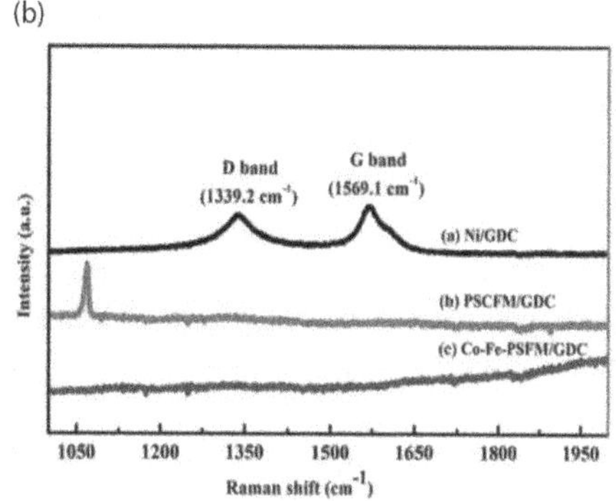

FIGURE 12.30 (a) Ni particles exsolved from $La_{0.52}Sr_{0.28}Ni_{0.06}Ti_{0.94}O_3$ (5% H_2/Ar, 1,000°C, 6 h) after coking showing limited coking; scale bars, 1 mm (overview), 100 nm (detail); (b) Raman spectra collected from the cathode surface of Ni/GDC, $Pr_{0.4}Sr_{0.6}Co_{0.2}Fe_{0.7}Mo_{0.1}O_{3-\delta}$/GDC and exsolved Co-Fe-$(Pr_{0.4}Sr_{0.6})_3(Fe_{0.85}Mo_{0.15})_2O_7$/GDC after testing in CO_2 electrolysis cell at the constant applied potential of 0.4 V (vs. OCV) at 850°C for 10 h. (Reprinted by permission from Macmillan Publishers Ltd. *Nat. Commun.*, Neagu, D. et al., 2015, Copyright 2015; Liu, S. et al., *J. Mater. Chem. A*, 4, 17521–17528, 2016. Reproduced by permission of The Royal Society of Chemistry.)

The generation of B-site exsolution is proposed to be driven by chemical/electrochemical reductions and controlled by bulk/surface defects as well as external conditions. When the lattice is reduced, surface defects such as oxygen vacancy formation becomes preferable, which destabilizes the lattice stoichiometry. Meanwhile, the generation of oxygen vacancies accompanied by the reduction of B-site transition metals can further lead to metal nucleation at the surface, where the nucleation barrier is lowered by crystal defects. These processes will eventually drive B-site cations to the surface from the inside bulk continuously and will cause the growth of exsolved particles until an equilibrium is reached. At present, additional research is still needed to enhance the activity and stability of hydrogen electrodes with exsolved metallic particles, and the results are expected to provide guidance for designing novel electrode materials for SOCs.

12.5 SURFACE ENGINEERING PROMOTES ORR/OER ACTIVITY FOR PEROVSKITE ELECTRODES

As reviewed in previous sections of this chapter, surface segregation plays an essential role in determining the performance of SOC electrodes. Much research focuses on suppressing detrimental segregation or promoting beneficial segregation to enhance the reactivity of SOCs electrode. This section will review some representative works that use surface engineering to control surface segregation and enhance the activity and stability of SOC electrodes. The most widely used approach is to decorate or cover existing materials with a new phase. These works will be introduced in detail and will be categorized according to the additional phase.

12.5.1 Surface Decoration with Alkaline Earth Metal Oxides

Alkaline earth elements are widely used as the A-site cations of perovskite-like electrode materials. As discussed in Section 12.4, the segregation of these cations, particularly Sr, may lead to the degradation of SOCs. However, it is also reported that intentional decoration of alkaline earth metal oxides on perovskite like oxides increases their ORR activity. Wagner et al. reported that the alkaline earth metal oxides (e.g., BaO or CaO) at the surface led to faster oxygen exchange kinetics in $SrTiO_3$.[134] Using ^{18}O isotope exchange depth profiling with ToF-SIMS, they observed that the surface exchange coefficient of $SrTiO_3$ films decorated by alkaline earth metal layers with 20-100 nm thicknesses was nearly an order of magnitude higher than that of the bare surface. Using EIS measurement, Shao-Horn et al. reported that, by covering with nanoscale "Sr coverage" (48% $Sr(OH)_2$, 35% SrO, and 17% $SrCO_3$), the ORR activity of $La_{0.8}Sr_{0.2}CoO_{3-\delta}$ increased by an order of magnitude.[135] They proposed that the reason for such enhanced performance could be attributed to the generation of a surface RP phase and the formation of highly activated heterointerface (Sr_xO_y/LSC_{113} or LSC_{113}/LSC_{214}).[31,47] Visco et al. reported that surface modification with Sr-nitrate could decrease the polarization resistance of LSM-YSZ components.[166] Moreover, Gorte et al. observed that the decoration of K_2O and CaO decreased the electrode impedance of LSM and LSF.[167]

Although the works cited above illustrate that decorating alkaline earth metal oxides may improve the performance of SOCs electrode, it should be also mentioned that with the increased thickness or extended operating time, the surface new phase could eventually turn to hinder the activity of electrodes. According to Shao-Horn et al., in SrO-decorated LSC thin films, the activated decoration layer would finally change into a completely inert phase with the increasing SrO amount.[135] In another in situ investigation on $La_{1-x}Sr_xCoO_{3-\delta}$ thin films, Ghislain et al. reported that tiny amounts of SrO (such as 4% of a monolayer) deposited on the surface could lead to severe deactivation effects.[136]

12.5.2 Surface Decoration of Transition Metal Cations

B-site cations in perovskite-based oxides are normally considered to be the active site for ORR/OER reaction. Thus, the decoration of transition metal cations on electrode materials is supposed to potentially enhance their catalytic performance. Tsvetkov et al. implemented a $TiCl_4$ solution to deposit Ti onto the surface of $La_{0.8}Sr_{0.2}CoO_3$ (LSC) thin films.[76] Using EIS measurement, they found much lower Sr segregation levels together with higher surface exchange kinetic and higher stability on Ti-modified films than bare LSC films at 530°C. Such activation became more obvious when surface Ti content increased from 2% (approximately two times) to 10% (approximately eight times). Nevertheless, if Ti content further increased to higher than 10%, it turned to hindering surface activity, suggesting that the excess Ti at the surface might be harmful to the electrochemical activity.

In additional research on LSC systems, they found that the concentration of oxygen vacancies ($V_O^{\bullet\bullet}$) would decrease by introducing any less reducible B-site cations (e.g., Hf^{4+}, Ti^{4+}, Zr^{4+}, Nb^{5+} or Al^{3+}, etc.).[44] Since oxygen vacancies have opposite charges with substitutional Sr defects (Sr_{La}), less oxygen vacancies on the surface

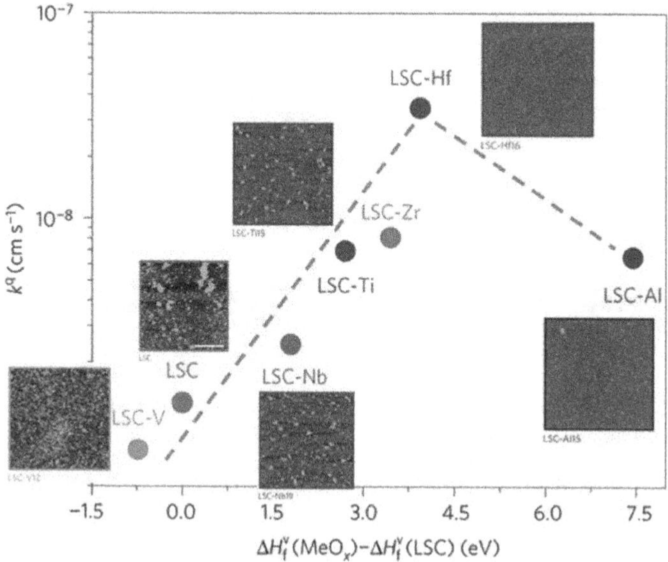

FIGURE 12.31 Surface exchange coefficient (represented by k^q) varies with the difference of $\Delta H_f^V(MeO_x)$ and $\Delta H_f^V(LSC)$ in V, Nb, Ti, Zr, Hf and Al modified LSC systems. (Reprinted by permission from Macmillan Publishers Ltd. Tsvetkov, N. et al., *Nat. Mater.*, 2016, Copyright 2016.)

led to the suppression of Sr migration towards the surface by weakened electronic attraction. As shown in Figure 12.31, there is a "volcano" relation of surface oxygen exchange consistent (k^q) with $\Delta H_f^V(MeO_x) - \Delta H_f^V(LSC)$, which is equivalent to oxygen vacancy formation enthalpy of substituted cations. As the difference of $\Delta H_f^V(MeO_x)$ and $\Delta H_f^V(LSC)$ increased, the enhancement of k^q reaches up to one to two orders of magnitude. Such improvement can be attributed to better stability against Sr segregation. However, if this value continues increasing, the decoration layer would turn to blocking surface activity due to the too low concentration of oxygen vacancies, which are thought to be reactive sites for ORR/OER.

Xia et al. found that $La_{0.6}Sr_{0.4}Co_{0.2}Fe_{0.8}O_{3-\delta}$ electrodes decorated with Ni by plasma glow discharge exhibited much faster oxygen surface exchange kinetics than that of an untreated one.[168] They attributed this improvement to the increase of surface area and the formation of a $NiFe_2O_4$ phase. Yamahara et al. reported that infiltrated Co particles on a LSM-YSZ electrode drastically enhanced its performance of in the range of 600 and 800°C.[169] Likewise, in another study on porous $La_{0.8}Sr_{0.2}FeO_{3-\delta}$ (LSF) electrode, Xia et al. found that surface infiltration by $Co(NO_3)_2$ solution significantly suppressed Sr segregation.[170] Using an in situ impedance spectroscopy during pulsed laser deposition (PLD), Ghislain et al. demonstrated that depositing Co_3O_4 on $La_{1-x}Sr_xCoO_{3-\delta}$ could lead to surface (re)activation.[136] This effect might be attributed to the exposure of B-site Co[3,171] as well as the elimination of surface segregated Sr,[3,30] as illustrated in Figure 12.32.

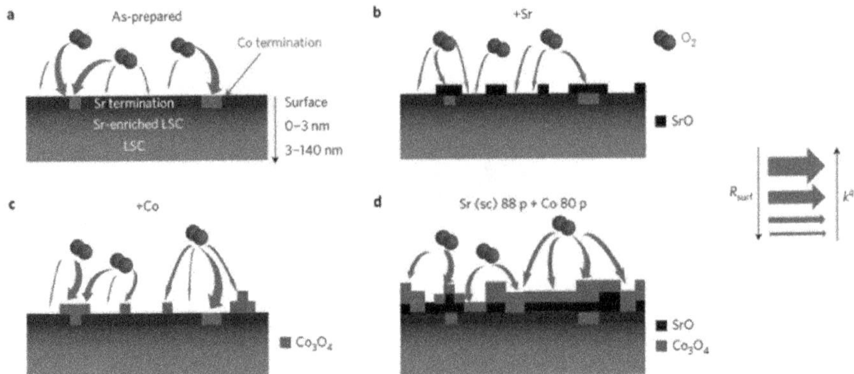

FIGURE 12.32 Illustration of the model of inhomogeneously active LSC surfaces for ORR. Four surface status including (a) bare surface, (b) Sr-decorated surface, (c) Co-decorated surface, and (d) surface covered by Sr and reactivated by Co are illustrated. (Reprinted by permission from Macmillan Publishers Ltd. *Nat. Mater.*, Rupp, G.M. et al., 2017, Copyright 2017.)

Although the works cited above show that the decoration of B-site cation can improve the electrode performance, there are some other works giving contradicted results. For instance, Gorte et al. found that additional 10 wt% CoO_x on a LSM-YSZ composite had no obvious effects on electrode performance under SOFC conditions.[172] Similarly, McIntosh et al. reported that $La_{0.6}Sr_{0.4}Co_{0.2}Fe_{0.8}O_{3-\delta}$ electrodes modified by cobalt nitrate solution via spray deposition exhibited no observable changes in surface exchange coefficient and bulk diffusion coefficient compared to the unmodified one.[173] Shim et al. modified $La_{0.6}Sr_{0.4}CoO_3$ (LSC) electrodes with CoO_x via atomic layer deposition.[174] They found that after annealing at 600°C for 4 hours in air, the CoO_x-treated films even showed lower power density than that of the bare cell. They attributed it to the transfer of CoO_x to Co_3O_4, which blocked O_2 adsorption and dissociation.

12.5.3 Surface Decoration by Secondary Perovskite-Based Phase

The formation of heterointerface by depositing secondary perovskite-based phase is reported to cause significant improved catalytic activity as well as higher structural stability for SOC electrodes.[6] One representative example of such a system is the A_2BO_4/ABO_3 heterostructure.[143,144,175–180] Sase et al. reported that deposition of $(La, Sr)_2CoO_4$ (LSC_{214}) onto PLD-layered $(La, Sr)CoO_3$ (LSC_{113}) films leaded to three-order-of-magnitude enhancement in oxygen exchange coefficient compared to the single LSC_{113} (Figure 12.33a).[143] Similar improved performance in LSC_{214}/LSC_{113} structure was observed later by Crumlin et al.,[178] Yashiro et al.,[179] Lee et al.,[144,180] and Hayd et al.[181] in their EIS measurements (Figure 12.33b).

Using ab initial calculation, Han et al. showed that lattice strain near the interface of LSC_{113} and LSC_{214} as well as the anisotropic oxygen transport path could be possible reasons that led to the fast oxygen exchange kinetic near the interface.[176] With in situ STM, Chen et al. observed the electronic activation effects near the

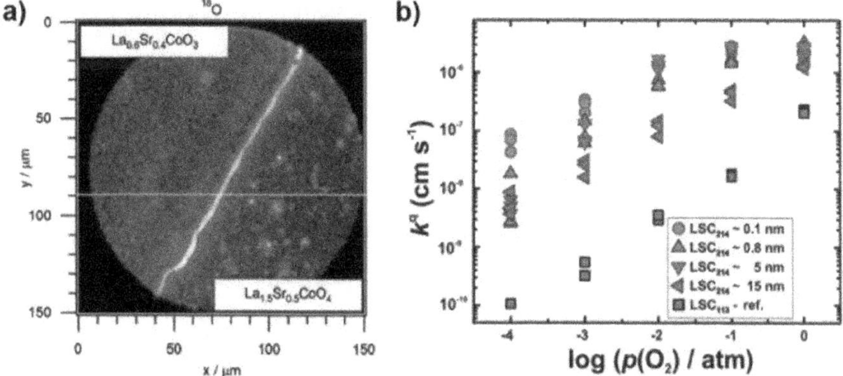

FIGURE 12.33 (a) ^{18}O intensity at the interface of $La_{1.5}Sr_{0.5}CoO_4/La_{0.6}Sr_{0.4}CoO_3$ after the diffusion annealing at 773 K for 180 s, in 0.2 bar $^{18}O_2$. Much higher ^{18}O intensity is presented at the interfacial region. (b) Surface exchange coefficients k^q as a function of oxygen partial pressure in various LSC_{113}/LSC_{214} films and single LSC_{113}. (Reprinted from Sase, M. et al., *J. Electrochem. Soc.*, 155, B793–B797, 2008. With permission from The Electrochemical Society, Copyright 2008.); Reprinted with permission from Crumlin, E.J. et al., *J. Phys. Chem. Lett.*, 1, 3149–3155, 2010. Copyright 2010 American Chemical Society.)

heterointerface of LSC_{214} and LSC_{113} and believed such effect could potentially lead to greatly enhanced performance of the heterostructure (Figure 12.34a).[175] Shao-Horn et al., on the other hand, attributed high performance of $LSC_{214/113}$ heterostructure to the stabilization of RP-phased decoration against decomposition into secondary phases than the single perovskite materials.[144,180] Similar improvement of electrode performance was reported by Lee et al. on LSM_{113}/LSC_{113} heterostructure prepared by PLD.[46] According to the analysis by AES, they attributed it to the Sr enrichment near the heterointerface. Based on density functional theory simulation, Gadre et al. found that Sr tended to replace La at the interface of LSC_{113} and LSC_{214}.[182] Combining the first principle-based simulation and coherent Bragg rod analysis (COBRA), Feng et al. found anomalous Sr enrichment at the interface of LSC_{113}/LSC_{214}, which they believed was the reason for the extremely enhanced oxygen surface exchange rate (Figure 12.34b).[78,183]

Besides the investigations in thin-film systems, several researchers indicated that the performance of porous perovskite-based materials could also be enhanced by surface engineering. Liu et al. reported that, by infiltration of the secondary perovskite-based phase, such as $La_{0.85}Sr_{0.15}MnO_{3-\delta}$ (LSM)[184] or $Sm_{0.5}Sr_{0.5}CoO_{3-\delta}$[185] onto $La_{1-x}Sr_xCo_yFe_{1-y}O_{3-\delta}$ (LSCF) systems, the electrochemical performance could be enhanced. Compared to the bare electrodes, LSM-infiltrated electrodes were found to exhibit apparently high-power density as well as enhanced stability at elevated temperatures (Figure 12.35). This difference could be explained by the fact that a favorable hybrid phase of LSM was formed at the interfacial region between LSM and LSCF after long-term operation, which would suppress surface Sr segregation and thus enhance its activity and stability during operation.[186]

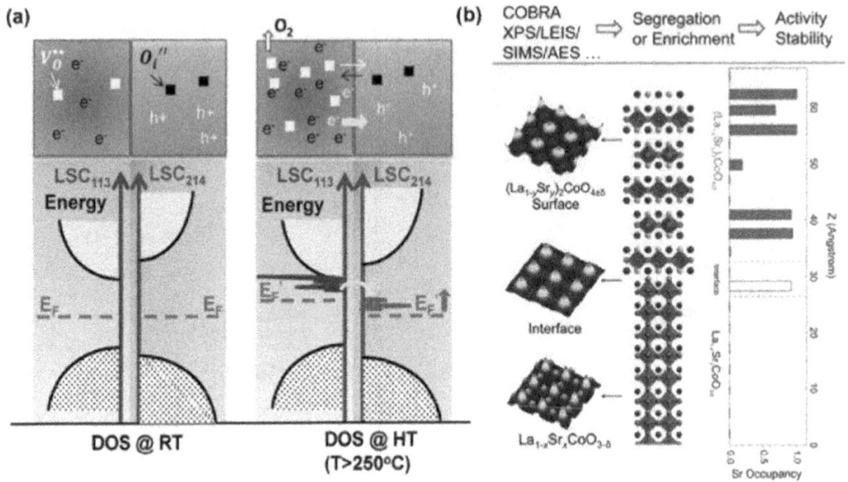

FIGURE 12.34 (a) Electronic structure alternation at the interface of LSC_{113} and LSC_{214} correlates to the creation of oxygen vacancies and electron injection from LSC_{113} to LSC_{214}, (b) Sr enrichment at the interface of LSC_{113} and LSC_{214} corresponds to strongly activity and stability enhancement. (From Gadre, M.J. et al.: Cation interdiffusion model for enhanced oxygen kinetics at oxide heterostructure interfaces. *Phys. Chem. Chem. Phys.*, 2012. 14. 2606–2616. Copyright Wiley-VCH Verlag GmbH & Co. KGaA. Reproduced with permission; Reprinted with permission from Feng, Z. et al., *Acc. Chem. Res.*, 49, 966–973, 2016. Copyright 2016 American Chemical Society.)

In another study of porous $La_{0.6}Sr_{0.4}Co_{0.2}Fe_{0.8}O_3$ (LSCF) electrode, Liu et al. modified its surface with $Pr_2(Ni, Mn)O_4$ decoration, which contained a conforming perovskite $Pr(Ni, Mn)O_3$ (PNM) layer and PrO_x nano particles, by a one-step infiltration process.[187] The anode-supported cells with this hybrid catalyst-coated LSCF cathode demonstrate remarkable peak power densities while maintaining

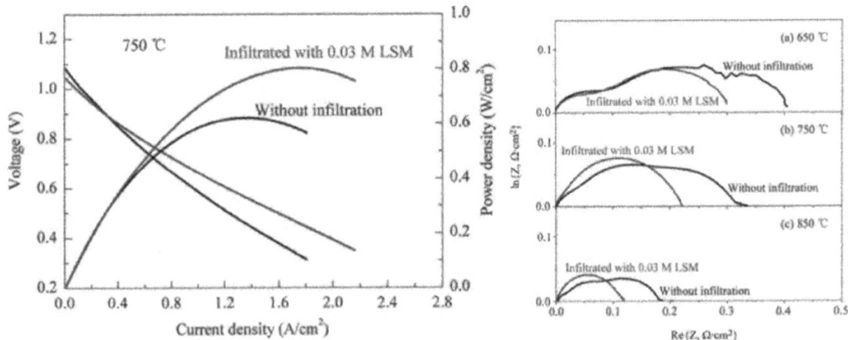

FIGURE 12.35 LSM-infiltrated LSCF cathodes appeared apparently high power density (under 750°C) as well as enhanced stability (under 650°C, 750°C, and 850°C) relative to LSCF without infiltration. (Reprinted from Liu, Z. et al., *J Energy Chem.*, 22, 555–559, 2013. With permission from Dalian Institute of Chemical Physics, Chinese Academy of Sciences. Copyright 2013.)

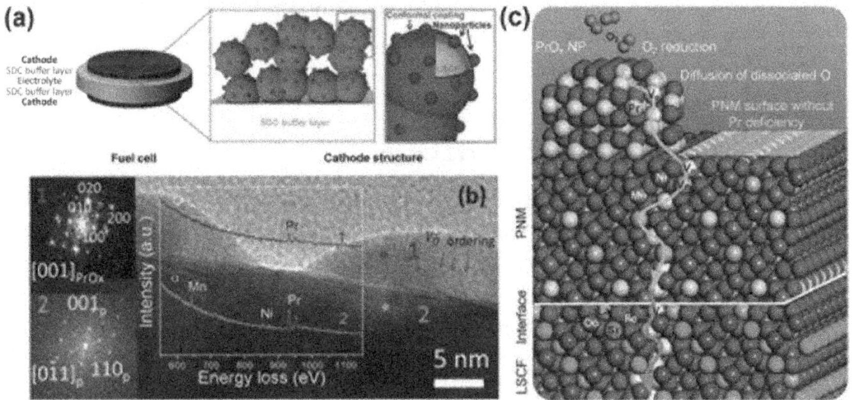

FIGURE 12.36 (a) An LSCF electrode decorated with PNM coating and exsoluted PrO_x nanoparticles, (b) high-resolution TEM image showing two PrO_x particles on a PNM coating deposited on an LSCF grain, (c) schematic representation of the lowest energy pathway for ORR on the hybrid catalyst (PrO_x/PNM) decorated LSC electrode and the enhanced bulk diffusion of oxygen vacancies in PNM by introducing Pr deficiency. (Chen, Y. et al., *Energ. Environ. Sci.*, 10, 964–971, 2017. Reproduced by permission of The Royal Society of Chemistry.)

excellent durability. Near ambient XPS and near edge X-Ray absorption fine structure (NEXAFS) analyses, together with density functional theory (DFT)–based simulation, showed that the perovskite PNM layer could significantly suppress Sr segregation and prevent performance degeneration. The PrO_x nano particles on the surface greatly accelerate the rate of electron transfer in the ORR, as shown in Figure 12.36. Other work using infiltration to improve performance and stability can be found in the review by Liu et al.[188]

12.5.4 Surface Decoration by Less Activated Phase

Some investigations focus on inhibiting detrimental segregation and enhancing surface stability of SOC electrodes by decorating non-electrode components. For instance, Huang et al. modified the surface of $La_{0.6}Sr_{0.4}CoO_{3-\delta}$ with ZrO_2 overcoats via atomic layer deposition (ALD).[189] According to EIS measurement and XPS characterization, its performance degradation rate (represented by polarization resistance) was much lower during 4000 h of operation. In another ALD study on $La_{0.6}Sr_{0.4}Fe_{0.8}Co_{0.2}O_{3-\delta}$-$Gd_{0.2}Ce_{0.8}O_{1.9}$ porous electrodes, they reported that a pristine electrode degraded approximately four times faster than a ZrO_2-overcoated one.[190] They proposed that Zr dopant at the surface of LSCF could introduce point defect with positive charge (e.g., $Zr^{\bullet}_{Fe(Co)}$), which could suppress surface Sr-segregation by reducing electronic attraction. Hwang et al. reported that in $La_{0.6}Sr_{0.4}CoO_3$-based electrodes modified by Al_2O_3 via atomic layer deposition, the apparent open-cell voltages and power densities decreased with the increasing thickness of the Al_2O_3 layer.[191] Such a decrease was due to the inert nature of Al_2O_3, which further led to a blocking effect when the overacted layer is too thick. These results show that modifying the surface with less active materials is a potential approach to achieve a balance of activity and stability for electrode materials.

12.6 CONCLUSION AND OUTLOOK

Under typical SOC operating conditions, variations in specific cation concentrations are often observed near the surfaces or interfaces of electrodes due to the accumulation (or depletion) of other charged defects such as oxygen vacancies or cation vacancies/interstitials. This compositional variation is further exacerbated by extreme electrical polarization or exposure to active/corrosive contaminants. In many cases, cation enrichment or depletion may lead to cation segregation or the formation of secondary phases on electrode surfaces or interfaces. Surface chemistry plays a vital role in determining the electrocatalytic activity and durability of many widely studied perovskite-based electrode materials for SOCs. Recent advancements in surface characterization techniques (especially in situ/operando tools) make it possible to probe the origins of cation segregation and the detailed chemical and structural features of electrodes associated with the process under operating conditions. According to existing investigations, the main driving forces for A-site cation enrichment or segregation are largely electrostatic and elastic interactions. Specifically, the degree of cation segregation is determined by both the nature of materials (such as cation deficiency, species of dopant ions, crystallinity, and lattice strain) and the environmental conditions (such as temperature and thermal history, atmosphere, and electrical polarization).

In general, a high operating temperature and a strong oxidizing atmosphere promotes A-site cation segregation in electrode materials, facilitating the formation of oxide particles on electrode surfaces. In contrast, a strong reducing atmosphere promotes B-site cation segregation, stimulating the formation of metallic particles on electrode surfaces. The different segregation modes have profound impacts on the electrocatalytic activity and stability of electrode materials. For example, Sr enrichment creates more oxygen vacancies, thus leading to considerable performance enhancements. However, when Sr segregation occurs due to excessive Sr enrichment (e.g., beyond the solubility limit of Sr), the formation of SrO (an inert compound) can degrade electrode performance. Furthermore, SrO may be more vulnerable to poisoning by contaminants or impurities such as H_2O, CO_2, or Cr (from Cr-containing interconnect), further degrading the electrocatalytic activity and durability of the electrodes. Still, studies have shown positive effects of some segregated AO_x oxides (such as PrO_x). In contrast, B-cation segregation under reducing conditions often results in the formation of highly active transition metal particles, generally enhancing surface conductivity, electrocatalytic activity, and stability of the electrodes under operating conditions.

A profound understanding of cation enrichment and segregation mechanisms as well as the correlation between the surface features (structure, chemistry, and nanostructure) and the electrochemical properties of SOC electrode materials provide guidance for the rational design of improved electrodes with enhanced electrocatalytic activity and stability. Since the electrochemical properties of electrodes depend sensitively on both A-site and B-site cation enrichment and segregation, it is critical to exert precise control of the synthesis processes in order to obtain electrodes with desired properties. Alternatively, the surface of the electrode may be modified using a thin, conformal coating of catalysts to minimize or eliminate the detrimental effect of cation enrichment and segregation of the electrode materials. The specific methods of surface modification include vapor phase deposition and solution infiltration of various catalyst materials. In addition, to enhancing long-term stability against detrimental segregation or contaminant poisoning, surface decoration by nanoparticles of some materials that are less electrocatalytically active has also proven effective.

Nevertheless, there is still much to learn about cation segregation. Some of the specific open issues to be addressed include how to track in real time the evolution of surface chemistry and structure and correlate these microscopic features with electrode performance under SOC operating conditions, how to enhance desired positive effects while suppressing negative effects associated with cation segregation under operating conditions, and how to achieve enhancing electrocatalytic activity while maintaining long-term durability for various perovskite-electrode systems. To address these scientific issues, powerful surface characterization tools (e.g., secondary ion mass spectrometry, high-temperature scanning probe microscopy, high-sensitivity low-energy ion scattering, etc.) must be used, either under well-controlled conditions or under operando conditions, to enhance our ability to study the origin and pinpoint the impact of surface cation segregation on electrode performance. These studies may provide critical insights into the mechanisms of cation segregation. In addition, advanced theoretical modeling and simulation of surface and interfacial processes are equally important. New theories and predictive models for the origin and overall impact of cation segregation must be developed and validated via carefully designed experiments. Once corroborated, these theories and models can in turn be used for the rational design of better electrode materials for various applications.

REFERENCES

1. Druce, J., et al. Surface termination and subsurface restructuring of perovskite-based solid oxide electrode materials. *Energy & Environmental Science* **7**, 3593–3599 (2014).
2. Shao, Z., Haile, S. M. A high-performance cathode for the next generation of solid-oxide fuel cells. *Nature* **431**, 170–173 (2004).
3. Rupp, G. M., et al. Surface chemistry of La0.6Sr0.4CoO3 d thin films and its impact on the oxygen surface exchange resistance. *Journal of Materials Chemistry A* **3**, 22759–22769 (2015).
4. Chen, D., Chen, C., Baiyee, Z. M., Shao, Z., Ciucci, F. Nonstoichiometric oxides as low-cost and highly-efficient oxygen reduction/evolution catalysts for low-temperature electrochemical devices. *Chemical Reviews* **115**, 9869–9921 (2015).
5. Chen, Y., et al. Advances in cathode materials for solid oxide fuel cells: Complex oxides without alkaline earth metal elements. *Advanced Energy Materials* **5**, 1500537 (2015).
6. Irvine, J. T. S., et al. Evolution of the electrochemical interface in high-temperature fuel cells and electrolysers. *Nature Energy* **1**, 15014 (2016).
7. Yang, G., Zhou, W., Liu, M., Shao, Z. Enhancing electrode performance by exsolved nanoparticles: A superior cobalt-free perovskite electrocatalyst for solid oxide fuel cells. *ACS Applied Materials & Interfaces* **8**, 35308–35314 (2016).

8. Sun, Y., et al. A-site deficient perovskite: The parent for in situ exsolution of highly active, regenerable nano- particles as SOFC anodes. *Journal of Materials Chemistry A* **3**, 11048–11056 (2015).
9. Kubicek, M., Limbeck, A., Mling, T. F., Hutter, H., Fleig, J. R. Relationship between cation segregation and the electrochemical oxygen reduction kinetics of $La_{0.6}Sr_{0.4}CoO_{3-\delta}$ thin film electrodes. *Journal of The Electrochemical Society* **6**, B727–B734 (2011).
10. Baqué, L. C., et al. Degradation of oxygen reduction reaction kinetics in porous $La_{0.6}Sr_{0.4}Co_{0.2}Fe_{0.8}O_{3-\delta}$ cathodes due to aging-induced changes in surface chemistry. *Journal of Power Sources* **337**, 166–172 (2017).
11. Du, Z., et al. High-performance anode material $Sr_2FeMo_{0.65}Ni_{0.35}O_{6-\delta}$ with in situ exsolved nanoparticle catalyst. *ACS Nano* **10**, 8660–8669 (2016).
12. Hua, B., Yan, N., Li, M., Sun, Y. F., Zhang, Y.Q. Anode-engineered protonic ceramic fuel cell with excellent performance and fuel compatibility. *Advanced Materials* **28**, 8922–8926 (2016).
13. Myung, J., Neagu, D., Miller, D. N., Irvine, J. T. S. Switching on electrocatalytic activity in solid oxide cells. *Nature* **537**, 528–531 (2016).
14. Li, Y., et al. Controlling cation segregation in perovskite-based electrodes for high electrocatalytic activity and durability. *Chemical Society Reviews* **46**, 6345–6378 (2017).
15. Chen, Y., et al. Segregated chemistry and structure on (001) and (100) surfaces of $(La_{1-x}Sr_x)_2CoO_4$ override the crystal anisotropy in oxygen exchange kinetics. *Chemistry of Materials* **27**, 5436–5450 (2015).
16. Téllez, H., Druce, J., Kilnerac, J. A., Ishihara, T. Relating surface chemistry and oxygen surface exchange in $LnBaCo_2O_{5+\delta}$ air electrodes. *Faraday Discussions* **182**, 145–157 (2015).
17. Téllez, H., Druce, J., Ju, Y., Kilner, J., Ishihara, T. Surface chemistry evolution in $LnBaCo_2O_{5+\delta}$ double perovskites for oxygen electrodes. *International Journal of Hydrogen Energy* **39**, 20856–20863 (2014).
18. Druce, J., Téllez, H., Simrick, N., Ishihara, T., Kilner, J. Surface composition of solid oxide electrode structures by laterally resolved low energy ion scattering (LEIS). *International Journal of Hydrogen Energy* **39**, 20850–20855 (2014).
19. Limbeck, A., et al. Dynamic etching of soluble surface layers with on-line inductively coupled plasma mass spectrometry detection—a novel approach for determination of complex metal oxide surface cation stoichiometry. *Journal of Analytical Atomic Spectrometry* **31**, 1638–1646 (2016).
20. Burriel, M., et al. Influence of crystal orientation and annealing on the oxygen diffusion and surface exchange of $La_2NiO_{4+\delta}$. *Journal of Physical Chemistry C* **120**, 17927–17938 (2016).
21. Cushman, C. V., et al. Low energy ion scattering (LEIS). A practical introduction to its theory, instrumentation, and applications. *Analytical Methods* **8**, 3419–3439 (2016).
22. Druce, J., Simrick, N., Ishihara, T., Kilner, J. "Imaging" LEIS of micro-patterned solid oxide fuel cell electrodes. *Nuclear Instruments and Methods in Physics Research Section B* **332**, 261–265 (2014).
23. Cavallaro, A., Harrington, G. F., Skinner, S. J., Kilner, J.A. Controlling the surface termination of NdGaO3 (110): The role of the gas atmosphere. *Nanoscale* **6**, 7263–7273 (2014).
24. Symianakis, E., et al. Electrochemical characterization and quantified surface termination obtained by low energy ion scattering and X-ray photoelectron spectroscopy of orthorhombic and rhombohedral $LaMnO_3$ powders. *Journal of Physical Chemistry C* **119**, 12209–12217 (2015).
25. Druce, J., Ishihara, T., Kilner, J. Surface composition of perovskite-type materials studied by low energy ion scattering (LEIS). *Solid State Ionics* **262**, 893–896 (2014).

26. Druce, J., Téllez, H., Ishihara, T., Kilner, J. A. Oxygen exchange and transport in dual phase ceramic composite electrodes. *Faraday Discussions* **182**, 271–288 (2015).
27. Tomkiewicz, A. C., Tamimi, M. A., Huq, A., McIntosh, S. Is the surface oxygen exchange rate linked to bulk ion diffusivity in mixed conducting Ruddlesden-Popper phases? *Faraday Discussions* **182**, 113–127 (2015).
28. Burriel, M., et al. Absence of Ni on the outer surface of Sr doped La_2NiO_4 single crystals. *Energy & Environmental Science* **7**, 311–316 (2014).
29. Druce, J., Kilner, J. A., Ishihara, T. Relating electrochemical performance measurements and surface analyses of $Pr_{2-x}La_xNiO_{4+\delta}$ electrode materials. *ECS Transactions* **57**, 3269–3276 (2013).
30. Cai, Z., Kubicek, M., Fleig, J., Yildiz, B. Chemical heterogeneities on $La_{0.6}Sr_{0.4}CoO_{3-\delta}$ thin films-correlations to cathode surface activity and stability. *Chemistry of Materials* **24**, 1116–1127 (2012).
31. Dulli, H., Dowben, P. A., Liou, S. H., Plummer, E. W. Surface segregation and restructuring of colossal-magnetoresistant manganese perovskites $La_{0.65}Sr_{0.35}MnO_3$. *Physical Review B* **62**, 14629–14632 (2000).
32. Borca, C. N., et al. The surface phases of the La0.65Pb0.35MnO3 manganese perovskite surface. *Surface Science* **512**, L346–L352 (2002).
33. Nenning, A., et al. Ambient pressure XPS study of mixed conducting perovskite-type SOFC cathode and anode materials under well-defined electrochemical polarization. *Journal of Physical Chemistry C* **120**, 1461–1471 (2016).
34. Crumlin, E. J., et al. In situ ambient pressure X-ray photoelectron spectroscopy of cobalt perovskite surfaces under cathodic polarization at high temperatures. *Journal of Physical Chemistry C* **117**, 16087–16094 (2013).
35. Stoerzinger, K. A., Hong, W. T., Crumlin, E. J., Bluhm, H., Shao-Horn, Y. Insights into electrochemical reactions from ambient pressure photoelectron spectroscopy. *Accounts of Chemical Research* **48**, 2976–2983 (2015).
36. Hong, W. T., et al. Near-ambient pressure XPS of high-temperature surface chemistry in $Sr_2Co_2O_5$ thin films. *Topics in Catalysis* **59**, 574–582 (2016).
37. Crumlin, E. J., Mutoro, E., Zhi, L., Grass, M. E., Biegalski, M. D. Surface strontium enrichment on highly active perovskites for oxygen electrocatalysis in solid oxide fuel cells. *Energy & Environmental Science* **5**, 6081–6088 (2012).
38. Vovk, G., Chen, X., Mims, C. A. In situ XPS studies of perovskite oxide surfaces under electrochemical polarization. *Journal of Physical Chemistry B* **109**, 2445–2454 (2005).
39. Wu, Q., Liu, M., Jaegermann, W. X-ray photoelectron spectroscopy of La0.5Sr0.5MnO3. *Materials Letters* **59**, 1980–1983 (2005).
40. Opitz, A. K., et al. Enhancing electrochemical water-splitting kinetics by polarization-driven formation of near-surface iron(0): An insitu XPS study on perovskite-type electrodes. *Angewandte Chemie International Edition* **54**, 2628–2632 (2015).
41. Cai, Z., Kuru, Y., Han, J. W., Chen, Y., Yildiz, B. Surface electronic structure transitions at high temperature on perovskite oxides: The case of strained $La_{0.8}Sr_{0.2}CoO_3$ thin films. *Journal of the American Chemical Society* **133**, 17696–17704 (2011).
42. Fulmer, A. T., Dondlinger, J., Langell, M. A. Passivation of the La_2NiMnO_6 double perovskite to hydroxylation by excess nickel, and the fate of the hydroxylated surface upon heating. *Applied Surface Science* **305**, 544–553 (2014).
43. Araki, W., Yamaguchi, T., Arai, Y., Malzbender, J. Strontium surface segregation in $La_{0.58}Sr_{0.4}Co_{0.2}Fe_{0.8}O_{3-\delta}$ annealed under compression. *Solid State Ionics* **268**, 1–6 (2014).
44. Tsvetkov, N., Lu, Q., Sun, L., Crumlin, E. J., Yildiz, B. Improved chemical and electrochemical stability of perovskite oxides with less reducible cations at the surface. *Nature Materials* **15**, 1010–1016 (2016).

45. Lee, W., Han, J. W., Chen, Y., Cai, Z., Yildiz, B. Cation size mismatch and charge interactions drive dopant segregation at the surfaces of manganite perovskites. *Journal of the American Chemical Society* **135**, 7909–7925 (2013).
46. Lee, D., et al. Enhanced oxygen surface exchange kinetics and stability on epitaxial $La_{0.8}Sr_{0.2}CoO_{3-\delta}$ thin films by $La_{0.8}Sr_{0.2}MnO_{3-\delta}$ decoration. *Journal of Physical Chemistry C* **118**, 14326–14334 (2014).
47. Feng, Z., et al. In situ studies of the temperature-dependent surface structure and chemistry of single-crystalline (001)-oriented $La_{0.8}Sr_{0.2}CoO_{3-\delta}$ perovskite thin films. *Journal of Physical Chemistry Letters* **4**, 1512–1518 (2013).
48. Katsiev, K., Yildiz, B., Balasubramaniam, K., Salvador, P. A. Electron tunneling characteristics on La0.7Sr0.3MnO3 thin-film surfaces at high temperature. *Applied Physics Letters* **95**, 092106 (2009).
49. Newell, D. T., Harrison, A., Silly, F., Castell, M. R. $SrTiO_3$ (001)-(root5 X root5)-R26.6 reconstruction: A surface resulting from phase separation in a reducing environment. *Physical Review B* **75**, 205429 (2007).
50. Huber, A., et al. In situ study of electrochemical activation and surface segregation of the SOFC electrode material $La_{0.75}Sr_{0.25}Cr_{0.5}Mn_{0.5}O_{3-\delta}$. *Physical Chemistry Chemical Physics* **14**, 751–758 (2012).
51. Lu, J., Hua, X., Long, Y. Recent advances in real-time and in situ analysis of an electrode–electrolyte interface by mass spectrometry. *Analyst* **142**, 691–699 (2017).
52. Téllez, H., et al. New perspectives in the surface analysis of energy materials by combined time-of-flight secondary ion mass spectrometry (ToF-SIMS) and high sensitivity low-energy ion scattering (HS-LEIS). *Journal of Analytical Atomic Spectrometry* **29**, 1361–1370 (2014).
53. Norrman, K., Hansen, K. V., Mogensen, M. Time-of-flight secondary ion mass spectrometry as a tool for studying segregation phenomena at nickel-YSZ interfaces. *Journal of the European Ceramic Society* **26**, 967–980 (2006).
54. Guerquin-Kern, J., Wu, T., Quintana, C., Croisy, A. Progress in analytical imaging of the cell by dynamic secondary ion mass spectrometry (SIMS microscopy). *Biochimica et Biophysica Acta (BBA)-General Subjects* **1724**, 228–238 (2005).
55. Kilner, J. A., Skinner, S. J., Brongersma, H. H. The isotope exchange depth profiling (IEDP) technique using SIMS and LEIS. *Journal of Solid State Electrochemistry* **15**, 861–876 (2011).
56. Kubicek, M., et al. Tensile lattice strain accelerates oxygen surface exchange and diffusion in $La_{1-x}Sr_xCoO_{3-\delta}$ thin films. *ACS Nano* **7**, 3276–3286 (2013).
57. Kessel, M., De Souza, R. A., Martin, M. Oxygen diffusion in single crystal barium titanate. *Physical Chemistry Chemical Physics* **17**, 12587–12597 (2015).
58. Kubicek, M., et al. Cation diffusion in $La_{0.6}Sr_{0.4}CoO_{3-\delta}$ below 800 and its relevance for Sr segregation. *Physical Chemistry Chemical Physics* 2715–2726 (2014).
59. Ravella, U. K., Liu, J., Corbel, G., Skinner, S. J., Lacorre, P. Cationic intermixing and reactivity at the $La_2Mo_2O_9/La_{0.8}Sr_{0.2}MnO_{3-\delta}$ solid oxide fuel cell electrolyte-cathode interface. *ChemSusChem* **9**, 2182–2192 (2016).
60. Huber, A., et al. In situ study of activation and de-activation of LSM fuel cell cathodes: Electrochemistry and surface analysis of thin-film electrodes. *Journal of Catalysis* **294**, 79–88 (2012).
61. Bucher, E., Gspan, C., Höschen, T., Hofer, F., Sitte, W. Oxygen exchange kinetics of $La_{0.6}Sr_{0.4}CoO_{3-\delta}$ affected by changes of the surface composition due to chromium and silicon poisoning. *Solid State Ionics* **299**, 26–31 (2017).
62. Kim, Y., Chen, X., Jiang, S. P., Bae, J. Chromium deposition and poisoning at $Ba_{0.5}Sr_{0.5}Co_{0.8}Fe_{0.2}O_{3-\delta}$ cathode of solid oxide fuel cells. *Electrochemical and Solid-State Letters* **4**, B41–B45 (2011).

63. Zhao, L., Drennan, J., Kong, C., Amarasinghe, S., Jiang, S. P. Insight into surface segregation and chromium deposition on $La_{0.6}Sr_{0.4}Co_{0.2}Fe_{0.8}O_{3-\delta}$ cathodes of solid oxide fuel cells. *Journal of Materials Chemistry A* **2**, 11114–11123 (2014).
64. Jiang, S. P., Zhang, S., Zhen, Y. D. Deposition of Cr species at (La, Sr)(Co, Fe)O_3 cathodes of solid oxide fuel cells. *Journal of The Electrochemical Society* **1**, A127–A134 (2006).
65. Rahmati, B., et al. Oxidation of reduced polycrystalline Nb-doped $SrTiO_3$: Characterization of surface islands. *Surface Science* **595**, 115–126 (2005).
66. Lenser, C., et al. Formation and movement of cationic defects during forming and resistive switching in $SrTiO_3$ thin film devices. *Advanced Functional Materials* **25**, 6360–6368 (2015).
67. Oh, D., Gostovic, D., Wachsman, E. D. Mechanism of $La_{0.6}Sr_{0.4}Co_{0.2}Fe_{0.8}O_3$ cathode degradation. *Journal of Materials Research* **27**, 1992–1999 (2012).
68. Lee, W., Yildiz, B. Factors that influence cation segregation at the surfaces of perovskite oxides. *ECS Transactions* **57**, 2115–2123 (2013).
69. Jalili, H., Han, J. W., Kuru, Y., Cai, Z., Yildiz, B. New insights into the strain coupling to surface chemistry, electronic structure, and reactivity of $La_{0.7}Sr_{0.3}MnO_3$. *Journal of Physical Chemistry Letters* **2**, 801–807 (2011).
70. Liang, Y., Bonnell, D. A. Structures and chemistry of the annealed $SrTiO_3$(001) surface. *Surface Science* **310**, 128–134 (1994).
71. Chen, Y., et al. Impact of Sr segregation on the electronic structure and oxygen reduction activity of $SrTi_{1-x}Fe_xO_3$ surfaces. *Energy & Environmental Science* **5**, 7979–7988 (2012).
72. Yang, C., et al. Sulfur-tolerant redox-reversible anode material for direct hydrocarbon solid oxide fuel cells. *Advanced Materials* **24**, 1439–1443 (2012).
73. Chang, K. C., Ingram, B., Hopper, M., Ilavsky, J., You, H. Ultra small angle X-ray scattering studies of solid oxide fuel cell cathode powders. *ECS Transactions* **50**, 111–115 (2013).
74. Hardy, J. S., Templeton, J. W., Edwards, D. J., Lu, Z., Stevenson, J. W. Lattice expansion of LSCF-6428 cathodes measured by in situ XRD during SOFC operation. *Journal of Power Sources* **198**, 76–82 (2012).
75. Richard, M. I., et al. In situ x-ray scattering study on the evolution of Ge island morphology and relaxation for low growth rate: Advanced transition to superdomes. *Physical Review B* **80**, 045313 (2009).
76. Tsvetkov, N., Lu, Q., Yildiz, B. Improved electrochemical stability at the surface of La0.8Sr0.2CoO3 achieved by surface chemical modification. *Faraday Discuss* **182**, 257–269 (2015).
77. Ma, W., et al. Vertically aligned nanocomposite $La_{0.8}Sr_{0.2}CoO_3/(La_{0.5}Sr_{0.5})_2CoO_4$ cathodes–electronic structure, surface chemistry and oxygen reduction kinetics. *Journal of Materials Chemistry A* **3**, 207–219 (2015).
78. Feng, Z., et al. Catalytic activity and stability of oxides: The role of near-surface atomic structures and compositions. *Accounts of Chemical Research* **49**, 966–973 (2016).
79. Su, D. S., Zhang, B., Robert, S. Electron microscopy of solid catalysts-transforming from a challenge to a toolbox. *Chemical reviews* **115**, 2818–2882 (2015).
80. Xin, H. L., et al. Revealing the atomic restructuring of Pt–Co nanoparticles. *Nano Letters* **14**, 3203–3207 (2014).
81. Han, B., et al. Nanoscale structural oscillations in perovskite oxides induced by oxygen evolution. *Nature Materials* **16**, 121–126 (2017).
82. Zwiebler, M., et al. Electronic depth profiles with atomic layer resolution from resonant soft x-ray reflectivity. *New Journal of Physics* **17**, 083046 (2015).
83. Fenter, P., Park, C., Nagy, K. L., Sturchio, N. C. Resonant anomalous X-ray reflectivity as a probe of ion adsorption at solid–liquid interfaces. *Thin Solid Films* **515**, 5654–5659 (2007).

84. Perret, E., et al. Resonant X-ray scattering studies of epitaxial complex oxide thin films. *Journal of Applied Crystallography* **46**, 76–87 (2013).
85. Zhao, L., Zhang, I., Becker, T., Jiang, S. P. Raman spectroscopy study of chromium deposition on $La_{0.6}Sr_{0.4}Co_{0.2}Fe_{0.8}O_{3-\delta}$ cathode of solid oxide fuel cells. *Journal of the Electrochemical Society* **161**, F687–F693 (2014).
86. McIntyre, M. D., Neuburger, D. M., Walker, R. A. In situ optical studies of carbon accumulation with different molecular weight alkanes on solid oxide fuel cell Ni anodes. *ECS Transactions* **66**, 11–19 (2015).
87. Li, X., et al. High-temperature surface enhanced Raman spectroscopy for in situ study of solid oxide fuel cell materials. *Energy & Environmental Science* **7**, 306–310 (2014).
88. Pomfret, M. B., et al. Hydrocarbon fuels in solid oxide fuel cells: In situ Raman studies of graphite formation and oxidation. *Journal of Physical Chemistry C* **112**, 5232–5240 (2008).
89. Yu, Y., et al. Chemical characterization of surface precipitates in $La_{0.7}Sr_{0.3}Co_{0.2}Fe_{0.8}O_{3-\delta}$ as cathode material for solid oxide fuel cells. *Journal of Power Sources* **333**, 247–253 (2016).
90. Fister, T. T., et al. In situ characterization of strontium surface segregation in epitaxial $La_{0.7}Sr_{0.3}MnO_3$ thin films as a function of oxygen partial pressure. *Applied Physics Letters* **93**, 151904 (2008).
91. Neagu, D., Tsekouras, G., Miller, D. N., Ménard, H., Irvine, J. T. S. In situ growth of nanoparticles through control of non-stoichiometry. *Nature Chemistry* **5**, 916–923 (2013).
92. Tsekouras, G., Neagu, D., Irvine, J. T. S. Step-change in high temperature steam electrolysis performance of perovskite oxide cathodes with exsolution of B-site dopants. *Energy & Environmental Science* **6**, 256–266 (2013).
93. Jin, T., Lu, K. Surface and interface behaviors of $(La_{0.8}Sr_{0.2})_xMnO_3$ air electrode for solid oxide cells. *Journal of Power Sources* **196**, 8331–8339 (2011).
94. Abernathy, H., et al. Examination of the mechanism for the reversible aging behavior at open circuit when changing the operating temperature of $(La_{0.8}Sr_{0.2})_{0.95}MnO_3$ electrodes. *Solid State Ionics* **272**, 144–154 (2015).
95. Davis, J. N., et al. Hard X-ray fluorescence measurements of heteroepitaxial solid oxide fuel cell cathode materials. *ECS Transactions* **41**, 19–24 (2012).
96. Kwon, H., Lee, W., Han, J. W. Suppressing cation segregation on lanthanum-based perovskite oxides to enhance the stability of solid oxide fuel cell cathodes. *RSC Advances* **6**, 69782–69789 (2016).
97. Estradé, S., et al. Effects of thickness on the cation segregation in epitaxial (001) and (110) La2/3Ca1/3MnO3 thin films. *Applied Physics Letters* **95**, 072507 (2009).
98. Sharma, V., et al. Effects of moisture on (La, A)MnO_3 (A = Ca, Sr, and Ba) solid oxide fuel cell cathodes: A first-principles and experimental study. *Journal of Materials Chemistry A* **4**, 5605–5615 (2016).
99. Sharma, V., Mahapatra, M. K., Singh, P., Ramprasad, R. Cationic surface segregation in doped $LaMnO_3$. *Journal of Materials Science* **50**, 3051–3056 (2015).
100. Galvez, M. E., et al. Physico-chemical changes in Ca, Sr and Al-doped La-Mn-O perovskites upon thermochemical splitting of CO_2 via redox cycling. *Physical Chemistry Chemical Physics* **17**, 6629–6634 (2015).
101. Araki, W., Miyashita, M., Arai, Y. Strontium surface segregation in $La_{0.6}Sr_{0.4}Co_{0.2}Fe_{0.8}O_{3-\delta}$ subjected to mechanical stress. *Solid State Ionics* **290**, 18–23 (2016).
102. Szot, K., Speier, W. Surfaces of reduced and oxidized $SrTiO_3$ from atomic force microscopy. *Physical Review B* **60**, 5909–5926 (1999).
103. Han, J. W., Jalili, H., Kuru, Y., Cai, Z., Yildiz, B. Strain effects on the surface chemistry of $La_{0.7}Sr_{0.3}MnO_3$. *ECS Transactions* **35**, 2097–2104 (2011).

104. Ding, H., Virkar, A. V., Liu, M., Liu, F. Suppression of Sr surface segregation in $La_{1-x}Sr_xCo_{1-y}Fe_yO_{3-\delta}$: a first principles study. *Physical Chemistry Chemical Physics* **15**, 489–496 (2013).
105. Lee, H., Park, C., Park, H. Effect of La^{3+} substitution with Gd3+ on the resistive switching properties of $La_{0.7}Sr_{0.3}MnO_3$ thin films. *Applied Physics Letters* **104**, 191604 (2014).
106. Ravkina, O., Yaremchenko, A., Feldhoff, A. Phase separation in BSCF perovskite under elevated oxygen pressures ranging from 1 to 50 bar. *Journal of Membrane Science* **520**, 76–88 (2016).
107. Caillol, N., Pijolat, M., Siebert, E. Investigation of chemisorbed oxygen, surface segregation and effect of post-treatments on $La_{0.8}Sr_{0.2}MnO_3$ powder and screen-printed layers for solid oxide fuel cell cathodes. *Applied Surface Science* **253**, 4641–4648 (2007).
108. Xie, H., et al. Effects of annealing on structure and composition of LSMO thin films. *Physica B: Condensed Matter* **477**, 14–19 (2015).
109. Neagu, D., et al. Nano-socketed nickel particles with enhanced coking resistance grown in situ by redox exsolution. *Nature Communications* **6**, 8120 (2015).
110. Tselev, A., et al. Surface control of epitaxial manganite filmsvia oxygen pressure. *ACS Nano* **9**, 4316–4327 (2015).
111. Darvish, S., Asadikiya, M., Hu, B., Singh, P., Zhong, Y. Thermodynamic prediction of the effect of CO_2 to the stability of $(La_{0.8}Sr_{0.2})_{0.98}MnO_{3\pm\delta}$ system. *International Journal of Hydrogen Energy* **41**, 10239–10248 (2016).
112. Bucher, E., Sitte, W., Klauser, F., Bertel, E. Oxygen exchange kinetics of $La_{0.58}Sr_{0.4}Co_{0.2}Fe_{0.8}O_3$ at 600°C in dry and humid atmospheres. *Solid State Ionics* **191**, 61–67 (2011).
113. Chen, K., Jiang, S. P. Review—materials degradation of solid oxide electrolysis cells. *Journal of the Electrochemical Society* **163**, F3070–F3083 (2016).
114. Chen, K., et al. Chromium deposition and poisoning of $La_{0.8}Sr_{0.2}MnO_3$ oxygen electrodes of solid oxide electrolysis cells. *Faraday Discussion* **182**, 457–476 (2015).
115. Mutoro, E., et al. Reversible compositional control of oxide surfaces by electrochemical potentials. *Journal of Physical Chemistry Letters* **3**, 40–44 (2012).
116. Zhu, L., et al. Electrochemically driven deactivation and recovery in $PrBaCo_2O_{5+\delta}$ oxygen electrodes for reversible solid oxide fuel cells. *ChemSusChem* **9**, 2443–2450 (2016).
117. Pan, Z., Liu, Q., Zhang, L., Zhang, X., Chan, S. H. Study of activation effect of anodic current on $La_{0.6}Sr_{0.4}Co_{0.2}Fe_{0.8}O_{3-\delta}$ air electrode in solid oxide electrolyzer cell. *Electrochimica Acta* **209**, 56–64 (2016).
118. Baumann, F. S., et al. Strong performance improvement of $La_{0.6}Sr_{0.4}Co_{0.8}Fe_{0.2}O_3$ - SOFC cathodes by electrochemical activation. *Journal of the Electrochemical Society* **152**, A2074–A2079 (2005).
119. Lin, Y., et al. Electron-induced Ti-rich surface segregation on $SrTiO_3$ nanoparticles. *Micron* **68**, 152–157 (2015).
120. Chen, D., Bishop, S. R., Tuller, H. L. Nonstoichiometry in oxide thin films operating under anodic conditions: A chemical capacitance study of the praseodymium-cerium oxide system. *Chemistry of Materials* **26**, 6622–6627 (2014).
121. Chen, D., Tuller, H. L. Voltage-controlled nonstoichiometry in oxide thin films: $Pr_{0.1}$ $Ce_{0.9}O_{2-\delta}$ case study. *Advanced Functional Materials* **24**, 7638–7644 (2014).
122. Waller, D., Lane, J. A., Kilner, J. A., Steele, B. C. H. The structure of and reaction of A-site deficient $La_{0.6}Sr_{0.4-x}Co_{0.2}Fe_{0.8}O_{3-\delta}$ perovskites. *Materials Letters* **27**, 225–228 (1996).
123. Tzoumanekas, C., Kelires, P. C. Segregation, clustering, and suppression of phase separation in amorphous silicon–germanium alloys. *Journal of Non-Crystalline Solids* **266**, 670–674 (2000).
124. Murugavel, P., et al. Effects of oxygen annealing on the physical properties and surface microstructures of $La_{0.8}Ba_{0.2}MnO_3$ films. *Journal of Physics D: Applied Physics* **35**, 3166–3170 (2002).

125. Yin, Z., et al. Oxygen in-diffusion at room temperature in epitaxial $La_{0.7}Ca_{0.3}MnO_{3-\delta}$ thin films. *Journal of Physics D-Applied Physics* **42**, 125002 (2009).
126. Wang, H., et al. Mechanisms of performance degradation of (La, Sr)(Co, Fe)$O_{3-\delta}$ solid oxide fuel cell cathodes. *Journal of The Electrochemical Society* **163**, F581–F585 (2016).
127. Wang, W., Jiang, S. A mechanistic study on the activation process of (La, Sr)MnO_3 electrodes of solid oxide fuel cells. *Solid State Ionics* **177**, 1361–1369 (2006).
128. Ohly, C., Hoffmann-Eifert, S., Szot, K., Waser, R. Electrical conductivity and segregation effects of doped SrTiO3 thin films. *Journal of the European Ceramic Society* **21**, 1673–1676 (2001).
129. Ridder, M. D., Vervoort, A. G. J., Welzenis, R. G. V., Brongersma, H. H. The limiting factor for oxygen exchange at the surface of fuel cell electrolytes. *Solid State Ionics* **156**, 255–262 (2003).
130. Pan, Z., Liu, Q., Zhang, L., Zhang, X., Chan, S. H. Effect of Sr surface segregation of $La_{0.6}Sr_{0.4}Co_{0.2}Fe_{0.8}O_{3-\delta}$ electrode on its electrochemical performance in SOC. *Journal of the Electrochemical Society* **162**, F1316–F1323 (2015).
131. Wei, B., Chen, K., Zhao, L., Lu, Z., Jiang, S. P. Chromium deposition and poisoning at $La_{0.6}Sr_{0.4}Co_{0.2}Fe_{0.8}O_{3-\delta}$ oxygen electrodes of solid oxide electrolysis cells. *Physical Chemistry Chemical Physics* **17**, 1601–1609 (2015).
132. Jung, W. C., Tuller, H. L. Investigation of surface Sr segregation in model thin film solid oxide fuel cell perovskite electrodes. *Energy & Environmental Science* **5**, 5370–5378 (2012).
133. la O', G. J., Savinell, R. F., Shao-Horn, Y. Activity enhancement of dense strontium-doped lanthanum manganite thin films under cathodic polarization: A combined AES and XPS study. *Journal of The Electrochemical Society* **156**, B771–B781 (2009).
134. Wagner, S. F., et al. Enhancement of oxygen surface kinetics of SrTiO3 by alkaline earth metal oxides. *Solid State Ionics* **177**, 1607–1612 (2006).
135. Mutoro, E., Crumlin, E. J., Biegalski, M. D., Christen, H. M., Yang, S. H. Enhanced oxygen reduction activity on surface-decorated perovskite thin films for solid oxide fuel cells. *Energy & Environmental Science* **4**, 3689–3696 (2011).
136. Rupp, G. M., Opitz, A. K., Nenning, A., Limbeck, A., Fleig, J. Real-time impedance monitoring of oxygen reduction during surface modification of thin film cathodes. *Nature Materials* **16**, 640–645 (2017).
137. Rothschild, A., Menesklou, W., Tuller, H. L., Ivers-Tiffée, E. Electronic structure, defect chemistry, and transport properties of $SrTi_{1-x}Fe_xO_{3-y}$ solid solutions. *Chemistry of Materials* **18**, 3651–3659 (2006).
138. Liu, S., Liu, Q., Luo, J. CO_2-to-CO conversion on layered perovskite with in situ exsolved Co–Fe alloy nanoparticles: An active and stable cathode for solid oxide electrolysis cells. *Journal of Materials Chemistry A* **4**, 17521–17528 (2016).
139. Zhou, Y., Lü, Z., Xu, S., Xu, D., Wei, B. Investigation of a solid oxide fuel cells catalyst $LaSrNiO_4$: Electronic structure, surface segregation, and oxygen adsorption. *International Journal of Hydrogen Energy* **41**, 21497–21502 (2016).
140. Suntivich, J., May, K. J., Gasteiger, H. A., Goodenough, J. B., Shao-Horn, Y. A perovskite oxide optimized for oxygen evolution catalysis from molecular orbital principles. *Science* **334**, 1383–1385 (2011).
141. Lee, Y. L., Kleis, J., Rossmeisl, J., Yang, S. H., Morgan, D. Prediction of solid oxide fuel cell cathode activity with first-principles descriptors. *Energy & Environmental Science* **4**, 3966–3970 (2011).
142. Hu, B., Keane, M., Mahapatra, M. K., Singh, P. Stability of strontium-doped lanthanum manganite cathode in humidified air. *Journal of Power Sources* **248**, 196–204 (2014).
143. Sase, M., et al. Enhancement of oxygen surface exchange at the hetero-interface of (La, Sr)CoO_3/(La, Sr)$_2CoO_4$ with PLD-layered films. *Journal of the Electrochemical Society* **155**, B793–B797 (2008).

144. Lee, D., Lee, Y., Wang, X. R., Morgan, D., Shao-Horn, Y. Enhancement of oxygen surface exchange on epitaxial $La_{0.6}Sr_{0.4}Co_{0.2}Fe_{0.8}O_{3-\delta}$ thin films using advanced hetero-structured oxide interface engineering. *MRS Communications* **6**, 204–209 (2016).
145. Zhu, Z., Qian, J., Wang, Z., Dang, J., Liu, W. High-performance anode-supported solid oxide fuel cells based on nickel-based cathode and $Ba(Zr_{0.1}Ce_{0.7}Y_{0.2})O_{3-\delta}$ electrolyte. *Journal of Alloys and Compounds* **581**, 832–835 (2013).
146. Niwa, E., Uematsu, C., Miyashita, E., Ohzeki, T., Hashimoto, T. Conductivity and sintering property of $LaNi_{1-x}Fe_xO_3$ ceramics prepared by Pechini method. *Solid State Ionics* **201**, 87–93 (2011).
147. Kiselev, E. A., Cherepanov, V.A. $p(O_2)$-stability of $LaFe_{1-x}Ni_xO_{3-\delta}$ solid solutions at 1100°C. *Journal of Solid State Chemistry* **183**, 1992–1997 (2010).
148. Sugita, S., et al. Cathode contact optimization and performance evaluation of intermediate temperature-operating solid oxide fuel cell stacks based on anode-supported planar cells with $LaNi_{0.6}Fe_{0.4}O_3$ cathode. *Journal of Power Sources* **185**, 932–936 (2008).
149. Chiba, R., Yoshimura, F., Sakurai, Y. An investigation of $LaNi_{1-x}Fe_xO_3$ as a cathode material for solid oxide fuel cells. *Solid State Ionics* **124**, 281–288 (1999).
150. Falcón, H., Goeta, A. E., Punte, G., Carbonio, R.E. Crystal Structure Refinement and Stability of $LaFe_xNi_{1-x}O_3$ Solid Solutions. *Journal of Solid State Chemistry* **133**, 379–385 (1997).
151. Stodolny, M. K., Boukamp, B. A., Blank, D. H. A., van Berkel, F. P. F. Cr-poisoning of a LaNi0.6Fe0.4O3 cathode under current load. *Journal of Power Sources* **209**, 120–129 (2012).
152. Komatsu, T., Arai, H., Chiba, R., Nozawa, K., Arakawa, M. Long-term chemical stability of $LaNi_{1-x}Fe_xO_3$ as a cathode material in solid oxide fuel cells. *Journal of the Electrochemical Society* **154**, B379–B382 (2007).
153. Stodolny, M. K., Boukamp, B. A., Blank, D. H. A., van Berkel, F. P. F. Impact of Cr-poisoning on the conductivity of $LaNi_{0.6}Fe_{0.4}O_3$. *Journal of Power Sources* **196**, 9290–9298 (2011).
154. Escudero, M. J., Fuerte, A., Daza, L. $La_2NiO_{4+\delta}$ potential cathode material on $La_{0.9}Sr_{0.1}Ga_{0.8}Mg_{0.2}O_{2.85}$ electrolyte for intermediate temperature solid oxide fuel cell. *Journal of Power Sources* **196**, 7245–7250 (2011).
155. Pérez-Coll, D., Aguadero, A., Escudero, M. J., Daza, L. Effect of DC current polarization on the electrochemical behaviour of $La_2NiO_{4+\delta}$ and $La_3Ni_2O_{7+\delta}$-based systems. *Journal of Power Sources* **192**, 2–13 (2009).
156. Kim, G. T., Wang, S., Jacobson, A. J., Yuan, Z., Chen, C. Impedance studies of dense polycrystalline thin films of $La_2NiO_{4+\delta}$. *Journal of Materials Chemistry* **17**, 1316–1320 (2007).
157. Zhou, X. D., et al. Electrochemical performance and stability of the cathode for solid oxide fuel cells: V. high performance and stable Pr_2NiO_4 as the cathode for solid oxide fuel cells. *Electrochimica Acta* **71**, 44–49 (2012).
158. Kobsiriphat, W., et al. Nickel- and ruthenium-doped lanthanum chromite anodes: Effects of nanoscale metal precipitation on solid oxide fuel cell performance. *Journal of the Electrochemical Society* **157**, B279 (2010).
159. Jardiel, T., et al. New SOFC electrode materials: The Ni-substituted LSCM-based compounds $(La_{0.75}Sr_{0.25})(Cr_{0.5}Mn_{0.5-x}Ni_x)O_{3-\delta}$ and $(La_{0.75}Sr_{0.25})(Cr_{0.5-x}Ni_xMn_{0.5})O_{3-\delta}$. *Solid State Ionics* **181**, 894–901 (2010).
160. Qin, Q., et al. Perovskite titanate cathode decorated by in-situ grown iron nanocatalyst with enhanced electrocatalytic activity for high-temperature steam electrolysis. *Electrochimica Acta* **127**, 215–227 (2014).
161. Li, Y., et al. Efficient carbon dioxide electrolysis based on perovskite cathode enhanced with nickel nanocatalyst. *Electrochimica Acta* **153**, 325–333 (2015).
162. Liu, S., Liu, Q., Luo, J. Highly stable and efficient catalyst with in situ exsolved Fe–Ni alloy nanospheres socketed on an oxygen deficient perovskite for direct CO_2 electrolysis. *ACS Catalysis* **6**, 6219–6228 (2016).

163. Wei, H., et al. In situ growth of Ni_xCu_{1-x} alloy nanocatalysts on redox-reversible rutile (Nb, Ti)O_4 towards high-temperature carbon dioxide electrolysis. *Scientific Reports* **4**, 5156 (2015).
164. Xie, K., et al. Composite cathode based on redox-reversible $NbTi_{0.5}Ni_{0.5}O_4$ decorated with in situ grown Ni particles for direct carbon dioxide electrolysis. *Fuel Cells* **14**, 1036–1045 (2014).
165. Qi, W., et al. Reversibly in-situ anchoring copper nanocatalyst in perovskite titanate cathode for direct high-temperature steam electrolysis. *International Journal of Hydrogen Energy* **39**, 5485–5496 (2014).
166. Lee, Y., et al. Conditioning effects on $La_{1-x}Sr_xMnO_3$-yttria stabilized zirconia electrodes for thin-film solid oxide fuel cells. *Journal of Power Sources* **115**, 219–228 (2003).
167. Bidrawn, F., Kim, G., Aramrueang, N., Vohs, J. M., Gorte, R. J. Dopants to enhance SOFC cathodes based on Sr-doped $LaFeO_3$ and $LaMnO_3$. *Journal of Power Sources* **195**, 720–728 (2010).
168. Zhang, Y., et al. Plasma glow discharge as a tool for surface modification of catalytic solid oxides: A case study of $La_{0.6}Sr_{0.4}Co_{0.2}Fe_{0.8}O_{3-\delta}$ perovskite. *Energies* **9**, 786 (2016).
169. Imanishi, N., et al. LSM-YSZ cathode with infiltrated cobalt oxide and cerium oxide nanoparticles. *Fuel Cells* **9**, 215–221 (2009).
170. Li, M., Zheng, M., Hu, B., Zhang, Y., Xia, C. Improving electrochemical performance of lanthanum strontium ferrite by decorating instead of doping cobaltite. *Electrochimica Acta* **230**, 196–203 (2017).
171. Pavone, M., Ritzmann, A. M., Carter, E. A. Quantum-mechanics-based design principles for solid oxide fuel cell cathode materials. *Energy & Environmental Science* **4**, 4933–4937 (2011).
172. Huang, Y., Vohs, J. M., Gorte, R. J. An examination of LSM-LSCo mixtures for use in SOFC cathodes. *Journal of the Electrochemical Society* **153**, A951–A955 (2006).
173. Cox-Galhotra, R. A., McIntosh, S. Unreliability of simultaneously determining kchem and Dchem via conductivity relaxation for surface-modified $La_{0.6}Sr_{0.4}Co_{0.2}Fe_{0.8}O_{3-\delta}$. *Solid State Ionics* **181**, 1429–1436 (2010).
174. Choi, H. J., Bae, K., Jang, D. Y., Kim, J. W., Shim, J. H. Performance degradation of lanthanum strontium cobaltite after surface modification. *Journal of the Electrochemical Society* **162**, F622–F626 (2015).
175. Chen, Y., et al. Electronic activation of cathode superlattices at elevated temperatures—source of markedly accelerated oxygen reduction kinetics. *Advanced Energy Materials* **3**, 1221–1229 (2013).
176. Han, J. W., Yildiz, B. Mechanism for enhanced oxygen reduction kinetics at the (La, Sr)$CoO_{3-\delta}$/(La, Sr)$_2CoO_{4+\delta}$ hetero-interface. *Energy & Environmental Science* **5**, 8598–8607 (2012).
177. Sase, M., et al. Enhancement of oxygen exchange at the hetero interface of (La, Sr) CoO_3/(La, Sr)$_2CoO_4$ in composite ceramics. *Solid State Ionics* **178**, 1843–1852 (2008).
178. Crumlin, E. J., et al. Oxygen reduction kinetics enhancement on a heterostructured oxide surface for solid oxide fuel cells. *Journal of Physical Chemistry Letters* **1**, 3149–3155 (2010).
179. Yashiro, K., et al. Composite cathode of perovskite-related oxides, (La, Sr)$CoO_{3-\delta}$/(La, Sr)$_2CoO_{4-\delta}$, for solid oxide fuel cells. *Electrochemical and Solid-State Letters* **12**, B135–B137 (2009).
180. Lee, D., et al. Oxygen surface exchange kinetics and stability of (La, Sr)$_2CoO_{4\pm\delta}$/$La_{1-x}Sr_xMO_{3-\delta}$ (M = Co and Fe) hetero-interfaces at intermediate temperatures. *Journal of Materials Chemistry A* **3**, 2144–2157 (2015).

181. Hayd, J., Yokokawa, H., Ivers-Tiffée, E. Hetero-interfaces at nanoscaled (La, Sr)$CoO_{3-\delta}$ thin-film cathodes enhancing oxygen surface-exchange properties. *Journal of the Electrochemical Society* **160**, F351–F359 (2013).
182. Gadre, M. J., Lee, Y. L., Morgan, D. Cation interdiffusion model for enhanced oxygen kinetics at oxide heterostructure interfaces. *Physical Chemistry Chemical Physics* **14**, 2606–2616 (2012).
183. Feng, Z., et al. Anomalous interface and surface strontium segregation in $(La_{1-y}Sr_y)_2CoO_{4\pm\delta}$/ $La_{1-x}Sr_xCoO3\text{-}\delta$ heterostructured thin films. *Journal of Physical Chemistry Letters***5**, 1027–1034 (2014).
184. Liu, Z., Liu, M., Lei, Y., Liu, M. LSM-infiltrated LSCF cathodes for solid oxide fuel cells. *Journal of Energy Chemistry* **22**, 555–559 (2013).
185. Lou, X., Wang, S., Liu, Z., Yang, L., Liu, M. Improving $La_{0.6}Sr_{0.4}Co_{0.2}Fe_{0.8}O_{3-\delta}$ cathode performance by infiltration of a $Sm_{0.5}Sr_{0.5}CoO_{3-\delta}$ coating. *Solid State Ionics* **180**, 1285–1289 (2009).
186. Lynch, M. E., et al. Enhancement of $La_{0.6}Sr_{0.4}Co_{0.2}Fe_{0.8}O_{3-\delta}$ durability and surface electrocatalytic activity by $La_{0.85}Sr_{0.15}MnO_{3\pm\delta}$ investigated using a new test electrode platform. *Energy & Environmental Science* **4**, 2249–2258 (2011).
187. Chen, Y., et al. A robust and active hybrid catalyst for facile oxygen reduction in solid oxide fuel cells. *Energy & Environmental Science* **10**, 964–971 (2017).
188. Ding, D., Li, X., Lai, S. Y., Gerdes, K. Liu, M. Enhancing SOFC cathode performance by surface modification through infiltration. *Energy & Environmental Science* **7**, 552–575 (2014).
189. Gong, Y., et al. Stabilizing nanostructured solid oxide fuel cell cathode with atomic layer deposition. *Nano Letters* **13**, 4340–4345 (2013).
190. Gong, Y., et al. Atomic layer deposition functionalized composite SOFC cathode $La_{0.6}Sr_{0.4}Fe_{0.8}Co_{0.2}O_{3-\delta}$-$Gd_{0.2}Ce_{0.8}O_{1.9}$: Enhanced long-term stability. *Chemistry of Materials* **25**, 4224–4231 (2013).
191. Kim, E., et al. Degradation of $La_{0.6}Sr_{0.4}CoO_{3-\delta}$ based cathode performance in solid oxide fuel cells due to the presence of aluminum oxide deposited through atomic layer deposition. *Ceramics International* **40**, 7817–7822 (2014).

13 High-Temperature Electrochemical Process of CO_2 Conversion with SOCs 8

Cell and Stack Design, Fabrication, and Scale-Up

13.1 SOEC COMPONENT/CELL/STACK STRUCTURE, FABRICATION, AND SCALE-UP

13.1.1 Component/Cell/Stack Structure

The main components of single-cell solid oxide electrolysis cells (SOECs) include the fuel electrode, the electrolyte, and the oxygen electrode. In regard to cell configurations, individual SOECs can be planar, tubular, or flat tubular, as shown in Figure 13.1a.[1] Early SOEC systems developed by Doenit et al. [2] and Feduska et al.[3] used tubular cells as their geometry. When compared to planar configurations, higher mechanical and thermal stability can be achieved with tubular configurations. Sealing of the cell is also much easier in tubular configurations than that of flat-plate configurations because the sealing area is significantly reduced. However, planar designs have still been widely adopted due to their much shorter current collection paths and significantly higher volumetric density.

In order to improve the chemical output of a single-cell SOEC system, the active cell area must be enlarged. Simply increasing the area of a single-cell SOEC comes with limitations, however. First, it is difficult to control the temperature across large cell areas; second, it is challenging to manufacture large and low-cost ceramic films. The 1.4 V voltage of a single cell is too low to operate in a co-electrolysis system. In order to circumvent these issues, a stack consisting of multiple cells is necessary. The structure of SOEC from stack to single cell is shown in Figure 13.1b (in case of planar cell). Obviously, interconnects must be added in a stack; however, a difficult challenge for the interconnects is to provide adequate electrical connections between the oxygen electrode of one single cell to the fuel electrode of the adjacent one without gas permeation.

In regard to SOEC stacks, durability is one of the most challenging topics for researchers,[4–8] Currently, we will focus on the basic structures of SOEC stacks.

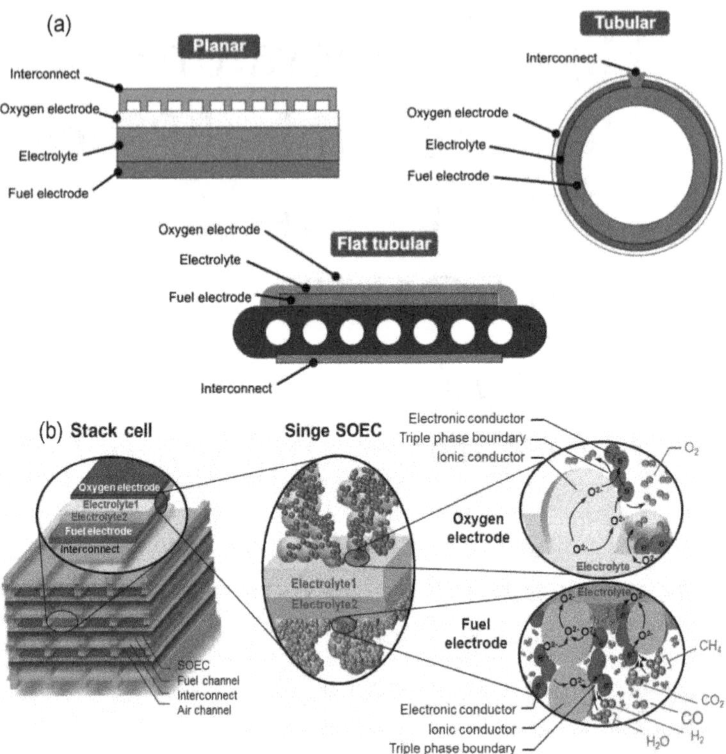

FIGURE 13.1 (a) Various SOEC cell configurations (Reprinted from Elder, R. et al., High temperature electrolysis (Chapter 11), in Styring, P. et al. (Eds.), *Carbon Dioxide Utilisation: Closing the Carbon Cycle*, pp. 183–209, Elsevier, Amsterdam, the Netherlands, Copyright 2015, with permission from Elsevier.), (b) Structure of SOEC from stack to single cell. (Reproduced from Wachsman, E.D. and Lee, K.T., *Science*, 334, 935–939, 2011, with permission from AAAS, Copyright 2011.)

Fundamentally, the stacks can be classified into three large categories: planar,[9–14] tubular,[15–24] and flat tubular[25] geometries.

1. Planar SOEC. Planar SOECs may be in the form of a square plate or a circular disc.[26] Both 10-cell and 5-cell SOEC stacks for square plate SOECs are presented in Figure 13.2. In addition to 10 planar SOECs, other components such as interconnects, seal gaskets, flow channels, and end plates are also shown in Figure 13.2a. Materials and Systems Research Inc. (MSRI) stacks consisting of five single Ni/YSZ-supported SOECs along with interconnects, meshes, and gaskets are shown in Figure 13.2c. Aside from the exploded views of stack structures, three-dimensional (3D) views of the test fixture for the two stacks are also shown in Figure 13.2b and d. Obviously, the detailed structures of the stacks depend on different demands.

   Ebbesen et al.[7] proposed multi-cells stacks with planar Ni/YSZ-supported solid oxide cells for steam electrolysis or CO_2/H_2O co-electrolysis, as shown

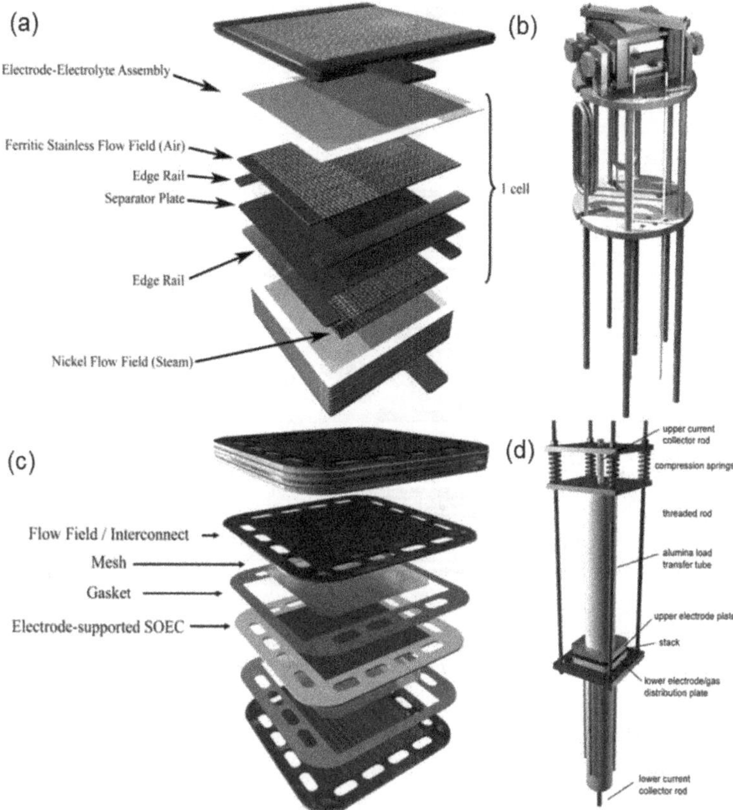

FIGURE 13.2 (a) Exploded views of a 10-cell stack (Ceramatec Inc.), (c) 5-cell stack (Materials and Systems Research Inc.), 3D views of the test fixtures for (b) the 10-cell stack and (d) the 5-cell stack. (Reprinted from *Int. J. Hydrog. Energy*, 38, Zhang, X. et al., Improved durability of SOEC stacks for high temperature electrolysis, 20–28, Copyright 2012, with permission from Elsevier.)

in Figure 13.3. The stacks consist of either six (for H_2O electrolysis) or 10 (for co-electrolysis) repeating units that contain the cell and the interconnect. A single repeat unit of the SOEC stack with circular discs presented by Boëdec et al.[28] is shown in Figure 13.4. In addition to the main components such as electrolyte and electrodes, the interconnects (flexible, cathodic, and anodic interconnects), seals, and cell supports are also important.

2. Tubular SOEC. Tubular stacks are developed due to their higher mechanical and thermal stabilities. In the past, multiple-cell tubular stacks were proposed by Spacil and Tedmon[29] for water vapor electrochemical dissociation. Two general designs for multiple-cell stacks are shown in Figures 13.5 and 13.6. The electrolyte in the first design is continuous (Figure 13.5), with one advantage being that the electrolyte can provide a convenient structural element. The series interconnection between the fuel electrode and the oxygen electrode is made by a lead such as a wire. The electrolyte is not continuous

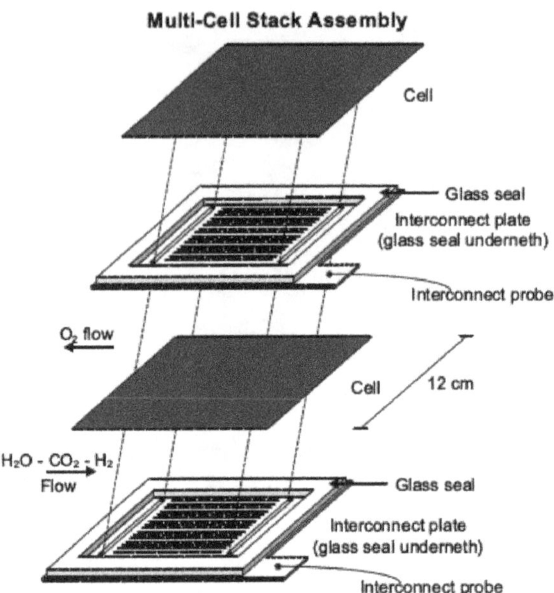

FIGURE 13.3 Assembly of two repeating units in stacks for H_2O electrolysis or CO_2/H_2O co-electrolysis. (Reprinted from *Int. J. Hydrog. Energy,* 36, Ebbesen, S.D. et al., Durable SOC stacks for production of hydrogen and synthesis gas by high temperature electrolysis, 7363–7373, Copyright 2011, with permission from Elsevier.)

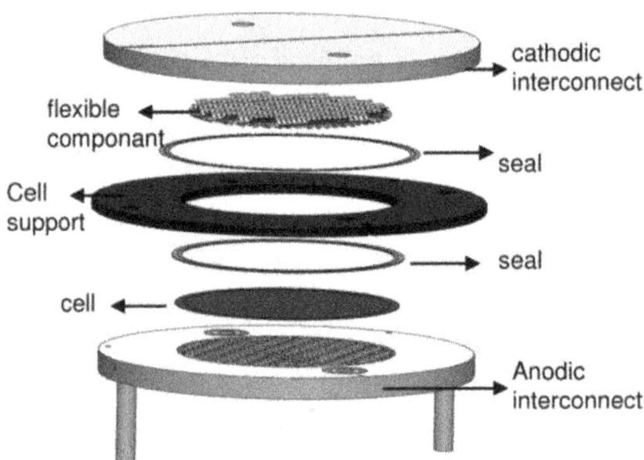

FIGURE 13.4 Structure of circular disc SOEC stack. (Boëdec, T. et al., A new stack to validate technical solutions and numerical simulations, *Fuel Cells*, 2012, 12, 239–247. Copyright Wiley-VCH Verlag GmbH & Co. KGaA. Reproduced with permission.)

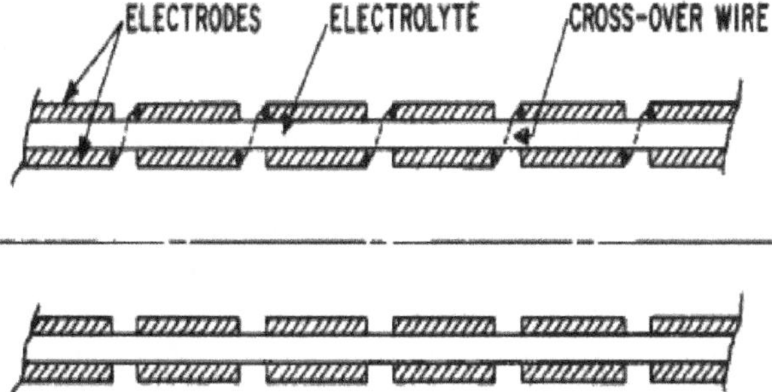

FIGURE 13.5 Schematic diagram of the tubular SOECs with a continuous electrolyte. (Reproduced from Spacil, H.S. and Tedmon, C.S., *J. Electrochem. Soc. Electrochem. Sci.*, 116, 1627–1633, 1969, with permission from the ECS, Copyright 1969.)

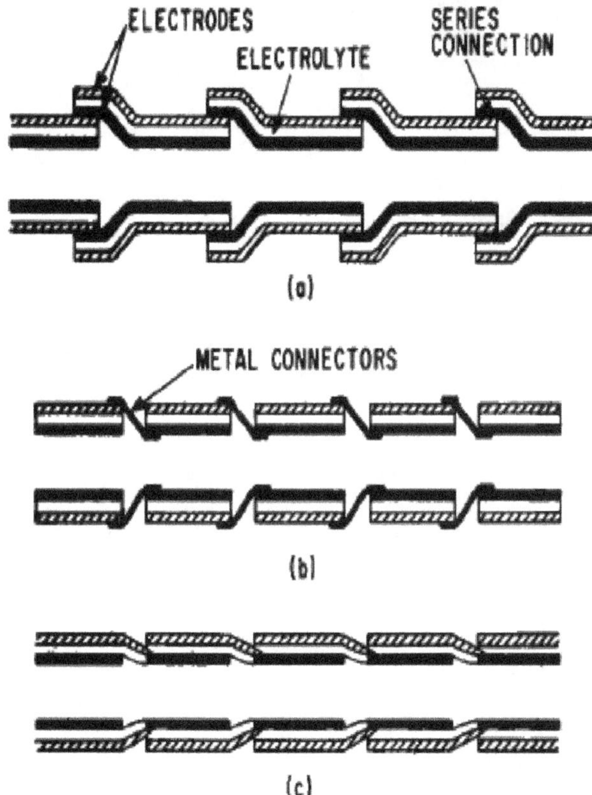

FIGURE 13.6 Schematic diagrams of several possible configurations for tubular stacks with discontinuous electrolyte. (a) and (c) Anode-cathode series connection directly; (b) By means of an intermediate metal connectors (Reproduced from Spacil, H.S. and Tedmon, C.S., *J. Electrochem. Soc. Electrochem. Sci.*, 116, 1627–1633, 1969. With permission from the ECS, Copyright 1969.)

in the second design (Figure 13.6) and the series connection between the two electrodes is made directly, as shown in Figures 13.5a and c, or the connection is achieved by an intermediate metal ring or collar, as indicated in Figure 13.6b. Novel designs have been developed in recent years. Figure 13.7 shows a solid oxide natural gas–assisted steam electrolyzer.[15] Compared to conventional SOECs, adding a system for heat recovery that consists of a catalytic reactor and heat exchangers can form a high-efficiency production system. Figure 13.7a shows a tubular SOEC stack with four tubes. The natural gas flows inside the tubes, while the steam/hydrogen mixture exits in the outside compartment of the tubes and the inside of the metal vessel. Figure 13.7b shows the experimental facility for testing.

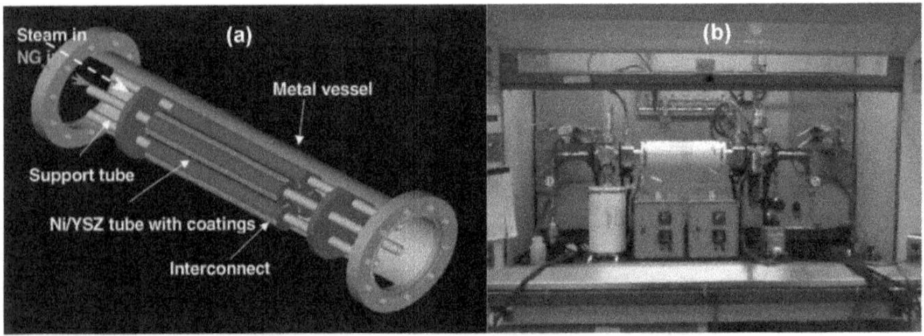

FIGURE 13.7 (a) Structure of a tubular SOEC stack with four tubes and (b) an experimental facility for a 100 W stack. (Reprinted from *Int. J. Hydrog. Energy*, 28, Martinez-Frias, J. et al., A natural gas-assisted steam electrolyzer for high-efficiency production of hydrogen, 483–490, Copyright 2003, with permission from Elsevier.)

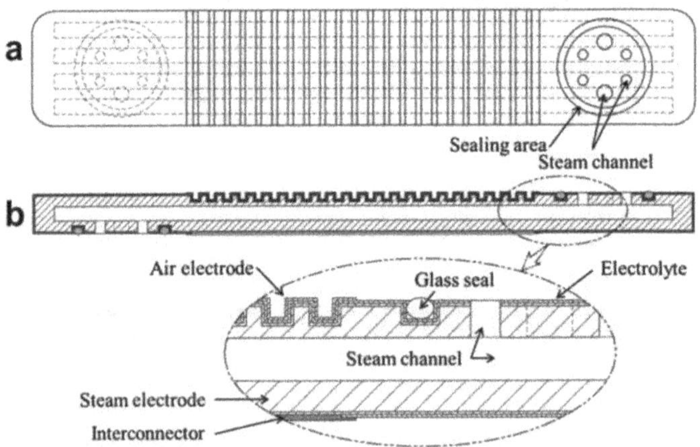

FIGURE 13.8 Flat-tubular SOECs in (a) top and (b) side views. (Reprinted from *Int. J. Hydrog. Energy*, 37, Kim, S. et al., Hydrogen production performance of 3-cell flat-tubular solid oxide electrolysis stack, 78–83, Copyright 2012, with permission from Elsevier.)

3. Flat tubular SOEC. For flat tubular SOECs (Figure 13.8),[25] the upper side of the single cell is used as the air/oxygen electrode, while the bottom side is used as the steam electrode and the interconnector. As shown in Figure 13.8b, the steam channels are located in both perpendicular and parallel directions, and the air channels are perpendicular to the steam channels. The thickness of a single cell is 4.9 mm and the area of the oxygen electrode is 30 cm^2. In addition, the area of the channels is equal to the area of non-channels. In order to investigate the influence of operating conditions on the electrolysis performance, button cells with an active area of 0.5 cm^2 were fabricated in this work.

13.1.2 Fabrication

The fabrication process selected for each SOEC cell/stack design depends on the various configurations of the cells in the stack, with the fabrication of the dense electrolyte being the key step in any selected process. There are two main approaches to ceramic fabrication: the particulate method and the deposition method. The particulate method involves the compaction of ceramic powder into the components of SOEC cells and then densification at high temperatures,[30] such as tape casting and tape calendaring. The deposition method for manufacturing of cell components on a support involves processes such as plasma spraying, chemical vapor deposition (CVD), and spray pyrolysis.

13.1.2.1 Particulate Method

There are currently three main particulate processes for the fabrication of SOECs: tape casting, tape calendering, and extrusion. The first two processes are often used in the fabrication of planar SOECs, whereas the third is used for tubular SOECs.

Tape casting is a common manufacturing process for thin, flat sheets of ceramic. This method has been used for the fabrication of various components for planar SOECs. The steps of tape casting include (1) using a doctor blade to make a layer of slip (e.g., ceramic powder suspended in a liquid), (2) drying this layer on a temporary support, and (3) stripping the layer from the support and then firing. Multilayer tapes can be manufactured by one continuous casting layer on top of another.

Tape calendering is also often used for the fabrication of various components for planar SOECs.[30] The steps of tape calendering include (1) mixing the ceramic powder and organic binder in a high shear mixer to form a plastic mass, (2) rolling the mass into a tape of desired thickness, and (3) firing the tape at elevated temperatures. Multiple tapes can be achieved by laminating with single layers for rolling operation again.

Extrusion is widely applied for the fabrication of tubular SOECs. The steps of extrusion conclude (1) mixing electrode powder and organic binder by mixer with certain temperature and rotational speed, and (2) extruding the ceramic slurry to form tubular framework through a single screw extruder. Particularly, the stability of ceramic slurry with plasticizer is the key issue for fabrication of large-density-value tubes.

Other particulate processes such as pressing[31] and extrusion[32,33] have also been developed for the fabrication of SOECs. The extrusion method is often used in fabricating tubular or flat tubular SOECs.

13.1.2.2 Deposition Method

Deposition techniques are widely used for the fabrication of both planar and tubular SOECs. Select deposition processes are introduced as follows:

Sputtering. This process involves depositing an YSZ film on a substrate using an electrical discharge in an argon/oxygen mixture atmosphere.

Dip coating. This process involves immersing porous substrates in a YSZ slurry of colloidal size particles, and then drying and firing the deposited films.

Spray pyrolysis. This process involves spraying a solution consisting of powdered precursors and/or the last particle ingredient onto a substrate at high temperatures and then sintering it to allow for densification of the deposited layer.

Plasma spraying. This process involves injecting a powder into a plasma jet where the powder is accelerated, melted, and deposited on a substrate.

Other deposition processes such as electrophoretic deposition, vapor phase electrolytic deposition, and so on, are also explored and developed for the fabrication of SOECs.

13.1.3 Scale-Up

Modular designs of CO_2/H_2O co-electrolysis systems are necessary for meeting the different scale demands of hydrogen and/or syngas production due to the limited size of a single cell. Normally, modular designs arrange single cells in uniform size together to form a SOEC stack. Then several of these stacks can be assembled to make a basic module. Obviously, the more modules there are, the larger the surface area, which can result in higher production capacity. This modular design allows for scale-up in sizes, which includes the increase in the size of module subassemblies and the development of subsystems required for system control operations. It also increases the reliability and flexibility of the stack system.

In the above section, the SOEC stacks presented have no more than 10 cells. For practical implementation and commercialization, scale-up is necessary. Therefore, large-scale SOEC stacks or even modules have been developed. For example, a low-weight and cost-effective stack with enhanced performance was designed and proposed by the Commissariat à l'Énergie Atomique (CEA).[34–36] This stack proved to be scalable from three to 25 cells. In particular, two stacks with 10 and 25 cells were tested in both steam electrolysis and CO_2/H_2O co-electrolysis modes by Reytier et al.,[36] with H_2 production rates of 0.6 and 1.7 Nm^3/h achieved, respectively, and steam conversion efficiencies of around 50% at 800°C below 1.3 V (Figure 13.9). A cost analysis was also performed, and the results supported the economic potential of this technology. Through their approach, the scale-up of

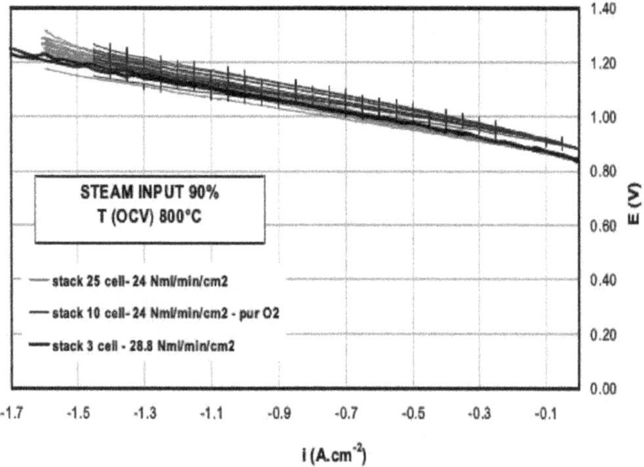

FIGURE 13.9 Performance of multiple cell stack (3 to 25 cell) (800°c, 90% H_2O/10% H_2). (Reprinted from *Int. J. Hydrog. Energ.*, 41, Odukoya, A. et al., Progress of the IAHE nuclear hydrogen division on international hydrogen production programs, 7878–7891, Copyright 2015, with permission from Elsevier.)

SOEC systems for practical application is validated for both steam electrolysis and H_2O/CO_2 co-electrolysis.

A 4 kW high-temperature steam electrolysis unit was presented for H_2 production on a large scale, which consisted of two 40-cell SOEC stacks.[37] The theoretical H_2 production rate was approximately 23 L/min for long-term operation (830 h, 0.41 A cm^{-2}). Furthermore, a large-scale planar stack was also fabricated by Stoots et al.,[38] and the feasibility of using SOECs for high-temperature steam electrolysis for H_2 production in large-scale was validated. The investigation of CO_2/H_2O co-electrolysis for syngas production was also performed. As shown in Figure 13.10, multiple stacks consisting of 720 cells, with each single cell possessing a 10 cm × 10 cm active area, was tested in a newly developed test facility of 15 kW integrated laboratory scale. Operation parameters such as operating voltage, gas composition, and operating temperature were varied during the testing. Results to date indicate that this technique is promising for hydrogen and syngas production.[38]

Recently, high-efficiency, low-cost hydrogen production SOEC electrolyzers were presented by SunFire GmbH, a company focusing on developing a competitive process to produce synthetic renewable fuels (e.g., petrol, diesel, kerosene, and methane) from carbon dioxide (CO_2) and water (H_2O) using renewable electrical energy.[39] Their multi-module SOEC electrolyzer installation is shown in Figure 13.11, and its technical specifications are listed in Table 13.1. The applications of this system include (1) industry: metallurgy, float glass, electronics, chemicals; (2) mobility: refineries, fueling stations; and (3) energy storage: island grids, peak power. The rated electrical power of hydrogen generation and power generation (in solid fuel cell mode) can reach 150 and 25 kW, respectively.

FIGURE 13.10 (a) Final installation of Integrated Laboratory Scale (ILS) module, (b) 15 kW hot zone with 3 module (720 cells) installed. (Reprinted from *Int. J. Hydrog. Energy*, 35, Stoots, C.M. et al., High-temperature electrolysis for large-scale hydrogen production from nuclear energy—Experimental investigations, 4861–4870, Copyright 2009, with permission from Elsevier.)

FIGURE 13.11 Multi-module reversible solid oxide cell electrolyzer installation. (Reproduced from Sunfire, RSOC electrolyzer high-efficency and low-cost hydrogrn production, 2016. With permission from Sunfire, Copyright 2016.)

TABLE 13.1
Technical Specifications of the Multi-Module Reversible Solid Oxide Cell Electrolyzer Installation

	Hydrogen Generation	Power Generation
Power (per module)		
Rated electrical power AC	150 kW	25 kW
Load variation		−100%–100%
Thermal output		20 kWth
Electrical efficiency AC based on LHV	82%	>50%
Total system efficiency		Up to 85% LHV
Specific electric power AC	3.7 kWh/Nm^3	
Production		
H_2 production	40 Nm^3/h	
H_2 pressure	10 bar (g)	
H_2 purity (after gas cleaning)	99.999% Atmospheric dew point temperature: −60°C	
O_2	On request	
Gas input	Saturated steam: Mass flow: 40 kg/h Pressure: 2–3 bar (g)	Natural gas Biogas Sewage gas
Emissions		NO_x < 40 mg/kWh CO < 50 mg/kWh
Installation		
Electrical interface		Three-phase, 380/400/480 VAC; 50 Hz/60 Hz
Noise	<60 dB @ 3 m distance	<70 dB @ 3 m distance
Ambient air temperature		−20°C–+45°C
Communication		Communication solution for remote monitoring and control.

According to SunFire GmbH, the competitive advantages of this installation from are (1) cost efficiency: high system efficiency through heat utilization (η > 80% LHV), (2) productivity: promising low hydrogen production costs compared to legacy technologies (<5 €/kg), (3) reversibility: reversible operations in electrolysis as well as in fuel cell mode, (4) modularity: scalable subsystem design to meet customer hydrogen demands, and (5) purity: renewable electricity and steam for clean hydrogen.

In summary, although the installations of co-electrolysis have not been commercialized, it is believed that the development of steam electrolysis with SOECs will be feasible and that both the technical and economic goals will be attainable in the near future.

13.2 SOEC SYSTEMS

13.2.1 System Management

SOEC stacks are the key components for high-temperature electrolysis. Similar to fuel cell systems, however, high-temperature electrolysis systems also include many subsystems besides the SOEC stack itself. These subsystems include heat management, power management, steam management, gas transfer/purification, data acquisition and control, and security.

Efficient heat management can improve the overall efficiency of high-temperature electrolysis processes by reducing energy consumption and recovering some heat contained in the outlet products of the electrolysis.[40] The chemical production rate of SOEC systems can be flexibly changed with power supply management. SOEC systems are more complex compared to solid oxide fuel cell (SOFC) systems because of steam condensation, which results in SOEC stacks cracking, as well as fluctuations in operating conditions. Therefore, steam management subsystems can guarantee the effective use and accurate control of steam. The main function of gas transfer/purification subsystems is to control the composition of the inlet gas and to purify the product. Data acquisition and control subsystems can monitor and control the SOEC modules. In a high-temperature electrolysis system, safety precautions are essential for handling high-temperature H_2 and O_2 effectively.

13.2.2 Lab-Scale SOEC Systems

In SOEC systems, electricity, heat, and steam with CO_2 enters the SOEC stack, and fuel products together with exhaust gases exit the system. Therefore, in addition to SOEC stack construction, other balance-of-plant (BOP) components are also required. BOP equipment may differ for each application depending on the size of the system and the operating parameters, including operating pressure.[1,21,41,42] Next, we present a case related to lab-scaled SOEC system.

A schematic of an electrolysis/co-electrolysis bench-scale system with a 10-cell stack is shown in Figure 13.12. In addition to the SOEC stack, there are several BOP components. Here, N_2 is used for the inert carrier gas, and CO_2 is used only during co-electrolysis. The components and their functions include (1) mass flow controllers, which establish the flow rates of gases; (2) the shop air system, which supplies airflow to the stack; (3) a heated humidifier, which mixes the $N_2/H_2/CO_2$ gas mixture with steam; (4) computerized feedback control, which maintains the temperature of the humidifier water; (5) absolute pressure transducers, which measure the pressure; (6) thermocouples, which monitor gas line temperatures; (6) a condenser, which removes the surplus steam in the exhaust after condensation; and (7) the detecting system, which measures the electrical current, steam/CO_2 concentrations (sensors), and products (GC). A large SOEC system with a 720-cell stack is also shown in Figure 13.13, its major BOP components such as hot zone enclosure lid, power supply and instrument racks and so forth are also presented here. The date acquisition and instrument control can be achieved by means of a custom lab view (National Instruments) program.

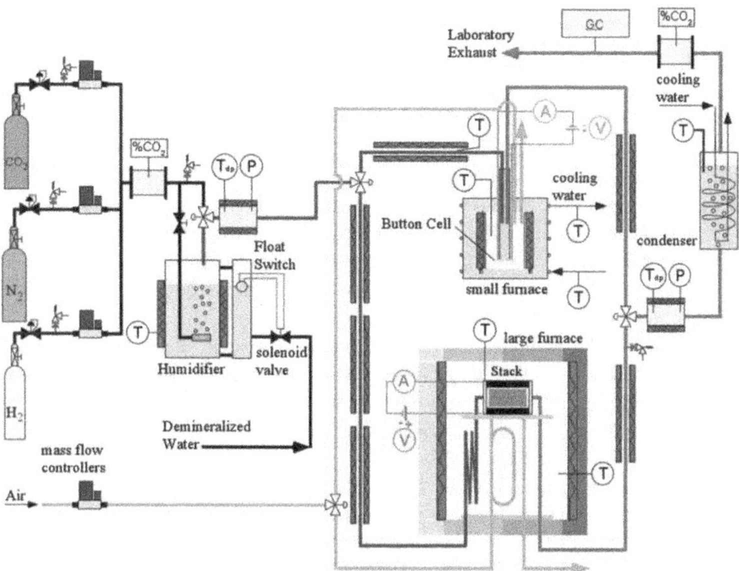

FIGURE 13.12 Schematic of INL electrolysis/co-electrolysis bench-scale system (10-cell stack). (Reproduced from *Int. J. Hydrog. Energy*, 35, Stoots, C.M. et al., High-temperature electrolysis for large-scale hydrogen production from nuclear energy—Experimental investigations, 4861–4870, Copyright 2009, with permission from Elsevier.)

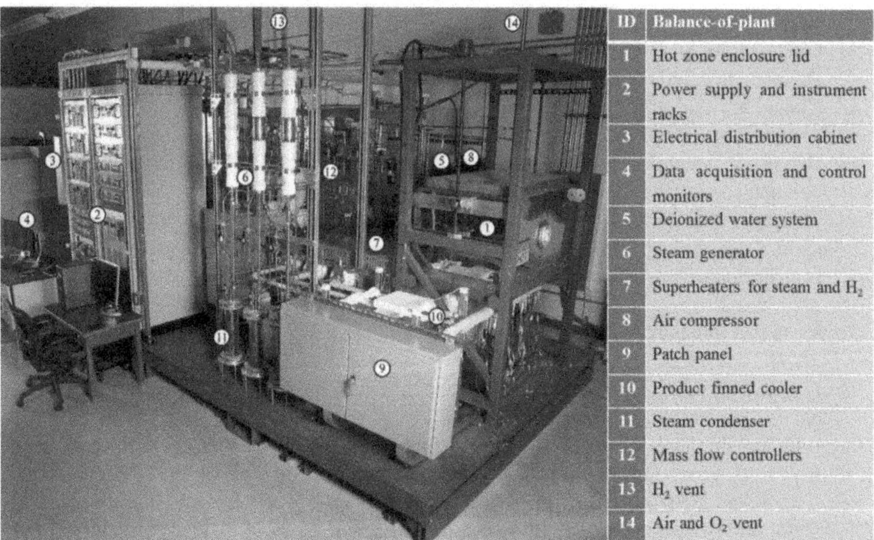

ID	Balance-of-plant
1	Hot zone enclosure lid
2	Power supply and instrument racks
3	Electrical distribution cabinet
4	Data acquisition and control monitors
5	Deionized water system
6	Steam generator
7	Superheaters for steam and H_2
8	Air compressor
9	Patch panel
10	Product finned cooler
11	Steam condenser
12	Mass flow controllers
13	H_2 vent
14	Air and O_2 vent

FIGURE 13.13 View of INL 15 k W high temperature electrolysis/co-electrolysis facility in integrated laboratory scale (720 cells). (Reproduced from *Int. J. Hydrog. Energy*, 35, Stoots, C.M. et al. High-temperature electrolysis for large-scale hydrogen production from nuclear energy—Experimental investigations, 4861–4870, Copyright 2009, with permission from Elsevier.)

REFERENCES

1. Elder, R., Cumming, D., Mogensen, M. B. High Temperature Electrolysis (Chapter 11). In Styring, P., Quadrelli, A., Armstrong, K. (Eds.), *Carbon Dioxide Utilisation: Closing the Carbon Cycle*, pp. 183–209 (Elsevier, Amsterdam, The Netherlands, 2015).
2. Doenitz, W., Schmidberger, R. S. Concepts and design for scaling up high temperature water vapour electrolysis. *International Journal of Hydrogen Energy* **7**, 321–330 (1982).
3. Feduska, W., Isenberg, A. O. High-temperature solid oxide fuel cell—Technical status. *Journal of Power Sources* **10**, 89–102 (1983).
4. Zheng, Y., et al. Comparison of performance and degradation of large-scale solid oxide electrolysis cells in stack with different composite air electrodes. *International Journal of Hydrogen Energy* **40**, 2460–2472 (2015).
5. Fang, Q., Blum, L., Menzler, N. H. Performance and degradation of solid oxide electrolysis cells in stack. *Journal of The Electrochemical Society* **162**, F907–F912 (2015).
6. Fang, Q., et al. Durability test and degradation behavior of a 2.5 kW SOFC stack with internal reforming of LNG. *International Journal of Hydrogen Energy* **38**, 16344–16353 (2013).
7. Ebbesen, S. D., Høgh, J., Nielsen, K. A., Nielsen, J. U., Mogensen, M. Durable SOC stacks for production of hydrogen and synthesis gas by high temperature electrolysis. *International Journal of Hydrogen Energy* **36**, 7363–7373 (2011).
8. Menzler, N. H., et al. Studies of material interaction after long-term stack operation. *Fuel Cells* **7**, 356–363 (2007).
9. O'Brien, J. E., Stoots, C. M., Herring, J. S., Hartvigsen, J. J. Performance of planar high-temperature electrolysis stacks for hydrogen production from nulcear energy. *Nuclear Technology* **158**, 118–131 (2007).
10. O'Brien, J. E., Stoots, C. M., Herring, J. S., Hartvigsen, J. J. Hydrogen production performance of a 10-cell planar solid-oxide electrolysis stack. *Journal of Fuel Cell Science and Technology* **3**, 213–219 (2006).
11. Chen, M., et al. High temperature co-electrolysis of steam and CO_2 in an SOC stack: Performance and durability. *Fuel Cells* **13**, 638–645 (2013).
12. Zhang, X., et al. Improved durability of SOEC stacks for high temperature electrolysis. *International Journal of Hydrogen Energy* **38**, 20–28 (2013).
13. Brisse, A. Schefold, J. High temperature electrolysis at EIFER, main achievements at cell and stack level. *Energy Procedia* **29**, 53–63 (2012).
14. Nguyen, V. N., Fang, Q., Packbier, U., Blum, L. Long-term tests of a Jülich planar short stack with reversible solid oxide cells in both fuel cell and electrolysis modes. *International Journal of Hydrogen Energy* **38**, 4281–4290 (2013).
15. Martinez-Frias, J., Pham, A. Q., Aceves, S. M. A natural gas-assisted steamelectrolyzer for high-efficiency production of hydrogen. *International Journal of Hydrogen Energy* **28**, 483–490 (2003).
16. Luo, Y., Shi, Y., Li, W., Cai, N. Dynamic electro-thermal modeling of co-electrolysis of steam and carbon dioxide in a tubular solid oxide electrolysis cell. *Energy* **89**, 637–647 (2015).
17. Luo, Y., Shi, Y., Li, W., Cai, N. Comprehensive modeling of tubular solid oxide electrolysis cell for co-electrolysis of steam and carbon dioxide. *Energy* **70**, 420–434 (2014).
18. Lee, S., et al. Electrochemical performance of H_2O–CO_2 coelectrolysis with a tubular solid oxide coelectrolysis (SOC) cell. *International Journal of Hydrogen Energy* **41**, 7530–7537 (2016).
19. Dipu, A. L., Ujisawa, Y., Ryu, J., Kato, Y. Electrolysis of carbon dioxide for carbon monoxide production in a tubular solid oxide electrolysis cell. *Annals of Nuclear Energy* **81**, 257–262 (2015).

20. Kleiminger, L., Li, T., Li, K., Kelsall, G. H. Syngas (CO-H_2) production using high temperature micro-tubular solid oxide electrolysers. *Electrochimica Acta* **179**, 565–577 (2015).
21. Dipu, A. L., Ujisawa, Y., Ryu, J., Kato, Y. Carbon dioxide reduction in a tubular solid oxide electrolysis cell for a carbon recycling energy system. *Nuclear Engineering and Design* **271**, 30–35 (2014).
22. Chen, L., Chen, F., Xia, C. Direct synthesis of methane from CO_2–H_2O co-electrolysis in tubular solid oxide electrolysis cells. *Energy Environmental & Science* **7**, 4018–4022 (2014).
23. Shao, L., Wang, S., Qian, J., Ye, X., Wen, T. Optimization of the electrode-supported tubular solid oxide cells for application on fuel cell and steam electrolysis. *International Journal of Hydrogen Energy* **38**, 4272–4280 (2013).
24. Laguna-Bercero, M. A., Campana, R., Larrea, A., Kilner, J. A., Orera, V. M. Steam electrolysis using a microtubular solid oxide fuel cell. *Journal of the Electrochemical Society* **157**, B852–B855 (2010).
25. Kim, S., Yu, J., Seo, D., Han, I., Woo, S. Hydrogen production performance of 3-cell flat-tubular solid oxide electrolysis stack. *International Journal of Hydrogen Energy* **37**, 78–83 (2012).
26. Schefold, J., Brisse, A., Tietz, F.. Nine thousand hours of operation of a solid oxide cell in steam electrolysis mode. *Journal of the Electrochemical Society* **159**, A137–A144 (2012).
27. Wachsman, E. D., Lee, K. T. Lowering the temperature of solid oxide fuel cells. *Science* **334**, 935–939 (2011).
28. Boëdec, T., et al. A new stack to validate technical solutions and numerical simulations. *Fuel Cells* **12**, 239–247 (2012).
29. Spacil, H. S., Tedmon, C. S. Electrochemical dissociation of water vapor in solid oxide electrolyte cells II. Materials, fabrication, and properties. *Journal of the Electrochemical Society: Electrochemical Science* **116**, 1627–1633 (1969).
30. Ormerod, R. M. Solid oxide fuel cells. *Chemical Society Reviews* 32, 17–28 (2003).
31. Xin, X., Lü, Z., Zhu, Q., Huang, X., Su, W. Fabrication of dense YSZ electrolyte membranes by a modified dry-pressing using nanocrystalline powders. *Journal of Materials Chemistry* **17**, 1627 (2007).
32. Kim, S. D., Hyun, S. H., Moon, J., Kim, J., Song, R. H. Fabrication and characterization of anode-supported electrolyte thin films for intermediate temperature solid oxide fuel cells. *Journal of Power Sources* **139**, 67–72 (2005).
33. Orera, V. M., Laguna-Bercero, M. A., Larrea, A. Fabrication methods and performance in fuel cell and steam electrolysis operation modes of small tubular solid oxide fuel cells: A review. *Frontiers in Energy Research* **2**, 22 (2014).
34. Odukoya, A., et al. Progress of the IAHE nuclear hydrogen division on international hydrogen production programs. *International Journal of Hydrogen Energy* **41**, 7878–7891 (2016).
35. Reytier, M., Cren, J., Petitjean, M., Chatroux, A., Gousseau, G. Development of a cost-efficient and performing high temperature steam electrolysis stack. *Electrochemical Society Transactions* **57**, 3151–3160 (2013).
36. Reytier, M., et al. Stack performances in high temperature steam electrolysis and co-electrolysis. *International Journal of Hydrogen Energy* **40**, 11370–11377 (2015).
37. Zhang, X., O'Brien, J. E., Tao, G., Zhou, C., Housley, G. K. Experimental design, operation, and results of a 4 kW high temperature steam electrolysis experiment. *Journal of Power Sources* **297**, 90–97 (2015).

38. Stoots, C. M., O'Brien, J. E., Condie, K. G., Hartvigsen, J. J. High-temperature electrolysis for large-scale hydrogen production from nuclear energy—Experimental investigations. *International Journal of Hydrogen Energy* **35**, 4861–4870 (2010).
39. Products and technology in Sunfire GmbH, RSOC electrolyzer: High-efficiency and low-cost hydrogen production, 2016. http://www.sunfire.de/en/.
40. Mansilla, C., Sigurvinsson, J., Bontemps, A., Maréchal, A., Werkoff, F. Heat management for hydrogen production by high temperature steam electrolysis. *Energy* **32**, 423–430 (2007).
41. Salzano, F. J., Skaperdas, G., Mezzina, A. Water vapor electrolysis at high temperature—Systems considerations and benefits. *International Journal of Hydrogen Energy* **10**, 801–809 (1985).
42. Döenitz, W., Schmidberger, R., Steinheil, E., Streicher, R. Hydrogen production by high temperature electrolysis of water vapour. *International Journal of Hydrogen Energy* **5**, 55–63 (1980).

14 High-Temperature Electrochemical Process of CO_2 Conversion with SOCs 9

Degradation Issues

It is well known that degradation is one of the most important issues in high temperature co-electrolysis (HTCE) using solid oxide cell (SOC).[1] Under practical operation in solid oxide electrolysis cell (SOEC) mode, the lowest degradation is 1.7%/1000 h during 3600 h at –1 A/cm^2, as reported in the literature,[2] which is twice as high as that in solid oxide fuel cell (SOFC) mode. The degradation rates of a 5000-h successful operation with a five-cell SOEC stack were approximately 15%/1000 h (0–2000 h) and 6%/1000 h degradation (2000–5000 h) for two kinds of feed-gas composition. Another point is that the 18% loss in H_2 production was observed in the 1000 h-test, which was performed by Idaho National Laboratory with a 25-cell SOEC stack, as shown in Figure 14.1.[3]

Therefore, in order to reduce degradation and prolong the lifetime of SOEC stacks, a deeper understanding of the degradation mechanism is critical. Generally, the approaches to deal with this issue can be classified into two large groups: altering operating conditions and modifying cell architecture by optimizing or replacing SOEC materials, especially electrode or electrolyte materials. The degradation of SOEC materials is much more challenging when the cell or stack is cycled between fuel cell and electrolysis cell modes.[4] Some degradation issue will be discussed in this chapter.

14.1 DELAMINATION OF OXYGEN ELECTRODE

Compared with that of the electrolyte or the fuel electrode, the degradation of the oxygen electrode appears to be a more significant factor in cell/stack failure, particularly at high currents.[5–9] The delamination of the electrode should be considered first,[3] and the most widely observed phenomenon is delamination between the interface of the oxygen electrode and the electrolyte.[10–13] Various models were developed in the literature to understand and explain the origins of delamination.[1,12,14,15] According to Virkar's approach,[12] high oxygen pressure in oxygen electrolytes is one of the main causes of delamination. This model was also presented by Jacobsen and Mogensen.[15] Rashkeev et al.[16] studied the mechanisms of oxygen electrode delamination in atomic-scale using both density functional theory (DFT) calculation and

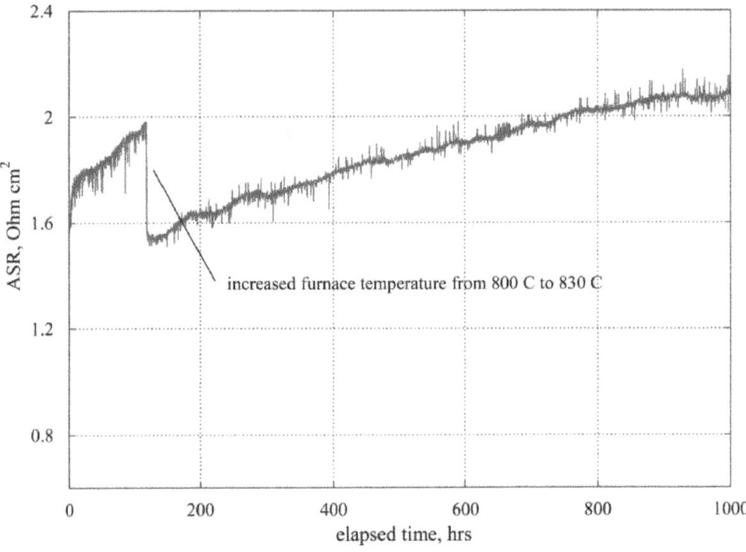

FIGURE 14.1 Area-specific resistance (ASR) of a 25-cell stack with time for a 1000 h-test. (Reprinted from Sohal, M.S. et al., *J. Fuel Cell Sci. Tech.*, 9, 0110171, 2012. With permission from ASME (American Society of Mechanical Engineers), Copyright 2012.)

thermodynamic modeling. It was found that the stability and structure of oxygen electrodes were affected greatly by interdiffusion of various atoms across the electrode/electrolyte interface. The strategy such as putting an additional thin layer between the electrode and electrolyte to suppress the interdiffusion was proposed.

Regarding one of the most common oxygen electrodes, the strontium-doped lanthanum manganites (LSM, $La_{1-x}SrxMnO_3$) electrode, studies from Chen et al.[17,18] showed that the disintegration of LSM particles and formation of LSM nanoparticles at the interface of the electrode/electrolyte were the causes for the delamination of LSM oxygen electrodes. Under the electrolysis polarization conditions, the lattice shrinkage-induced localized internal stress in the LSM grains contacted with electrolyte could lead to the breaking of LSM particles and the formation of LSM nanoparticles.[17] For instance, the boron deposition and poisoning might lead to the delamination of oxygen electrodes, which was investigated by Chen et al.[18–20] The presence of volatile boron species could greatly deteriorate the activity and stability of oxygen electrodes and damage their microstructure, as well as accelerate the delamination of an oxygen electrode from electrolyte, as shown in Figure 14.2. Similar delamination of oxygen electrodes could also be caused by impurities such as volatile chromium or sulphur.

To solve this problem, some related investigations has been performed.[7,8] It is found that the cathodic polarization, namely, operation in SOFC mode, could suppress the electrode degradation caused in SOEC mode. Hughes et al.[8] presented the durability testing of solid oxide cell electrodes in revering-current and found that the reversing-current operation could lower the delamination of oxygen electrodes and increase the lifetime of solid oxide cells. Four thousand hours of reversible operation

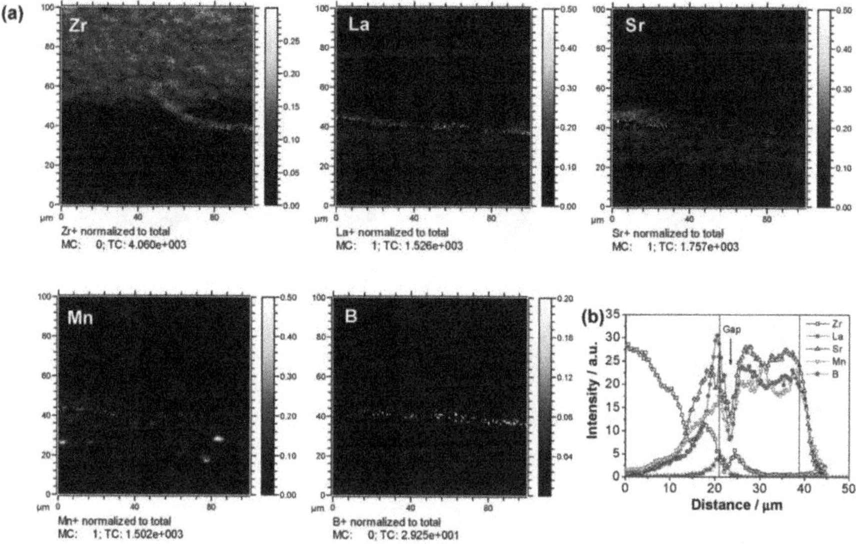

FIGURE 14.2 (a) TOF-SIMS elemental images (b) and line scan of the cross section of LSM oxygen electrode after polarization. (Reprinted from *Int. J. Hydrog. Energy*, 41, Chen, K. et al., Boron deposition and poisoning of $La_{0.8}Sr_{0.2}MnO_3$ oxygen electrodes of solid oxide electrolysis cells under accelerated operation conditions, 1419–1431, Copyright 2016, with permission from Elsevier.)

performed by Graves et al.[7] demonstrated that the electrolysis-induced degradation, which could lead to severe microstructure deterioration near the interface of the electrode/electrolyte, could be completely eliminated by reversibly cycling between SOEC and SOFC modes.

To interpret those favorable results, Hughes et al.[8] proposed that there might be a characteristic incubation time before the delamination of oxygen electrodes occurs. It is known that some unfavorable phenomena, that is, the formation of oxygen bubbles, cation migration, nucleation of zirconate phase, and so on, might be interrupted by shortening cycle times. Therefore, the degradation process could correspondingly be interrupted if the current-cycle period was decreased to no more than the incubation time. Graves et al.[7] proposed that the periodic release of high oxygen pressure in SOFC mode could avoid the formation of oxygen bubbles and nanopores or even reverse those processes to prevent degradation. More recently, Chen et al.[21] investigated the microstructural behavior of LSM oxygen electrodes under cyclic cathodic and anodic polarization, presenting an interpretation for the above reversible operation (Figure 14.3): the self-healing and regeneration of the interface under SOFC mode. As a consequence, in addition to adding a diffusion barrier between the oxygen electrode and electrolyte, or controlling some conditions such as adjusting oxygen pressure and avoiding impurity sources, reversing-current operation was also a promising approach to solve the problems related to the delamination of the oxygen electrode.

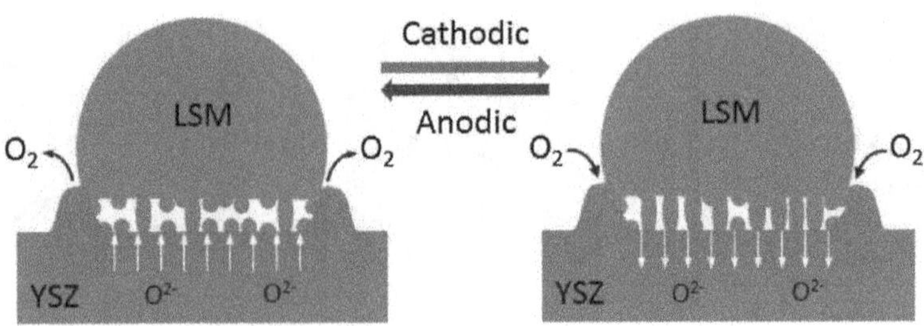

FIGURE 14.3 Self-healing and regeneration process of LSM oxygen electrode under reversible operation of solid oxide cells. (Reproduced from Chen, K. et al., *Phys. Chem. Chem. Phys.*, 17, 31308–31315, 2015. With permission from PCCP Owner Societies, Copyright 2015.)

14.2 Cr POISONING OF THE OXYGEN ELECTRODE

Aside from delamination of electrodes, the poisoning of the oxygen electrode in SOFC was reported in many research papers. The presence of volatile contaminants such as Cr, B, and S generated from a Fe-Cr alloy interconnect, borosilicate glass sealant, and air stream, respectively, has a serious harmful effect on oxygen electrode activity and performance stability of SOC.[18–20,22–32] Particularly, the Cr poisoning of the SOFC cathodes has been extensively investigated by researchers.[29–42] There are three main hypotheses to explain the degradation mechanism, which are summarized by Bilge Yildiz[4]:

1. The first hypothesis suggests that the process of poisoning occurs primarily via the formation of Cr^{6+}-containing gaseous species including CrO_3 or $CrO_2(OH)_2$, which is from the oxidation of chromium oxide on the interconnect. Then those Cr species are reduced at triple phase boundaries (TPB) (electrode/electrolyte/gas) to form Cr-rich phases such as Cr_2O_3. As a result, the electrochemistry of electrode will be impaired and polarization losses will increase.
2. The suggestion from the second hypothesis is that both chemical dissociation of the electrode material and solid-state diffusion of the Cr-containing species into the oxygen electrode are the potential causes for Cr deposition.
3. The third hypothesis was proposed by Yongda Zhen[43] and Xinbing Chen.[34] It suggests that the Cr deposition process is thermodynamically driven and kinetically limited by a nucleation reaction between the Cr species.

Compared to the substantial body of reported work on SOFCs, there is only a few studies of the Cr poisoning of the oxygen electrode under SOEC modes.[11,27,28,44] In 2010, Sharma and Yildiz[44] studied the degradation mechanism of $La_{0.8}Sr_{0.2}CoO_3$ (LSC) on oxygen electrodes in SOEC mode. They presented a mechanism for Cr deposition, and this governing process is shown in Figure 14.4a and b. More exact mechanisms about La-Cr-O phase formations, long-range transport of Sr and Co cations, and so on, need to be further quantified. More recently, Chen[27] investigated the impacts of Fe-Cr alloy

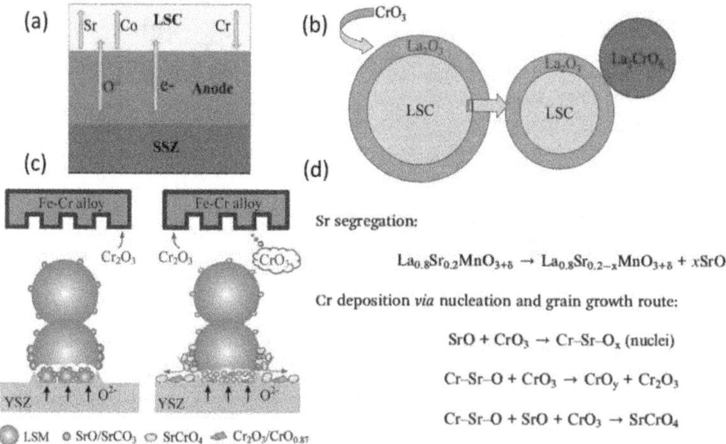

FIGURE 14.4 (a) Cation diffusion and charge transport in SOEC. (Reproduced from Sharma V.I. and Yildiz, B., *J. Electrochem. Soc.*, 157, B441–B448, 2010. With permission from ECS, Copyright 2010.) (b) A possible reaction pathway for Cr deposition from volatile Cr species on LSM oxygen electrode. (Reproduced from Sharma V.I. and Yildiz, B., *J. Electrochem. Soc.*, 157, B441–B448, 2010. With permission from ECS, Copyright 2010.) (c) Left: LSM nanoparticle formation, right: $Co_2O_3/CrO_{0.87}$ and $SrCrO_4$ formation. (Chen, K. et al., *Faraday Discuss.*, 182, 457–476, 2015. Reproduced by permission of The Royal Society of Chemistry.) (d) Possible reactions of Sr segregation, nucleation, and grain route. (From Jiang, S.P. and Chen, X., *Int. J. Hydrog. Energy*, 39, 505–531, 2014.)

metallic interconnect on the degradation and delamination process of $La_{0.8}Sr_{0.2}MnO_3$ (LSM) oxygen electrode first in SOEC mode at 800°C. According to the nucleation theory,[45] the accelerated migration and segregation of SrO from the bulk to the free surface and subsequent reaction between SrO and gaseous Cr species are shown in Figure 14.4c and d. The results demonstrated that, in essence, the deposition was a chemical process induced by the nucleation and grain growth reaction between the gaseous Cr species and segregated SrO on LSM oxygen electrodes in SOEC mode.

14.3 SIO_2 POISONING OF THE FUEL ELECTRODE

With regard to cell degradation, impurities play a key role in the degradation of fuel electrode (especially the Ni-YSZ electrode) in HTCE with SOEC.[46–51] Studies about model[52–56] and real[54,55] systems have reported that the locations of Si-containing impurity phase were very critical and the impurities also have a serious harmful effect on the electrochemical property of fuel electrodes and the performance of the system.[56] By contrast, the investigations of fuel electrode degradation in SOFC is much more sufficient than that of SOEC. A cell voltage degradation below 2%/1000 h could be achieved in SOFC, whereas the work of fuel electrode degradation in SOEC is still in the primary stage.[57]

Anne Hauch and Mogens Mogensen[57,58] from Risø National Laboratory studied a silica-related poisoning of the fuel electrodes (or hydrogen electrodes) in SOEC.

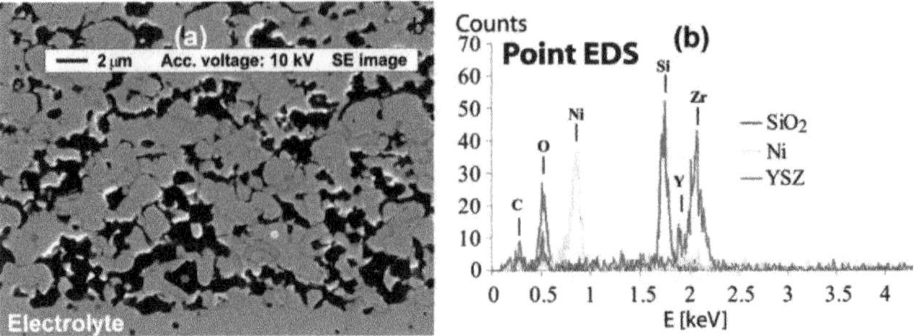

FIGURE 14.5 Postmortem SEM secondary electron (SE) image: (a) three colored points are shown on the SE image; (b) energy dispersive spectroscopy (EDS) for each point is shown in the same color with the SEM image. (Reproduced from Hauch, A. et al., *J. Electrochem. Soc.*, 154, A619–A626, 2007. With permission from ECS, Copyright 2007.)

The results showed that the initial passivation of SOEC resulted from Si-containing species, which was from the applied glass sealing and existed in SiO_2 form on TPB of the Ni/YSZ fuel electrode. The results of energy dispersive spectroscopy (EDS) are shown in Figure 14.5a and b. It was found that the high steam partial pressures were the cause of SiO_2 deposition in SOEC mode. The $Si(OH)_4$ (g) from the sealing is the main Si species source in SOEC mode; then the equilibrium ($Si(OH)_4(g) \rightleftarrows SiO_2(s) + 2H_2O(g)$) was shifted to the formation of $SiO_2(s)$ in the regions of the cell where the most steam was reduced to H_2. Obviously, the degradation can be avoided by removing the albite glass sealing, which contained Si species.

The degradation of SOEC through modeling of electrochemical potential profiles was proposed by Chatzichristodoulou in 2016.[59] The process of SiO_2 deposition at the Ni-YSZ TPB is shown in Figure 14.6a–c. It is well established that silica exists as an impurity in the Ni-YSZ fuel electrode materials.[56,57]

1. In highly humidified atmospheres, Si evaporates from the glass seal and is transformed into $Si(OH)_4(g)$ and subsequent $SiO_2(s)$ at the TPB of the Ni-YSZ fuel electrode, as shown in Figure 14.6a.
2. Under high cathodic polarization at the TPB, Si may further dissolve into Ni particle, as shown in Figure 14.6b by the yellow dotted arrow. According to the basic knowledge of physical chemistry, it is known that the maximum solubility of Si in Ni is 14 mil% at 850°C; in other words, Si can exist in the metallic form.
3. When the oxygen activity is high enough, the dissolved Si is re-oxidized into $SiO_2(s)$ at a later stage, as shown in Figure 14.6c.

In addition, the mechanism described above is further supported by thermodynamic calculations as the phase diagram of Ni-Si-O_2 at 850°C, which is shown in Figure 14.6d.

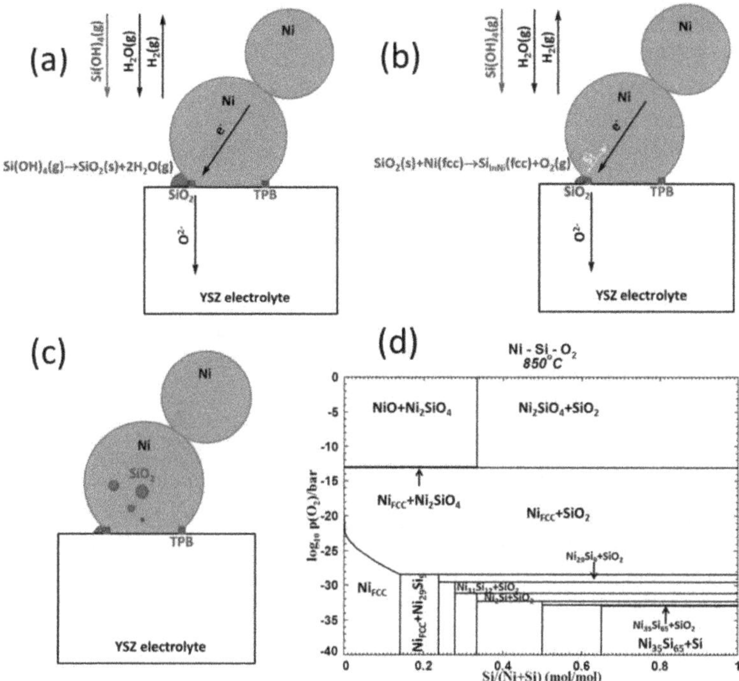

FIGURE 14.6 (a–c) Schematic illustration for deposition of SiO_2 at the TPB of Ni-YSZ fuel electrode and the generation of Si and SiO_2, and (d) phase diagram for Ni-Si-O_2 at 850°C (d). (Reprinted from *Electrochim. Acta*, 189, Chatzichristodoulou, C. et al., Understanding degradation of solid oxide electrolysis cells through modeling of electrochemical potential profiles, 265–282, Copyright 2016, with permission from Elsevier.)

14.4 REDOX STABILITY OF THE FUEL ELECTRODE

For fuel electrodes in SOEC, aside from impurity poisoning, another issue of reduction and oxidation (redox) can also cause the instability of fuel electrode (such as Ni-YSZ electrode). Redox of Ni catalyst particles can result in the volume changes of Ni-YSZ electrodes, and the volume changes can then lead to mechanical stresses and consequently to microstructural changes in the electrode. As a result, the activity of fuel electrode will be destroyed.

Hauch and Mogensen[57] investigated the microstructure and degradation of Ni-YSZ electrodes systematically and scrupulously. It was found that some significant microstructure was degraded in the Ni-YSZ electrode under the conditions of high-steam partial pressure and SOEC mode operation. The degradation would be more serious with high current densities. Electrolysis testing at conditions including high current density, high temperature, a high partial pressure of steam, and 2 A/cm,$^{-2}$ 950°C, $p(H_2O) = 0.9$ atm was presented. The result was that the microstructures at electrode/

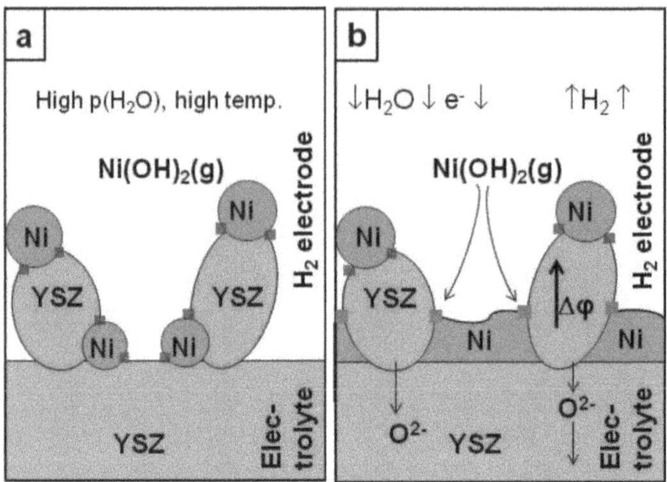

FIGURE 14.7 Possible mechanism in the microstructure at the YSZ-Ni/YSZ interface: (a) the YSZ-Ni/YSZ interface at OCV; (b) the YSZ-Ni/YSZ interface upon degradation at high current density electrolysis testing; the TPBs are marked as squares. (Reproduced from Hauch, A. et al., *J. Electrochem. Soc.*, 155, B1184–B1193, 2008. With permission from ECS, Copyright 2008.)

electrolyte interface were significantly changed, and it was ascribed to the relocation of Ni as the mechanism (Figure 14.7). The Ni could exist in the form of $Ni(OH)_2$ (g) at the conditions above, as shown in Figure 14.7a; then a dense layer of Ni on electrolyte (YSZ) was formed according to the reaction of $Ni(OH)_2$ (g) + 4e− → Ni (s) + $2O^{2-}$ + H_2 (g), as shown in Figure 14.7b. Such relocation of Ni would consequently block the oxygen transfer from the cathode into the electrolyte.

REFERENCES

1. Moçoteguy, P., Brisse, A. A review and comprehensive analysis of degradation mechanisms of solid oxide electrolysis cells. *International Journal of Hydrogen Energy* **38**, 15887–15902 (2013).
2. Schefold, J., Brisse, A., Tietz, F. Nine thousand hours of operation of a solid oxide cell in steam electrolysis mode. *Journal of The Electrochemical Society* **159**, A137–A144 (2012).
3. Sohal, M. S., et al. Degradation issues in solid oxide cells during high temperature electrolysis. *Journal of Fuel Cell Science and Technology* **9**, 0110171–10 (2012).
4. Yildiz, B. Reversible solid oxide electrolytic cells for large scale energy storage–challenges and opportunities. *Functional Materials for Sustainable Energy Applications* **35**, 149–178 (2012).
5. Elangovan, S., et al. Materials for solid oxide electrolysis cells. *ECS Transactions* **35**, 2875–2882 (2011).
6. Zhang, X., et al. Durability evaluation of reversible solid oxide cells. *Journal of Power Sources* **242**, 566–574 (2013).
7. Graves, C., Ebbesen, S. D., Jensen, S. H., Simonsen, S. B., Mogensen, M. B. Eliminating degradation in solid oxide electrochemical cells by reversible operation. *Nature Materials* **14**, 239–244 (2014).

8. Hughes, G. A., Yakal-Kremski, K., Barnett, S. A. Life testing of LSM-YSZ composite electrodes under reversing-current operation. *Physical Chemistry Chemical Physics* **15**, 17257–17262 (2013).
9. Daroukh, A., et al. Post-test analysis of electrode-supported solid oxide electrolyser cells. *Ionics* **21**, 1039–1043 (2015).
10. Momma, A., et al. Polarization behavior of high temperature solid oxide electrolysis cells (SOEC). *Journal of the Ceramic Society of Japan* **105**, 369–373 (1997).
11. Mawdsley, J. R., Carter, J. D., Kropf, A. J., Yildiz, B., Maroni, V. A. Post-test evaluation of oxygen electrodes from solid oxide electrolysis stacks. *International Journal of Hydrogen Energy* **34**, 4198–4207 (2009).
12. Virkar, A. V. Mechanism of oxygen electrode delamination in solid oxide electrolyzer cells. *International Journal of Hydrogen Energy* **35**, 9527–9543 (2010).
13. Knibbe, R., et al. Solid oxide electrolysis cells: Degradation at high current densities. *Journal of the Electrochemical Society* **157**, B1209–B1217 (2010).
14. Zhang, Y., Chen, K., Xia, C., Jiang, S. P., Ni, M. A model for the delamination kinetics of La0.8Sr0.2MnO3 oxygen electrodes of solid oxide electrolysis cells. *International Journal of Hydrogen Energy* **37**, 13914–13920 (2012).
15. Jacobsen, T., Mogensen, M. The course of oxygen partial pressure and electric potentials across an oxide electrolyte cell. *ECS Transactions* **13**, 259–273 (2008).
16. Rashkeev, S. N., Glazoff, M. V. Atomic-scale mechanisms of oxygen electrode delamination in solid oxide electrolyzer cells. *International Journal of Hydrogen Energy* **37**, 1280–1291 (2012).
17. Jiang, S. P. Challenges in the development of reversible solid oxide cell technologies: A mini review. *Asia-Pacific Journal of Chemical Engineering* **11**, 386–391 (2016).
18. Chen, K., et al. Boron deposition and poisoning of $La_{0.8}Sr_{0.2}MnO_3$ oxygen electrodes of solid oxide electrolysis cells under accelerated operation conditions. *International Journal of Hydrogen Energy* **41**, 1419–1431 (2016).
19. Chen, K., Ai, N., Zhao, L., Jiang, S. P. Effect of volatile boron species on the electrocatalytic activity of cathodes of solid oxide fuel cells I: (La, Sr)MnO_3 based electrodes. *Journal of the Electrochemical Society* **160**, F183–F190 (2013).
20. Chen, K., Ai, N., Zhao, L., Jiang, S. P. Effect of volatile boron species on the electrocatalytic activity of cathodes of solid oxide fuel cells II: (La, Sr)(Co, Fe)O_3 based electrodes. *Journal of the Electrochemical Society* **160**, F301–F308 (2013).
21. Chen, K., Liu, S., Ai, N., Koyama, M., Jiang, S. P. Why solid oxide cells can be reversibly operated in solid oxide electrolysis cell and fuel cell modes *Physical Chemistry Chemical Physics* **17**, 31308–31315 (2015).
22. Liu, H., et al. Reducing the reaction between boron-containing sealing glass-ceramics and lanthanum-containing cathode: Effect of La_2O_3. *Journal of the European Ceramic Society* **36**, 1103–1107 (2016).
23. Chen, K., et al. A fundamental study of boron deposition and poisoning of $La_{0.8}Sr_{0.2}MnO_3$ cathode of solid oxide fuel cells under accelerated conditions. *Journal of the Electrochemical Society* **162**, F1282–F1291 (2015).
24. Liu, H. L., et al. Reducing the reaction between boron-containing sealing glass-ceramics and lanthanum-containing cathode: Effect of Bi_2O_3. *Journal of the European Ceramic Society* **34**, 4463–4468 (2014).
25. Chen, K., et al. Effect of volatile boron species on the electrocatalytic activity of cathodes of solid oxide fuel cells: III. $Ba_{0.5}Sr_{0.5}Co_{0.8}Fe_{0.2}O_{3-\delta}$ Electrodes. *Journal of the Electrochemical Society* **161**, F1163–F1170 (2014).
26. Chen, K., et al. New zinc and bismuth doped glass sealants with substantially suppressed boron deposition and poisoning for solid oxide fuel cells. *Journal of Materials Chemistry A* **2**, 18655–18665 (2014).

27. Chen, K., et al. Chromium deposition and poisoning of $La_{0.8}Sr_{0.2}MnO_3$ oxygen electrodes of solid oxide electrolysis cells. *Faraday Discussions* **182**, 457–476 (2015).
28. Bo, W., et al. Chromium deposition and poisoning at $La_{0.6}Sr_{0.4}Co_{0.2}Fe_{0.8}O_{3-d}$ oxygen electrodes of solid oxide electrolysis cells. *Physical Chemistry Chemical Physics* **17**, 1601–1609 (2015).
29. Wang, C. C., et al. Effect of temperature on the chromium deposition and poisoning of $L_{a0.6}Sr_{0.4}Co_{0.2}Fe_{0.8}O_{3-\delta}$ cathodes of solid oxide fuel cells. *Electrochimica Acta* **139**, 173–179 (2014).
30. Kim, Y., et al. Effect of strontium content on chromium deposition and poisoning in $Ba_{1-x}Sr_xCo_{0.8}Fe_{0.2}O_{3-\delta}$ ($0.3 \leq x \leq 0.7$) cathodes of solid oxide fuel cells. *Journal of the Electrochemical Society* **159**, B185–B194 (2012).
31. Horita, T., et al. Chromium poisoning and degradation at (La, Sr)MnO_3 and (La, Sr) FeO_3 cathodes for solid oxide fuel cells. *Journal of the Electrochemical Society* **157**, B614–B620 (2010).
32. Stanislowski, M., et al. Reduction of chromium vaporization from SOFC interconnectors by highly effective coatings. *Journal of Power Sources* **164**, 578–589 (2007).
33. Matsuzaki, Y., Yasuda, I. Dependence of SOFC cathode degradation by chromium-containing alloy on compositions of electrodes and electrolytes. *Journal of the Electrochemical Society* **148**, A126–A131 (2001).
34. Xinbing, C., et al. Chromium deposition and poisoning on $(La_{0.6}Sr_{0.4-x}Ba_x)(Co_{0.2}Fe_{0.8})O_3$($0 \leq x \leq 0.4$) cathodes of solid oxide fuel cells. *Journal of the Electrochemical Society* **155**, B1093–B1101 (2008).
35. Yang, Z., Xia, G., Singh, P., Stevenson, J. W. Electrical contacts between cathodes and metallic interconnects in solid oxide fuel cells. *Journal of Power Sources* **155**, 246–252 (2006).
36. Jiang, S. P., Chen, X. Chromium deposition and poisoning of cathodes of solid oxide fuel cells-A review. *International Journal of Hydrogen Energy* **39**, 505–531 (2014).
37. Chen, X., Zhang, L., Liu, E., Jiang, S. P. A fundamental study of chromium deposition and poisoning at $(La_{0.8}Sr_{0.2})_{0.95}(Mn_{1-x}Cox)O_{3\pm\delta}$ ($0.0 \leq x \leq 1.0$) cathodes of solid oxide fuel cells. *International Journal of Hydrogen Energy* **36**, 805–821 (2011).
38. Konysheva, E., et al. Chromium poisoning of perovskite cathodes by the ODS alloy $Cr5Fe1Y_2O_3$ and the high chromium ferritic steel Crofer22APU. *Journal of the Electrochemical Society* **153**, A765–A773 (2006).
39. Hilpert, K., et al. Chromium vapor species over solid oxide fuel cell interconnect materials and their potential for degradation processes. *Journal of the Electrochemical Society* **143**, 3642–3647 (1996).
40. Zhao, L., et al. Insight into surface segregation and chromium deposition on $La_{0.6}Sr_{0.4}Co_{0.2}Fe_{0.8}O_{3-\delta}$ cathodes of solid oxide fuel cells. *Journal of Material Chemistry A* **2**, 11114–11123 (2014).
41. Fergus, J. Effect of cathode and electrolyte transport properties on chromium poisoning in solid oxide fuel cells. *International Journal of Hydrogen Energy* **32**, 3664–3671 (2007).
42. Quadakkers, W. J., et al. Compatibility of perovskite contact layers between cathode and metallic interconnector plates of SOFCs. *Solid State Ionics* **91**, 55–67 (1996).
43. Zhen Y., Tok A. I. Y., Boey F. Y. C., Jiang, S. P. Development of Cr-Tolerant cathodes of solid oxide fuel cells. *Electrochemical and Solid-State Letters* **11**, B42–B46 (2008).
44. Sharma V. I., Yildiz, B. Degradation mechanism in $La_{0.8}Sr_{0.2}CoO_3$ as contact layer on the solid oxide electrolysis cell anode. *Journal of the Electrochemical Society* **157**, B441–B448 (2010).
45. Badwal, S. P. S., Deller, R., Foger, K., Ramprakash, Y., Zhang, J. P. Interaction between chromia forming alloy interconnects and air electrode of solid oxide fuel cells. *Solid State Ionics* **99**, 297–310 (1997).

46. Tao, Y., Ebbesen, S. D., Mogensen, M. Degradation of solid oxide cells during co-electrolysis of H_2O and CO_2: Carbon deposition under high current densities. *ECS Transactions* **50**, 139–151 (2013).
47. Ebbesen, S. D., et al. Production of synthetic fuels by co-electrolysis of steam and carbon dioxide. *International Journal of Green Energy* **6**, 646–660 (2009).
48. Graves C., Ebbesen, S. D., Mogensen, M. Co-electrolysis of CO_2 and H_2O in solid oxide cells: Performance and durability. *Solid State Ionics* **192**, 398–403 (2011).
49. Ebbesen S. D., Mogensen M. Electrolysis of carbon dioxide in solid oxide electrolysis cells. *Journal of Power Sources* **193**, 349–358 (2009).
50. Ebbesen, S. D., Mogensen, M. Exceptional durability of solid oxide cells. *Electrochemical and Solid-State Letters* **13**, B106–B108 (2010).
51. Ebbesen, S. D., et al. Poisoning of solid oxide electrolysis cells by impurities. *Journal of the Electrochemical Society* **157**, B1419–B1429 (2010).
52. Jensen, K. Effect of impurities on structural and electrochemical properties of the Ni-YSZ interface. *Solid State Ionics* **160**, 27–37 (2003).
53. Jensen, K. V., Primdahl S., Chorkendorff, I., Mogensen M. Microstructural and chemical changes at the Ni/YSZ interface. *Solid State Ionics* **144**, 197–209 (2001).
54. Liu Y. L., Jiao C. G. Microstructure degradation of an anode/electrolyte interface in SOFC studied by transmission electron microscopy. *Solid State Ionics* **176**, 435–442 (2005).
55. Liu Y. L., Primdahl, S., Mogensen, M. Effects of impurities on microstructure in Ni/YSZ-YSZ half-cells for SOFC. *Solid State Ionics* **161**, 1–10 (2003).
56. Hansen, K. V., Norrman, K., Mogensen, M. TOF-SIMS studies of yttria-stabilised zirconia. *Surface and Interface Analysis* **38**, 911–916 (2006).
57. Hauch, A., et al. Solid oxide electrolysis cells: Microstructure and degradation of the Ni/Yttria-stabilized zirconia electrode. *Journal of the Electrochemical Society* **155**, B1184–B1193 (2008).
58. Hauch, A., et al. Silica segregation in the Ni/YSZ electrode. *Journal of the Electrochemical Society* **154**, A619–A626 (2007).
59. Chatzichristodoulou, C., Chen, M., Hendriksen, P. V., Jacobsen, T., Mogensen M. B. Understanding degradation of solid oxide electrolysis cells through modeling of electrochemical potential profiles. *Electrochimica Acta* **189**, 265–282 (2016).

15 Economic Analysis of CO_2 Conversion to Useful Fuels/Chemicals

The actual use of CO_2 for the production of chemicals/fuels and the predicted use in year 2030 are listed in Table 15.1.[1] The estimated global demand for CO_2 is around 200 Mt y^{-1}. This is only a small fraction of the estimated production from fossil fuels (24 Gt y^{-1}). Several common chemicals/fuels from CO_2 conversion are shown, with the main consumption of CO_2 being the production of inorganic carbonates (250 Mt y^{-1}), urea (180 Mt y^{-1}), and methanol (60 Mt y^{-1}). The products from CO_2 conversion can be divided into chemicals and fuels, with the global market of chemicals being 12–14 times smaller than that of fuels. However, current global yields of fuels from CO_2 conversion are much lower than that of chemicals. Therefore, new technologies need to be developed to fulfill expected market demands, especially for fuels. These developments need to be economically feasible, and economic analysis is of great importance for promoting further developments in CO_2 conversions.

Relative added value of products from CO_2 conversion, as defined in Equation (15.1), is a simple but effective criterion used to measure and evaluate the economic benefits and measure the profits. Here, m_P is the mass of the product (kg), $m_{E,i}$ is the mass of the reactant i (kg), W_P is the specific value of the product (€ kg^{-1}), $W_{E,i}$ is the specific value of the reactant i (€ kg^{-1}). Otto et al.[2] calculated the relative added value of various chemicals/fuels reacting with H_2 for different CO_2 prices using Equation (15.1). Their results are shown in Figure 15.1. The change in CO_2 prices has no significant impact on relative added values, and the calculated results of p-salicylic acid, oxalic acid, and dimethyl carbonate are the three highest among all 23 chemicals/fuels calculated. On the contrary, the relative added values of products such as dimethyl ether, propionic acid, formaldehyde, and methanol are negative as the value of products are lower than the cost of the reactants. Thus, the calculation of relative added value can be used for theoretical economic evaluations.

TABLE 15.1
CO_2 Use of and Market for CO_2 Conversion Products

Compound	Formula		Market 2016 Mt y^{-1}	CO_2 use Mt y^{-1}	Market 2030 Mt y^{-1}	CO_2 use Mt y^{-1}
Urea	$(H_2N)_2CO$	+4	**180**	132	**210**	154
Carbonate linear	$OC(OR)_2$	+4	>2	0.5	10	5
Carbonate cyclic	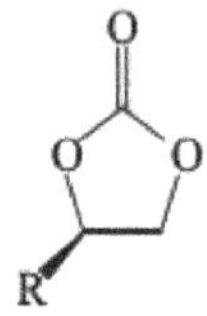	+4				
Polycarbonates	$-[OC(O)OCH_2CHR]_n-$	+4	5	1	9–10	2–3
Carbamates	RNH-COOR	+4	>6	1	11	*Ca.* 4
Acrylate	CH_2=CHCOOH	+3	5	1.5	8	5
Formic acid	HCOOH	+2	1	0.9	>10	>9
Inorganic carbonates	M_2CO_3, $M'CO_3$, $CaCO_3$	+4	**250**	70	**400**	100
Methanol	CH_3OH	−2	60	10	80	>28
Total				207		>332

Source: Aresta, M. et al., *J. Catal.*, 343, 2–45, 2016.

$$\text{Relative added value} = \frac{(m_{\text{p}} \cdot W_{\text{p}}) - \sum_{i}^{n} (m_{\text{E},i} \cdot W_{\text{E},i})}{\sum_{i}^{n} (m_{\text{E},i} \cdot W_{\text{E},i})} \tag{15.1}$$

With respect to the economic analysis of CO_2 conversions into useful fuels/chemicals in a specific production process or a chemical plant, the fixed capital costs, variable/fixed costs of production, product prices, and so on, of each product depends strongly on the product category, plant capacity, production processes, and even the difference between various countries and regions. Hence, it is difficult to compare systematically the costs of CO_2 conversion to different useful chemicals/fuels. Therefore, economic analysis for various typical chemicals/fuels will be reviewed in various cases in this section.

Note that all economic analysis in the literature is based on normal electricity costs. If redundant electricity from clean energy and/or nuclear sources is used, the results of the process economics can be different, allowing the process to be more profitable. In addition, the environmental benefits from CO_2 reduction is normally not considered in the following analysis. Aside from these two points, the environment benefits from CO_2 reduction is far more important than the economic benefits.

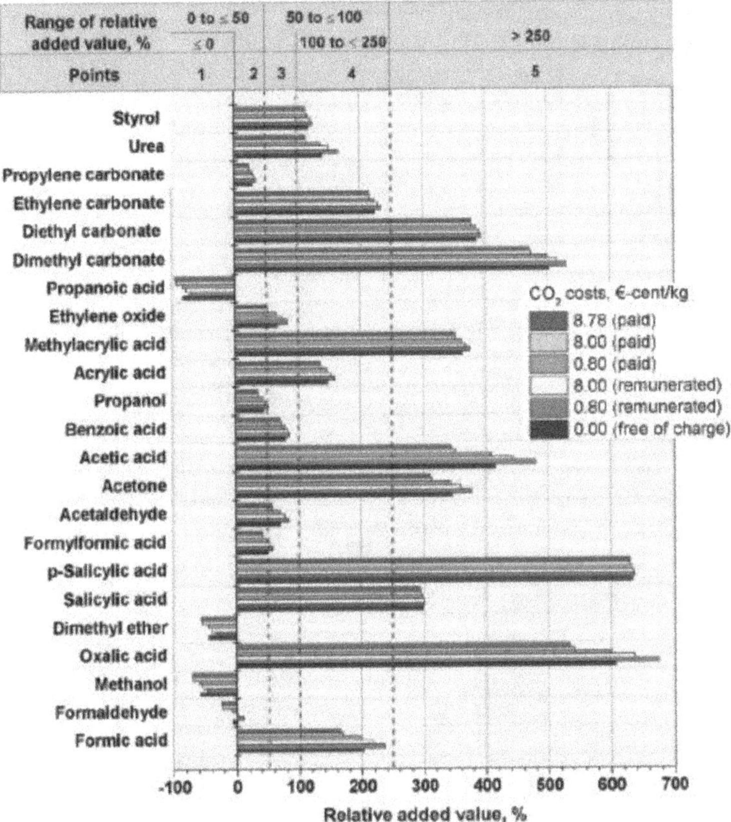

FIGURE 15.1 Relative added value of various chemicals/fuels for different CO_2 price. (Otto, A. et al., *Energ. Environ. Sci.*, 8, 3283–3297, 2015. Reproduced by permission of The Royal Society of Chemistry.)

15.1 METHANOL

Due to its versatility in the synthesis of several important chemicals such as acetic acid, chloromethane, alkyl halides, formaldehyde, and even higher hydrocarbons, methanol (CH_3OH) is a common feedstock and a building block in the chemical industry.[1,3–5] In 2014, around 60% of the produced methanol was used for the production of these chemicals, with the remaining 40% used for the production of fuels such as biodiesel, di-methyl-ether (DME), marine fuels, and so on.[6] Normally, the production of methanol is performed via a hydrogenation (CO_2+H_2) and bi-reforming (CO_2+natural gas) process. More recently, Kourkoumpas et al.[6] performed a techno-economic investigation into the implementation of "power to methanol (PtM)" using CO_2 from lignite power plants. Their concept consists of electricity generation, CO_2 capture and

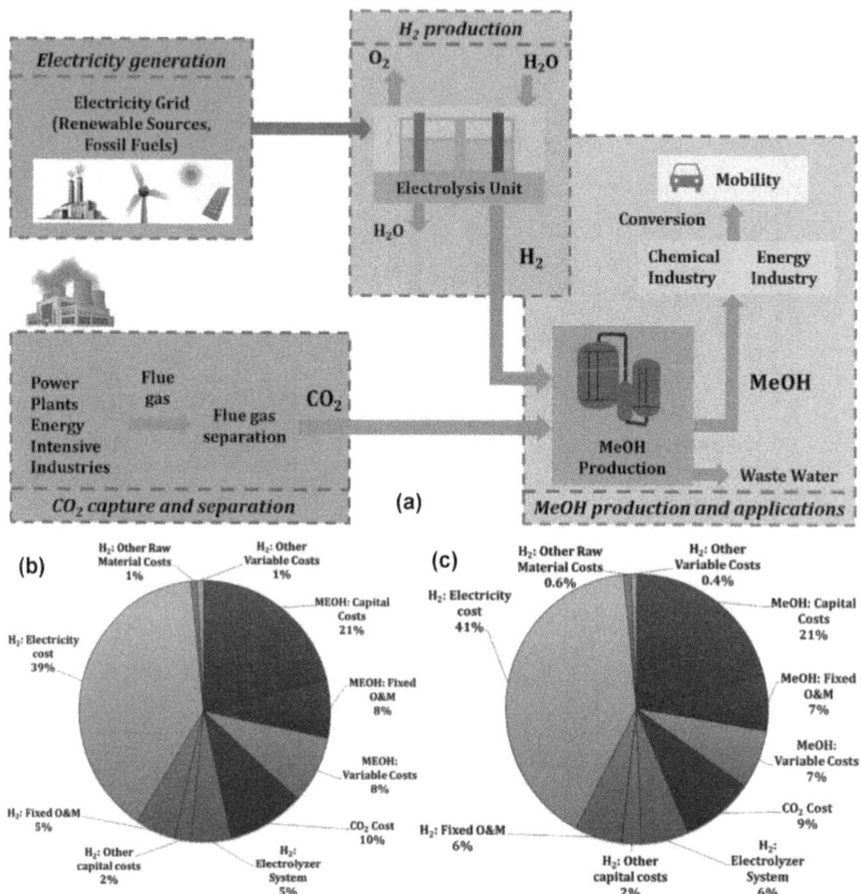

FIGURE 15.2 (a) Diagram of power to methanol (PtM), (b) the cost breakdown of two cases (the power plant owner case, and (c) and the private investor case. (Reprinted from *Int. J. Hydrog. Energy*, 41, Kourkoumpas, D.S. et al., Implementation of the Power to Methanol concept by using CO_2 from lignite power plants: Techno-economic investigation, 16674–16687, Copyright 2016, with permission from Elsevier.)

separation (CCS), H_2 production, as well as key stages of methanol production and application, as shown in Figure 15.2a. Two distinct case studies, a power plant owner case and a private investor case, were selected as cost breakdowns, shown in Figure 15.2b and c. In this figure, the methanol cost is estimated, and the results show that the cost for the power plant owner was 421 € ton^{-1} which was approximately 40% lower than that of the other case (580 € ton^{-1}) due to lower electricity costs and larger operating capacities of H_2 and methanol plants in the former. In addition, other economic analyses for conversion into methanol have been studied[7,8] and it is believed that methanol costs can be further reduced through the development of related technologies.

15.2 UREA

Urea (NH_2CONH_2) is usually commercialized as a solid, typically as granules or prills. There are four steps in the production of urea including the reaction of CO_2 and ammonia, low-pressure recirculation and purification, as well as granulation.[9] Bose et al.[10] proposed a coal-based poly-generation by first capturing CO_2 and then using the captured CO_2 to produce urea, a common fertilizer. The economic analysis of this proposed process is presented in Table 15.2 The utilization and recycling of waste is essential for sustainability, and utilizing waste can increase economic feasibility.

15.3 DIMETHYL CARBONATE (DMC)

Based on CO_2 conversion, dimethyl carbonate ($(CH_3O)_2CO$) can be synthesized through four proposed processes: (1) CO_2 and methanol directly, (2) urea, (3) propylene carbonate, and (4) ethylene carbonate. As a typical CO_2 conversion for dimethyl carbonate production, the techno-economic evaluation of this process using novel catalytic membrane reactors was studied,[11] especially for investment costs and operating costs, as listed in Tables 15.3 and 15.4, respectively (the production DMC is

TABLE 15.2
Data for Economic Analysis of the Proposed Process of Urea Production

Parameters	Value
Cost of urea plant	45 \$/MT
Cost of electricity	80 \$/MWh
Average cost of CO_2 transport and sequestration	14.1 \$/ton
Cost of utility heat	3.66 \$/GJ
Revenue earned through urea	570.5 \$/MT

Source: Bose, A. et al., *Clean Technol. Environ. Policy*, 17, 1271–1280, 2015.

TABLE 15.3
Overview of the Investment Costs for a typical CO_2 conversion for dimethyl carbonate production process

Equipment Type	Costs [M\$]	Costs [M\$]	Total Costs [M\$]
Membrane reactor	–	1.7	10.1
Flash vessels	0.2	0.3	1.8
Distillation columns	1.3	2.0	11.7
Compressors	1.1	1.5	9.2
Pumps	0.1	0.1	0.8
Heat exchangers	–	2.0	11.9
Total investment cost	2.7	7.6	45.5

Source: Kuenen, H.J. et al., *Comput. Chem. Eng.*, 86, 136–147, 2016.

TABLE 15.4
Overview of the Operating Costs for a typical CO_2 conversion for dimethyl carbonate production process

Parameters	Costs [My^{-1}]	Costs [My^{-1}]
Raw materials		6.63
Utilities		4.18
Electricity (0.06$/kWh)	0.83	
Chilled water (4$/GJ)	0.77	
Cooling water (0.05$/m^3)	0.40	
Hot water –95°C (0.05$/m^3)	0.004	
Steam 2 bar (5$/ton)	0.60	
Steam 5 bar (7$/ton)	0.54	
Steam 20 bar (12$/ton)	1.05	
Operations (labor related)		1.32
Maintenance		3.66
Operation overhead		0.74
Property taxes		0.91
Depreciation		4.07
Depreciation (8%)	3.50	
Membrane life time (3 years)	0.56	
Cost of manufacture		21.51
General expenses		2.31
Total production cost		23.82
Sales (DMC)		20.00
Revenue		–3.82

Source: Kuenen, H.J. et al., *Comput. Chem. Eng.*, 86, 136–147, 2016.

20 kton y^{-1}). Here, the DMC price is 1000$/ton, and the methanol price is 445$/ton, with the CO_2 cost being zero. It can be concluded that, under the performed conditions, the production process is unprofitable due to the high costs of depreciation, maintenance, and utilities.

15.4 FORMIC ACID

Formic acid (HCOOH) is a chemical that is widely used in textiles, food chemicals, and pharmaceuticals. It also has potential to be a hydrogen carrier and fuel that can be directly used for fuel cells. Formic acid is a high-value product because it has a concentrated, mature, and small market with low risk of substitution.[12] The process of CO_2 utilization for formic acid production is shown in Figure 15.3a, with the installed and operating costs of this plant also being depicted in Figure 15.3b and c. Forty-three percent of the total installed cost is for electrolysis, and the remaining costs are for the compression and separation processes. Utilities and consumables such as steam and electricity are the main contributors to operating costs.

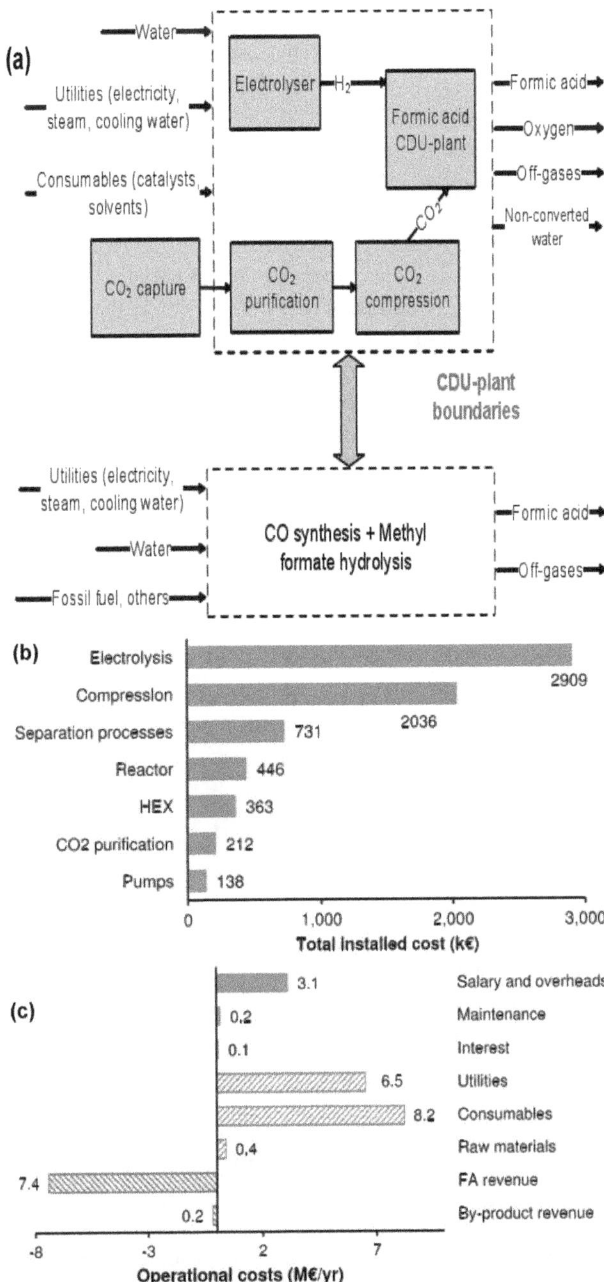

FIGURE 15.3 (a) Process of carbon dioxide utilization for formic acid production and the breakdown of (b) operating costs and (c) installed costs. (Reprinted from *Int. J. Hydrog. Energy*, 41, Pérez-Fortes, M. et al., Formic acid synthesis using CO_2 as raw material: Techno-economic and environmental evaluation and market potential, 16444–16462, Copyright 2016, with permission from Elsevier.)

The revenue of formic acid and other by-products are approximately 50% of the production costs. Therefore, more research, including R&D for electrolyzers, coupling with renewable energies, as well as developments of other related technologies to lower the cost for CO_2 conversion to produce formic acid, is needed.

15.5 SYNGAS ($CO+H_2$)

Compared to low-temperature electrochemical conversion of CO_2, the selectivity of products such as CO (electrolysis) and CO/H_2 (CO_2/H_2O co-electrolysis) at high temperatures using solid oxide cell (SOC) is preferable. In the synthetic natural gas (SNG) production of integrated high-temperature electrolysis (steam electrolysis or H_2O/CO_2 co-electrolysis using SOC) and methanation of CO_2, the effects of the following essential parameters were further investigated: CO_2 feedstock costs, cell degradation rates, and electrolysis stack costs. Figure 15.4 shows the share and specific costs of each category. It is evident that electricity is the most relevant cost.

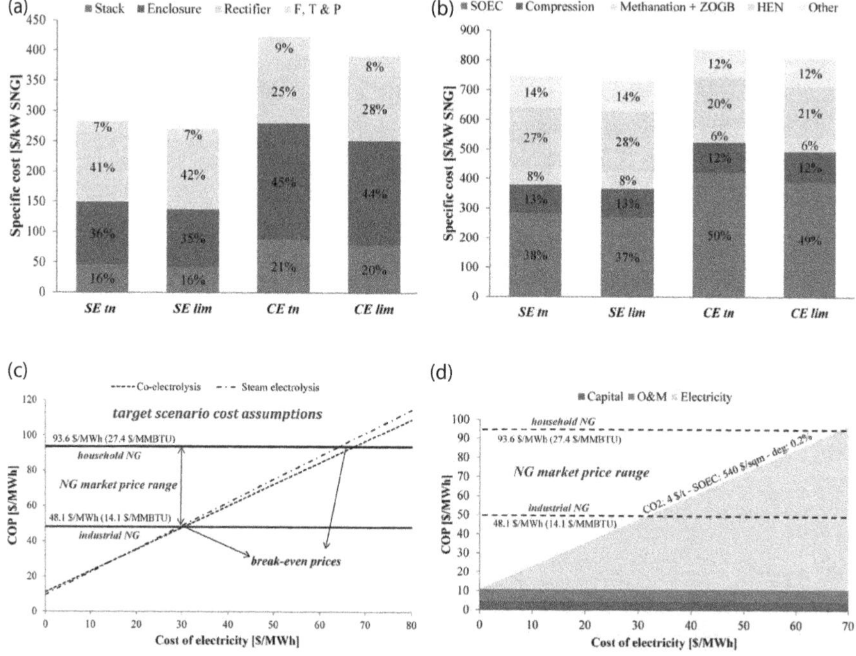

FIGURE 15.4 Specific share and cost for each category: (a) SOEC, (b) plant, (c) cost of production (COP) behavior of varied electricity feedstock prices, (d) SNG COP versus electricity costs for the co-electrolysis plant operating and limit voltage under target case assumptions. SE: steam electrolysis; CE: co-electrolysis; lim: limit; tn: thermoneutral; FT&P: foundations, transport, and placement; HEN: heat exchanger network; ZOGB: zinc oxide guard bed; O&M includes carbon dioxide cost, spare capacity installation, and stack replacement. (Reprinted from *J. Energ. Storage*, 2, Giglio, E. et al., Synthetic natural gas via integrated high-temperature electrolysis and methanation: Part II-Economic analysis, 64–79, Copyright 2015, with permission from Elsevier.)

The capital, maintenance, and operating costs are higher in co-electrolysis as well. For co-electrolysis, the calculated breakeven electricity price for target and state-of-the-art scenarios are 67$/MWh and 8 $/MWh, respectively.

In addition, the integration of Fischere-Tropsch and high-temperature H_2O/CO_2 co-electrolysis with SOC processes can efficiently produce syngas to store electricity. Fu et al.[13] studied its economic assessment; results show that excess nuclear power and wind power are the preferred electricity sources and that the cost of producing fuels was comparable to the cost of biomass-to-liquid processes.

15.6 SUMMARY

In this chapter, the economic analysis of CO_2 conversion and the relative added value of various products were discussed based on several typical production cases, including methanol, urea, dimethyl carbonate, formic acid, and syngas. Although the economic feasibility for the production of chemicals/fuels need further verification, it is believed that the costs will decrease further as related technologies are further developed.

REFERENCES

1. Aresta, M., Dibenedetto, A., Quaranta, E. State of the art and perspectives in catalytic processes for CO_2 conversion into chemicals and fuels: The distinctive contribution of chemical catalysis and biotechnology. *Journal of Catalysis* **343**, 2–45 (2016).
2. Otto, A., Grube, T., Schiebahn, S., Stolten, D. Closing the loop: Captured CO_2 as a feedstock in the chemical industry. *Energy & Environmental Science* **8**, 3283–3297 (2015).
3. Liao, F., Wu, X. P., Zheng, J., Li, M. J., Kroner, A. A promising low pressure methanol synthesis route from CO_2 hydrogenation over Pd@Zn core-shell catalysts. *Green Chemistry* (2016).
4. Rungtaweevoranit, B., et al. Copper nanocrystals encapsulated in Zr-based metal-organic frameworks for highly selective CO_2 hydrogenation to methanol. *Nano Letters* **16**, 7645–7649 (2016).
5. Kiss, A. A., Pragt, J. J., Vos, H. J., Bargeman, G., de Groot, M. T. Novel efficient process for methanol synthesis by CO_2 hydrogenation. *Chemical Engineering Journal* **284**, 260–269 (2016).
6. Kourkoumpas, D. S., et al. Implementation of the Power to Methanol concept by using CO_2 from lignite power plants: Techno-economic investigation. *International Journal of Hydrogen Energy* **41**, 16674–16687 (2016).
7. Wiesberg, I. L., de Medeiros, J. L., Alves, R. M. B., Coutinho, P. L. A., Araújo, O.Q.F. Carbon dioxide management by chemical conversion to methanol: HYDROGENATION and BI-REFORMING. *Energy Conversion and Management* **125**, 320–335 (2016).
8. Pérez-Fortes, M., Schöneberger, J. C., Boulamanti, A., Tzimas, E. Methanol synthesis using captured CO_2 as raw material: Techno-economic and environmental assessment. *Applied Energy* **161**, 718–732 (2016).
9. Pérez-Fortes, M., Bocin-Dumitriu, A., Tzimas, E. CO_2 utilization pathways: Techno-economic assessment and market opportunities. *Energy Procedia* **63**, 7968–7975 (2014).
10. Bose, A., Jana, K., Mitra, D., De, S. Co-production of power and urea from coal with CO_2 capture: Performance assessment. *Clean Technologies and Environmental Policy* **17**, 1271–1280 (2015).

11. Kuenen, H. J., Mengers, H. J., Nijmeijer, D. C., van der Ham, A. G. J., Kiss, A. A. Techno-economic evaluation of the direct conversion of CO_2 to dimethyl carbonate using catalytic membrane reactors. *Computers & Chemical Engineering* **86**, 136–147 (2016).
12. Pérez-Fortes, M., Schöneberger, J. C., Boulamanti, A., Harrison, G., Tzimas, E. Formic acid synthesis using CO_2 as raw material: Techno-economic and environmental evaluation and market potential. *International Journal of Hydrogen Energy* **41**, 16444–16462 (2016).
13. Fu, Q., et al. Syngas production via high-temperature steam/CO_2 co-electrolysis: An economic assessment. *Energy & Environmental Science* **3**, 1365–1608 (2010).
14. Giglio, E., Lanzini, A., Santarelli, M., Leone, P. Synthetic natural gas via integrated high-temperature electrolysis and methanation: Part II-Economic analysis. *Journal of Energy Storage* **2**, 64–79 (2015).

16 Summary and Possible Research Directions for CO_2 Conversion Technologies

To facilitate the research and development of CO_2 conversion, this book provides a comprehensive overview of advanced CO_2 conversion technologies for the production of useful fuels/chemicals. The fundamentals of CO_2, including molecule structure, thermodynamics, and kinetics, were summarized to explain its chemical stability, which is a direct obstacle for CO_2 conversion. Various conversion technologies as well as enzymatic, mineralization, photochemical/photoelectrochemical, thermochemical, and electrochemical processes were classified and reviewed. Particularly, electrochemical technologies at both low temperatures and high temperatures were considered and compared. Economic feasibilities were analyzed in light of various chemicals/fuels. Finally, technical and application challenges of CO_2 conversion to useful fuels/chemicals were proposed.

To overcome these challenges, the following future research directions are proposed:

1. *Further fundamental understanding of the CO_2 activation and conversion processes.* The mechanisms of CO_2 activation/conversion need to be further studied through both theoretical modeling and experimental testing (especially in electrochemical processes). For theoretical modeling, the transition states and elementary reactions in CO_2 conversions need to be further defined, and the rate-limiting step as well as the reaction pathways need to be controlled to overcome conversion barriers. For example, the mechanisms of CO_2 reduction and their relationship to reaction active site composition and structures should be fundamentally understood to guide new material development. To mitigate catalyst or electrode material degradation, further fundamental understanding of material behaviors and degradation mechanisms in CO_2 conversion processes is of great important. They can be further revealed by implementing some advanced techniques, for example, in situ time-resolved spectroscopy, X-ray computed tomography (XCT), variable temperature scanning tunneling microscope (VT-STM), low-energy ion scattering spectroscopy (LESI), ambient pressure-X-ray photoelectron spectroscopy (AP-XPS), environmental transmission electron microscopy (TEM), and secondary ion mass spectroscopy (SIMS). For instance, SIMS

in a three-dimensional (3D) model can be performed to display elemental 3D distributions with fast oxygen-incorporation paths and separation of electrode/electrolyte interface regions to analyze the causes for degradation and guide new designs of electrodes. To better control the reactions of CO_2 conversion kinetically, simulation techniques such as density functional theory (DFT) and molecular dynamics (MD) are needed to calculate energy barriers. With further understanding, it is possible to develop new methods to enhance activity and develop mitigation strategies for degradation.

2. *Enhancement of stability, activity, and product selectivity by optimizing or exploring advanced materials.* Many investigations have been performed to improve the activity, stability, and product selectivity of CO_2 conversion systems, with some progress achieved. However, this technological progress is still insufficient for practical applications. Breakthroughs are needed. Aside from further fundamental understanding and optimization of material performance, structural and morphological tailors of catalyst materials on the nanoscale and the exploration of appropriate innovative materials are also effective methods for enhancing activity, stability, and even product selectivity. For instance, the adsorption and activation of CO_2 onto the surface of a catalyst used in photocatalytic/photoelectrochemical processes can be optimized by morphological tailor. The enzymes used in the enzymatic process should be modified and immobilized to reduce the cost and improve activity and stability. With respect to high-temperature electrochemical processes, mixed ionic and electronic conductors (MIECs) and perovskites must be widely studied to improve oxygen electrode performances. For example, the heterostructured perovskite materials can exhibit remarkable enhancements in oxygen reduction reaction (ORR) kinetics, with three orders of magnitude greater than that of previously used materials.[1–4]
3. *Optimizing electrodes, system designs, and reactors for practical applications.* Aside from the materials themselves, the major limitations to activity and stability are partially related to operating environments and/or reaction processes, which largely depend on the design of reactors. In addition to improving material stability and activity, improving reactor designs and exploring novel appropriate reactors are important for performance enhancements. The CO_2 mineralization system mentioned above can be designed to generate electricity and produce industrial valuable products simultaneously, which is economically feasible and environmentally friendly. Regarding the electrochemical process, certain novel designs for solid oxide cell (SOC) configurations can lower ohmic and polarization resistances and enhance the power densities of SOC stacks. Another point is that certain designs for electrolyte membrane-based and gas diffusion electrodes (GDEs) electrochemical cells might be the right approaches to reduce internal resistance and improve reactant mass transfer.
4. *Development of CO_2 conversion at intermediate temperatures.* Although CO_2 conversions at low temperatures (<200°C) in aqueous and nonaqueous techniques, such as transition metal electrodes in liquid electrolytes or proton

exchange membranes, typically can be applied with simpler materials and rapid starts in milder conditions, the system is complex and most catalysts are costly (especially catalysts containing Pt). By contrast, CO_2 conversion techniques at high temperatures (>800°C) such as SOCs typically can perform with superior product selectivity and efficiency. However, requirements for materials, including electrodes, electrolytes, seals, and interconnects, are much higher, and the stack lifetime needs further verification due to degradation during high-temperature operations. Therefore, to enhance conversion efficiency, product selectivity, and stability, it is vitally important to draw on the strengths of low-temperature and high-temperature processes. It is believed that CO_2 conversion at intermediate temperatures (300–700°C) is an important and promising trend, but materials such as catalysts, electrodes, and electrolyte materials need to be developed to fulfill the requirements for CO_2 conversion process at intermediate temperatures.

5. *Developing the connection between CO_2 conversion systems and other clean energy sources.* If clean energy sources such as nuclear, solar, wind, geothermal, and so on, are used to run CO_2 conversion into useful chemicals/fuels, CO_2 emission and energy shortage issues can be relieved along with providing better economic feasibility. For instance, the conversion of CO_2 into valuable fuels (e.g., CH_4, CH_3OH, HCOOH, etc.) via artificial photosynthesis can be performed using solar energy on a large scale. The co-electrolysis of CO_2 and H_2O can be performed to produce useful fuels of CO and H_2 using solid oxide electrolysis cell stacks running on electricity and heat from redundant nuclear energy. Although efforts have been undertaken to accelerate these developments, wider and scalable applications have a long way to go, with coupling techniques between CO_2 conversion systems and other clean energy sources playing a key role. The implementation of this connection is difficult and requires worldwide government support.

REFERENCES

1. Ma, W., et al. Vertically aligned nanocomposite $La_{0.8}Sr_{0.2}CoO_3$ /$(La_{0.5}Sr_{0.5})_2CoO_4$ cathodes- electronic structure, surface chemistry and oxygen reduction kinetics. *Journal of Materials Chemistry A* **3**, 207–219 (2015).
2. Chen, Y., et al. Electronic activation of cathode superlattices at elevated temperatures–source of markedly accelerated oxygen reduction kinetics. *Advanced Energy Materials* **3**, 1221–1229 (2013).
3. Han J. W., Yildiz B. Mechanism for enhanced oxygen reduction kinetics at the (La, Sr) CoO_3/(La, Sr)$_2CoO_{4+\delta}$ heterointerface. *Energy & Environmental Science* **5**, 8598–8607 (2012).
4. Sase, M., et al. Enhancement of oxygen surface exchange at the hetero-interface of (La, Sr) CoO_3(La, Sr)$_2CoO_4$ with PLD-Layered Films. *Journal of The Electrochemical Society* **155**, B793–B797 (2008).

Index

Note: Page numbers in italic and bold refer to figures and tables, respectively.